SPREAD SPECTRUM CDMA SYSTEMS FOR WIRELESS COMMUNICATIONS

The Artech House Mobile Communications Series

John Walker, Series Editor

Advanced Technology for Road Transport: IVHS and ATT, Ian Catling, editor

An Introduction to GSM, Siegmund M. Redl, Matthias K. Weber, Malcolm W. Oliphant

CDMA for Wireless Personal Communications, Ramjee Prasad

Cellular Digital Packet Data, Muthuthamby Sreetharan and Rajiv Kumar

Cellular Mobile Systems Engineering, Saleh Faruque

Cellular Radio: Analog and Digital Systems, Asha Mehrotra

Cellular Radio Systems, D. M. Balston, R. C. V. Macario, editors

Cellular Radio: Performance Engineering, Asha Mehrotra

Digital Beamforming in Wireless Communications, John Litva, Titus Kwok-Yeung Lo

GSM System Engineering, Asha Mehrotra

Introduction to Radio Propagation for Fixed and Mobile Communications, John Doble

Land-Mobile Radio System Engineering, Garry C. Hess

Mobile Communications in the U.S. and Europe: Regulation, Technology, and Markets, Michael Paetsch

Mobile Antenna Systems Handbook, K. Fujimoto, J. R. James

Mobile Data Communications Systems, Peter Wong, David Britland

Mobile Information Systems, John Walker, editor

Personal Communications Networks, Alan David Hadden

RF and Microwave Circuit Design for Wireless Communications, Lawrence E. Larson, editor

Smart Highways, Smart Cars, Richard Whelan

Spread Spectrum CDMA Systems for Wireless Communications, Savo G. Glisic and Branka Vucetic

Understanding GPS: Principles and Applications, Elliott D. Kaplan, editor

Wireless Communications in Developing Countries: Cellular and Satellite Systems, Rachael E. Schwartz

Wireless Communications for Intelligent Transportation Systems, Scott D. Elliott, Daniel J. Dailey

Wireless Data Networking, Nathan J. Muller

Wireless: The Revolution in Personal Telecommunications, Ira Brodsky

For a complete listing of *The Artech House Telecommunications Library,* turn to the back of this book.

Spread Spectrum CDMA Systems for Wireless Communications

Savo Glisic
Branka Vucetic

Artech House, Inc.
Boston • London

Library of Congress Cataloging-in-Publication Data
Glisic, Savo.
  Spread spectrum CDMA systems for wireless communications / Savo Glisic.
    p.   cm.
  Includes bibliographical references and index.
  ISBN 0-89006-858-5 (alk. paper)
  1. Code division multiple access.   I. Vucetic, Branka   II. Title.
TK5103.45.G57   1997
621.3845—dc21   97-860
                                                              CIP

British Library Cataloguing in Publication Data
Glisic, Savo G.
  Spread spectrum CDMA systems for wireless communications
  1. Code division multiple access   2. Wireless communication systems.
  I. Title II. Vucetic, Branka
  621.3'82

  ISBN 0-89006-858-5

Cover design by Jennifer Makower.

© 1997 ARTECH HOUSE, INC.
685 Canton Street
Norwood, MA 02062

All rights reserved. Printed and bound in the United States of America. No part of this book may be reproduced or utilized in any form or by any means, electronic or mechanical, including photocopying, recording, or by any information storage and retrieval system, without permission in writing from the publisher.
  All terms mentioned in this book that are known to be trademarks or service marks have been appropriately capitalized. Artech House cannot attest to the accuracy of this information. Use of a term in this book should not be regarded as affecting the validity of any trademark or service mark.

International Standard Book Number: 0-89006-858-5
Library of Congress Catalog Card Number: 97-860

10 9 8 7 6 5 4 3 2 1

Contents

Preface   xi

Introduction   xiii

Chapter 1   Fundamentals of Spread Spectrum Systems   1
    1.1   Spread Spectrum System Concept   1
        1.1.1   Antijamming Capabilities   2
        1.1.2   Operation in Multipath Environment   4
        1.1.3   Code Division Multiple Access   6
        1.1.4   Capacity of a CDMA Network   7
    1.2   Examples of Spread Spectrum Systems   8
        1.2.1   Direct Sequence Spread Spectrum System   8
        1.2.2   Frequency Hopping Spread Spectrum System   13
        1.2.3   Hybrid DS/FH Schemes   14
        1.2.4   Chirp Signals for Pulse Compression   14
    1.3   Sequence Generation   18
        1.3.1   Correlation Functions   18
        1.3.2   Linear Recursions   19
        1.3.3   Sequence Generation by Using LFSR Long Division   22
        1.3.4   Sequence Representation by Using Elements of an Extension Galois Field   23

	1.3.5 m-Sequences	24
	1.3.6 The Trace Representation of m-Sequences	26
	1.3.7 Decimation of Sequences	27
	1.3.8 Crosscorrelation of Binary m-Sequences	27
	1.3.9 Composite Binary Sequences	29
	1.3.10 Gold Sequences	30
	1.3.11 The Small Set of Kasami Sequences	31
References		31
Selected Bibliography		32
Appendix A1 The Trace Function		33
Appendix B1 Binary Field and Vector Space		37
Chapter 2 Code Acquisition		**49**
2.1	Maximum Likelihood Parameter Estimation	50
2.2	ML-Motivated Code Synchronization	53
	2.2.1 Problem Definition	53
	2.2.2 Uncertainty Range of the Signal Parameters	54
	2.2.3 Detector Structures: DS Signal	55
	2.2.4 Detector Structures: FH Signal	56
	2.2.5 Parallel Versus Serial Implementation	57
	2.2.6 Two-Dimensional Search	58
	2.2.7 Serial Search	59
2.3	Performance Measures	60
2.4	Performance Analysis Tools	63
	2.4.1 Transform Domain Analysis	64
	2.4.2 Direct Method	70
	2.4.3 Modified Search Strategies	73
2.5	Automatic Decision Threshold Level Control	84
	2.5.1 Instantaneous Threshold Setting Algorithm	88
	2.5.2 Constant False Alarm Rate Algorithm	88
2.6	Sequential Detection	91
2.7	Lock Detectors	95
References		99
Selected Bibliography		101
Appendix A2 Code Acquisition Time Variance		102
Chapter 3 Code Tracking		**105**
3.1	Phase-Locked Loop Fundamentals	105
	3.1.1 Analog PLL	105
	3.1.2 The Linear Model	107
	3.1.3 Steady-State Phase Error	108
	3.1.4 Transient Response of the PLL Under Linear Conditions	110

	3.1.5	Loop Operation in the Presence of Noise	110
	3.1.6	Nonlinear Model of PLL	115
	3.1.7	Digital PLL	117
3.2	Code Tracking Loops		120
	3.2.1	Baseband Full-Time Early-Late Tracking Loop	120
	3.2.2	Full-Time Early-Late Noncoherent Tracking Loop	121
	3.2.3	Tau Dither Early-Late Noncoherent Tracking Loop	124
	3.2.4	Double Dither DLL	125
	3.2.5	Code Tracking Loop for Frequency Hopping Systems	128
	3.2.6	Time-Shared Two-Channel Noncoherent Code	129
	3.2.7	Discrete Tracking Process	131
	3.2.8	Discrete Tracking Systems for Slow FH	137
3.3	Code Tracking in Fading Channels		154
	3.3.1	Channel Model	156
	3.3.2	Joint Estimation of PN Code Delay and Multipath Using the Extended Kalman Filter	158
References			163
Selected Bibliography			163
Appendix A3 Parameters of FH System Code Tracking			166

Chapter 4	Spread Spectrum System Technology		175
4.1	CCD Technology		175
4.2	CCD PN-Matched Filter Imperfections		178
	4.2.1	DC Zero Offset	178
	4.2.2	Charge Transfer Inefficiency	180
	4.2.3	Additional Impairments	184
4.3	Reported Signal Processors Based on CCD Components		185
	4.3.1	Analog-Analog Correlators	185
	4.3.2	Analog-Binary Correlators	186
	4.3.3	Double Sampling	186
	4.3.4	Simultaneous/Parallel Search	187
	4.3.5	High-Speed Correlators	188
	4.3.6	Wide Bandwidth Devices	188
4.4	Receiver Configurations Based on Matched Filters		188
	4.4.1	Coherent Digital Demodulation of BPSK Data and BPSK Spread Spectrum Modulation	190
	4.4.2	BPSK Data and OQPSK Spread Spectrum Modulation	191
	4.4.3	Differentially Coherent Demodulation of QPSK Data Modulation	193

	4.4.4	Sum and Difference Demodulator	195
	4.4.5	Noncoherent Pulse Amplitude Modulation	195
	4.4.6	Coherent PAM	197
4.5	Surface Acoustic Wave Technology		198
	4.5.1	Convolution Function	198
	4.5.2	Fourier Transform	200
	4.5.3	Inverse Fourier Transform	201
	4.5.4	Transform Domain Signal Filtering	202
4.6	Reported Surface Acoustic Wave Devices		203
References			206
Selected Bibliography			209

Chapter 5 CDMA Systems — 211

5.1	A Cellular CDMA System		211
5.2	A Single-Cell CDMA Demodulator in a Gaussian Channel		215
5.3	Bit Error Probability		220
5.4	Capacity of a Single-Cell CDMA System		223
5.5	Multipath Propagation		225
	5.5.1	Fading and Doppler Shifts for Single Tone Transmission	225
	5.5.2	Frequency Selective Fading Channel	229
	5.5.3	Wideband Channel Models	230
5.6	Rake Receiver		236
	5.6.1	Performance of Rake Receiver	237
5.7	Capacity of a Multiple-Cell CDMA System		241
	5.7.1	Propagation Loss Model	241
	5.7.2	Power Control	242
	5.7.3	Reverse Link Capacity in Cellular CDMA	246
	5.7.4	Erlang Capacity of the Reverse Link in a Single-Cell CDMA	252
	5.7.5	Erlang Capacity of the Reverse Link in a Cellular CDMA	255
	5.7.6	Erlang Capacity of the Reverse Link in a Cellular CDMA System With Imperfect Power Control	256
	5.7.7	Forward Link Capacity in a Cellular CDMA System	258
References			261
Selected Bibliography			262
Appendix A5 Interim Standard 95			262

Chapter 6 Multiuser CDMA Receivers — 265

6.1	Introduction	265
6.2	The System Model for Multiuser Demodulation	267

	6.3	Optimum Receiver	269
	6.4	Suboptimum DS/CDMA Receivers	276
		6.4.1 Decorrelating Receiver	277
		6.4.2 Interference Canceler	281
		6.4.3 Adaptive MMSE Receivers	283
		6.4.4 The Adaptive Algorithm	286
	References		289
	Selected Bibliography		290
Chapter 7	Adaptive Interference Suppression and CDMA Overlay Systems		291
	7.1	Estimation Filters	293
		7.1.1 DS Spread Spectrum Receiver and Estimation Filter Structures	293
		7.1.2 Narrowband Interference Suppression by Estimation Filters	294
	7.2	Modeling of Narrowband Interference by Autoregressive Processes	300
	7.3	Examples of the Interfering Signal	302
	7.4	Bit Error Rate	304
	7.5	Adaptive Estimation Filters	306
		7.5.1 The Gradient and the Wiener Solution	307
		7.5.2 The Steepest Descent Algorithm	310
		7.5.3 The LMS Algorithm	311
		7.5.4 Decision Feedback Interference Suppression Filter	313
	7.6	Transform Domain Filters	314
	7.7	Adaptive Receivers	318
	7.8	CDMA Overlay for Personal Communication Networks	319
		7.8.1 CDMA Overlay	319
		7.8.2 Overlay Geometry	321
		7.8.3 Propagation Model	321
		7.8.4 Forward Link CDMA Network Capacity	324
		7.8.5 Reverse Link CDMA Network Capacity	329
	References		333
	Selected Bibliography		335
Chapter 8	Multiple-Access Protocols in Spread Spectrum Packet Radio Networks		337
	8.1	Packet Radio Networks—General Concepts	337
	8.2	Multiplexing and Multiple-Access Techniques	340
		8.2.1 Multiplexing	340
	8.3	Multiple-Access Schemes	341
		8.3.1 Frequency Division Multiple Access	342
		8.3.2 Time Division Multiple Access	344

8.4	Code Division Multiple Access		346
	8.4.1	Random Access Schemes	346
	8.4.2	Controlled Access Schemes	351
	8.4.3	CDMA Slotted ALOHA Schemes	354
References			365
Selected Bibliography			367
About the Authors			369
Index			371

PREFACE

This book is based on material used for postgraduate courses at the University of Oulu and the University of Sydney. This material has also been used for intensive courses organized by CEI-Europe in England, Italy, Spain, Sweden, and Switzerland, and is intended for audiences from both academia and industry. The main goal of these courses is to present the most current insight into the rather complex field of spread spectrum techniques, with an emphasis on code division multiple access (CDMA). CDMA has become extremely attractive for applications in mobile communications and wireless networks because of its high capacity and flexibility for multirate systems.

The book consists of eight chapters. The first three chapters cover inevitable topics: pseudonoise sequences, code acquisition, and code tracking. Although these topics have been covered in other books, we have tried to be different in the presentation and selection of the material. In the chapter on code tracking, for example, we emphasize the frequency hopping systems due to a recent renewed interest—even of time division multiple access systems—in this technique. Material covered in the remaining five chapters is not covered in any other book thus far, and in that sense represents the first systematic treatment of the most relevant topics in the field of code division multiple access.

In the organization of the material and its presentation, we owe a great deal to a number of colleagues. Special thanks to Professor Andreas Polydoros from the

University of Southern California, who wrote the introductory material to Chapter 2, and to Professor Peter Grant from the University of Edinburgh for his contribution to Chapter 4.

Brian Gardner, from Racal Messenger, reviewed the manuscript as it was created and provided very constructive criticism. This help is highly appreciated.

We are also indebted to a number of volunteers—David Clark, Matti Latva-aho, Jari Iinatti, and Graeme Woodward—who read parts of the book, corrected mistakes, and provided constructive criticism and suggestions for its improvement.

We also wish to thank Susanna Taggart of Artech House, for the constant care, encouragement, patience, and understanding throughout the process of creating the manuscript.

Special thanks to Pirjo Kumpumäki and Maree Beleli who, with great patience and skill, typed the entire manuscript and did the drawings.

Savo Glisic
Branka Vucetic
April 1997

INTRODUCTION

Spread spectrum is a signal transmission technique that has been around for more than a half century. Recent applications in mobile communications have renewed and considerably extended interest in the theory and practice of this technique.

In the field of radar systems, examples of frequency hopping signal patterns were patented already in 1938 [1]. Toward the end of World War II, the Germans were developing a linear-FM (chirp) radar system called Kugelschale. Even then they had been exploiting the fact that bandwidth expansion without pulse narrowing could also provide finer time resolution.

Antijam communications, based on spread spectrum techniques, have been studied for military applications for a long time. A frequency hopping system for the guidance of torpedoes was described in a patent application by Hedy Keisler Markey and George Anthiel in 1941. A comprehensive survey of spread spectrum origins can be found in [2]. Spread spectrum also provides better system performance in multipath propagation environments. The first experimental results were demonstrated by R. Price and P. Green, who developed the so-called Rake receiver structure [3]. The experiment was carried out via a teletype communications system over a transcontinental HF link.

The key elements of any spread spectrum system are pseudonoise sequences. It seems that the first extensive mathematical study of the subject was carried out by Sol Golomb during the 1950s, and his books [4,5] became the first textbooks in this field. The first ideas of *code division multiple access* (CDMA) are almost 50 years old [6]. During the early 1970s a number of papers already pointed out that CDMA can achieve higher capacity than *time division multiple access* (TDMA) system [7–12].

The first direct sequence spread spectrum systems were built during the 1950s. Examples of the early systems are ARC-50 by Magnavox [13] and satellite radios OM-55 and USC-28. Among the earliest papers in the open literature comparing different multiple access techniques such as TDMA, FDMA, and CDMA is reference [14]. Other examples of early military systems are the tactical communications satellite TATS [15] and the *Global Positioning System* (GPS) described in [16].

The first proposals for commercial application of CDMA came at the time when analog and later TDMA were going into system trials. In 1978 G. Cooper and R. Nettleton suggested CDMA for applications in cellular communication system in order to achieve higher system capacity [17]. This proposal came at the same time when the *American Mobile Phone System* (AMPS), based on analog technology, was going into system trials. Prior to the development of the GSM system in Europe, several proposals were put forth as trial systems to be used in selecting the access and channelization approaches for GSM. Among those are S900D [18] and SFH900 [19]. These systems were based on the *slow frequency hopping* (SFH) concept. Having accepted TDMA for the second generation of cellular mobile system, all major systems are now additionally introducing SFH combined with TDMA to improve the system performance in fading channel [20,21].

A separate direction of introducing the CDMA concept for cellular mobile communications has been traced out by FCC policy that encouraged solutions based on spread spectrum techniques. As a result *Interim Standard 95* (IS-95) was adopted by the *Telecommunications Industry Association* (TIA) in 1993. This standard is based on the hardware solution built up by Qualcomm Inc., a company that has a significant influence on shaping the way the modern cellular mobile communications are taking nowadays.

In the field of personal satellite communications, more and more systems are accepting CDMA as multiple-access scheme. Examples are Loral/Qualcomm's Globalstar with 48 satellites in LEO orbit, TRW's Odyssey with 12 satellites in MEO orbit, Constellation with 48 satellites in LEO orbit, Ellipso with 24 satellites in HEO orbit, and ESA's Archimedes.

Keeping in mind the importance of spread spectrum/CDMA techniques for future communications, the authors have tried to summarize the main issues relevant for understanding and applying this technique. The idea is to update and simplify the presentation of the material.

Chapter 1 introduces the basic ideas and motivation for using the spread spectrum technique. The main advantages of using techniques like antijam resistance, good performance in multipath propagation environment, and high capacity are discussed at the very beginning of the chapter. The main part of the chapter covers pseudonoise sequences, which are the key elements of each spread spectrum system.

Chapter 2 deals with code acquisition, which is a crucial step in the initialization of spread spectrum receiver operation. Besides a number of code acquisition strategies, different mathematical tools for their analysis are also described. Some additional topics like automatic decision threshold setting, sequential detector, and in lock/out of lock detectors are also discussed.

Chapter 3 discusses code tracking, which is a part of the overall synchronization process. As a first step, a short introduction into *phase-locked loop* (PLL) theory is presented because the PLL is the basic element of any code tracking loop. As a main part of this chapter, a comparison between different code tracking loops like delay-locked loop and tau dither loop is presented. Due to an exceptional interest in slow frequency hopping even in TDMA systems, discrete tracking loops for these systems are additionally elaborated. A brief description of multiple delay tracking in multipath channel based on Kalman filtering is also included.

Chapter 4 describes a number of receiver configurations based on using matched filters fabricated in *charge coupled devices* (CCD), *surface acoustic wave* (SAW), or standard *digital signal processing* (DSP) technology. A survey of the reported CCD and SAW matched filters is included along with a discussion of the main impairments in these components.

Chapter 5 describes the CDMA cellular radio network and covers the main issues related to system capacity. Discussion of uplink and downlink capacity, imperfect power control, and multipath propagation, is presented in detail. As a supplement, a short description of standard IS-95 is presented in an appendix.

Chapter 6 describes problems of multiuser detection. Near far problem and imperfect power control are inherent weaknesses of CDMA networks that can be further eliminated using the multiuser detection concept. This powerful and complex technique is discussed in detail in Chapter 8.

Chapter 7 deals with a special form of CDMA in the so-called overlay concept. These systems are designed to coexist with standard "narrowband" communications in the same frequency band. The whole idea is based on using very efficient interference suppression techniques in addition to the inherent characteristics of spread spectrum itself.

Chapter 8 covers packetized transmission in CDMA networks and discusses issues like throughput and packet delay. Protocols like ALOHA load sense and packet reservation are discussed in detail.

We believe that the structure of this book will provide insight into the current theoretical and practical results in the field of spread spectrum/CDMA techniques for a broad audience of engineers and students of modern telecommunications.

References

[1] Guanella, G., "Distance Determinig System," U.S. Patent 2,253,975, Aug. 26, 1941 (filed in U.S. on May 27, 1939; in Switzerland on Sept. 26, 1938).

[2] Scholtz, R. A., "The Origins of Spread-Spectrum Communications," *IEEE Trans. on Communications*, Vol. COM-30, May 1982, pp. 822–854.
[3] Price, R., and P. E. Green, Jr., "A Communication Technique for Multipath Channels," *Proc. IRE*, Vol. 46, March 1958, pp. 555–570.
[4] Golomb, S. W., *Shift Register Sequences*. San Francisco, CA: Holden-Day, 1967.
[5] Golomb, S. W. (ed.), *Digital Communications with Space Applications*. Englewood Cliffs, NJ: Prentice-Hall, 1964.
[6] Pierce, J. R., "Time Division Multiplex System with Erratic Sampling Times," Technical Memorandum 49-150-15, Bell Telephone Laboratories, June 15, 1949.
[7] Ahlswede, R., "Multi-way Communication Channels," *2nd Int. Symp. on Information Transmission*, USSR, 1971.
[8] Liao, H., "Multiple Access Channels," Ph.D. dissertation, Dept. of Electrical Engineering, University of Hawaii, Honolulu, Hawaii, 1972.
[9] Slepian, D., and J. K. Wolf, "A Coding Theorem for Multiple Access Channels with Correlated Sources," *Bell Syst. Tech. J.*, Sept. 1973.
[10] Wyner, A. D., "Recent Results in Shannon Theory," *IEEE Trans. on Information Theory*, Vol. IT-20, Jan. 1974, pp. 2–10.
[11] Cover, T. M., R. J. McEliece, and E. C. Posner, "Asynchronous Multiple-Access Channel Capacity," *IEEE Trans. on Information Theory*, Vol. IT-27, No. 4, July 1981, pp. 409–413.
[12] Hui, Y. J. N., and P. A. Humblet, "The Capacity Region of the Totally Asynchronous Multiple-Access Channel," *IEEE Trans. on Information Theory*, Vol. IT-31, No. 2, March 1985, pp. 207–216.
[13] Taylor, J. E., "Asynchronous Multiplexing," *AIEE Trans. (Communications and Electronics)*, Vol. 79, Jan. 1960, pp. 1054–1062.
[14] Schwartz, J. W., J. M. Aein, and J. Kaiser, "Modulation Techniques for Multiple Access to a Hard-Limiting Satellite Repeater," *Proc. IEEE*, Vol. 54, May 1966, pp. 763–777.
[15] Drouilhet, P. R., Jr., and S. L. Bernstein, "TATS—A Bandspread Modulation System for Multiple Access Tactical Satellite Communication," *EASCON Convention Record*, 1969.
[16] Spilker, J. J., Jr., "GPS Signal Structure and Performance Characteristics," *J. Institute of Navigation*, Vol. 25, No. 2, Summer 1978, pp. 121–146.
[17] Cooper, G. R., and R. W. Nettleton, "A Spread-Spectrum Technique for High-Capacity Mobile Communications," *IEEE Trans. on Vehicular Technology*, Vol. 27, No. 4, Nov. 1978, pp. 264–275.
[18] Ketterling, H. P., "The Digital Mobile Radio Telephone System Proposal S 900 D," *Proc. Second Nordic Seminar on Digital Land Mobile Radio Communication*, Stockholm, Sweden, Oct. 14–16, 1986, pp. 126–132.
[19] Dornstetter, J. L., and D. Verhulst, "The Digital Cellular SFH 900 System," *Proc. IEEE Int. Conf. on Communications*, June 1986, pp. 36.3.1–36.3.5.
[20] European Telecommunications Standards Institute, GSM Specifications, ETSI TC-SMG, Sophia Antipolis, France.
[21] Rasky, P. D., G. M. Chiasson, and D. E. Borth, "Hybrid Slow Frequency-hop/CDMA-TDMA as a Solution for High-Mobility, Widearea Personal Communications," *Proc. Fourth Winlab Workshop on Third Generation Wireless Information Networks*, East Brunswick, NJ, Oct. 19–20, 1993, pp. 195–215.

CHAPTER 1

FUNDAMENTALS OF SPREAD SPECTRUM SYSTEMS

1.1 SPREAD SPECTRUM SYSTEM CONCEPT

The traditional approach to digital communications is based on the idea of transmitting as much information as possible in as narrow a frequency bandwidth as possible. Therefore, a concept called *narrowband signal* (s_n) is used to yield narrowband systems. The most general concept of spread spectrum systems is presented in Figure 1.1.

Formally, the operation of both transmitter and receiver can be partitioned into two steps. In the first step, which we refer to as primary modulation, the narrowband signal s_n is formed. In the second step, or secondary modulation, the operation $\epsilon(\ )$ is applied, resulting in the expansion of the signal spectrum to a very wide frequency band. This signal will be denoted s_w.

At the receiver site, the first step is despreading, which is formally presented by the operation $\epsilon^{-1}(\ ) = \epsilon(\ )$. In other words, after despreading (which is identical to the spreading process) the wideband signal s_w is converted back to the original form s_n and standard methods for narrowband signal demodulation are used. What are the reasons for doing this? Let us start by explaining the very origins of the spread spectrum systems.

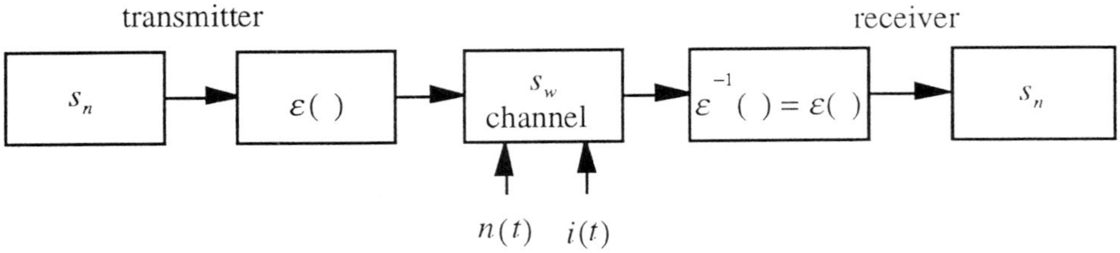

2 SPREAD SPECTRUM CDMA SYSTEMS FOR WIRELESS COMMUNICATIONS

Figure 1.1 Spread spectrum system concept.

1.1.1 Antijamming Capabilities

More than a half of century ago the spread spectrum concept was introduced to solve the problem of reliable communications in the presence of intensive jamming. If we assume that a signal s_w is received in the presence of a relatively narrowband and a much stronger jamming signal $i_n(t)$, then in the despreading process we have

$$\epsilon^{-1}(s_w + i_n) = \epsilon^{-1}(\epsilon(s_n)) + \epsilon^{-1}(i_n) = s_n + \epsilon(i_n) = s_n + i_w \tag{1.1}$$

In other words, the despreading process has converted the input signal into a sum of the narrowband useful and the wideband interfering signals. After narrowband filtering (operation $F(\ )$) with the bandpass filter of bandwidth B_n equal to the bandwidth of s_n, we have

$$F(s_n + i_w) \cong s_n + F(i_w) = s_n + i_{wr} \tag{1.2}$$

Only a small portion of the interfering signal energy will pass the filter and remain as residual interference i_{wr} because the bandwidth B_w of i_w is much larger than B_n. Figure 1.2 is a schematic representation of the process.

One can see from the figure that the power of the residual interference, $P(i_{wr}) = \eta_i B_n$, is related to the overall power of the interference signal, $P(i_w) = \eta_i B_w$, as

$$P(i_{wr}) = \frac{B_n}{B_w} P(i_w) = \frac{1}{G} P(i_w) \tag{1.3}$$

The parameter

$$G = \frac{B_w}{B_n} \tag{1.4}$$

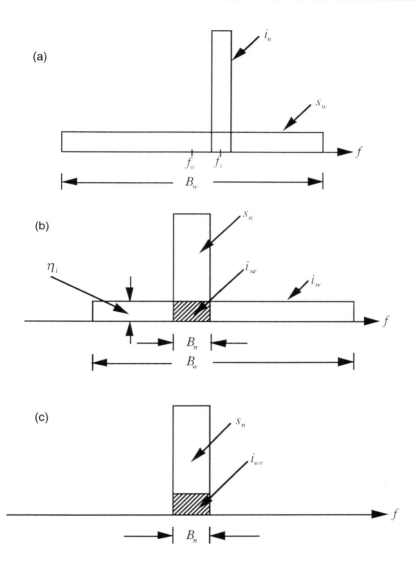

Figure 1.2 Despreading process in the presence of interference: (a) received signal, (b) result of $\epsilon^{-1}(\ )$ operation, and (c) result of $F(\ )$ operation.

that shows how much the interfering signal is suppressed in this process is called the processing gain. We will see later that for different implementations of the system, the statistics of i_{wr} will be different. The antijamming capability of spread spectrum systems has been exploited in military systems for a long time. The evolution of the practical solutions has been very dependent on available technology, and a comprehensive survey of the history of these systems can be found as an introduction to a number of books [1,2].

If the level of the interfering signal is too high, preliminary processing can be used to suppress interference prior to the $\epsilon^{-1}(\ )$ operation. These algorithms use different approaches, but the final effect is the same. A narrow notch in the frequency band occupied by the interference signal is formed using an adaptive algorithm that will be able to follow changes in the interfering signal parameters.

By forming the notch to suppress the interfering signals, a part of the useful signal in the same frequency band will also be removed. A detailed analysis, with which we will deal later, demonstrates that the signal degradation is acceptable as long as the bandwidth of the interfering signal is less than 20% of the signal bandwidth.

This approach has been used in civil applications too, for so-called *code division multiple access* (CDMA) overlay type networks. In these networks spread spectrum signals are used in the same frequency band where standard (narrowband) type users already exist. These users with much higher levels are suppressed with both preliminary suppression and spread spectrum processing so that their level in the spread spectrum receiver is tolerable. At the same time, the low spread spectrum signal density produces no excessive interference in the standard narrowband receivers.

1.1.2 Operation in Multipath Environment

As a result of multipath propagation, a transmitted signal will be received as a number of its mutually delayed replicas. A possible phasor representation is given in Figure 1.3. Most of the time, these signal components will act as interference to each other and the net result will be the degradation of the system performance. Now the question is: Is there any way that we can separate these components and combine them synchronously into one large signal vector that would provide good

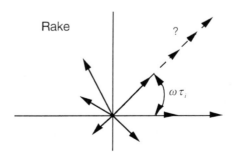

Figure 1.3 Multipath propagation.

signal demodulation conditions? The answer in the spread spectrum concept is the so-called Rake-receiver, which will be discussed in more detail later. For now, using the previously introduced notation, we can represent the received signal as

$$r(t) = \sum_{l=0}^{N\beta-1} \beta_l s_w(\tau_l) = \sum_{l=0}^{N\beta-1} \beta_l \epsilon_{\tau_l}(s_n(\tau_l)) \qquad (1.5)$$

where $\tau_0 = 0$ and τ_l is the relative delay (excess delay) experienced by the signal propagating through path l with respect to the signal propagation through the shortest path and β_l is the intensity coefficient of the corresponding path. Most of the time, this coefficient is modeled as a complex, zero mean Gaussian variable. In the receiver, the despreading operation $\epsilon_{\tau_s}^{-1}(\ ) = \epsilon_{\tau_s}(\ )$, synchronized to the signal with delay τ_s, will produce

$$\epsilon_{\tau_s}(r) = \sum_{l=0}^{L-1} \epsilon_{\tau_s}[\beta_l \epsilon_{\tau_l}(s_n(\tau_l))] \qquad (1.6)$$

The signal components will depend only on $\tau_s - \tau_l$, which can be represented as ϵ_{s-l} resulting in

$$\epsilon_{\tau_s}(r) = \sum_{l=1}^{L-1} \beta_l \epsilon_{s-l}[s_n(\tau_l)] \qquad (1.7)$$

This will result in

$$\epsilon_{s-l}[s_n(\tau_l)] \Rightarrow \begin{cases} s_n(\tau_l) & \text{if } \tau_s - \tau_l = 0 \\ s_w(\tau_l) & \text{if } \tau_s - \tau_l > \tau_r \end{cases} \qquad (1.8)$$

where τ_r is the range of ϵ operation. In other words, if the despreading operation is synchronized to the mth signal component ($\tau_s - \tau_m \cong 0$) and $\tau_s - \tau_l > \tau_r$ after despreading and bandpass filtering, we will have

$$F\{\epsilon_{\tau_m}(r)\} = \beta_m s_n(\tau_m) + \sum_{\substack{l \neq m \\ l=0}}^{L-1} \beta_l s_{wr}(\tau_l) \qquad (1.9)$$

$$= \beta_m s_n(\tau_m) + R_m$$

where $s_{wr}(\tau_l)$ are residuals of the multipath signal components included in R_m. Now if a number of parallel $\epsilon_{\tau_i}^{-1}(\ )$ ($i = 0, \ldots, L-1$) operations are performed, all signal components can be separated and then combined coherently into a signal

$$\Sigma = \sum_{m=0}^{L-1} [w_m \beta_m s_n(\tau_m) + w_m R_m] \tag{1.10}$$

where w_m is the combining coefficient. If $\beta_l s_{wr}(\tau_l)$ are independent and if the signal power is P_s, then the residual R_m will have the power

$$P_{rm} = \frac{P_s}{G} \sum_{\substack{l=0 \\ l \neq m}}^{L-1} |\beta_l|^2 \tag{1.11}$$

It can be shown that for the maximum signal-to-noise ratio at the output of the combiner (maximum ratio combining), $w_m \cong \beta_m^*$. In order to get an initial insight into the system performance we will assume equal gain combining $w_m = 1$ and multipath propagation with $\beta_l = \beta$ so that

$$P_{rm} = P_s(L-1)|\beta|^2/G \tag{1.12}$$

Therefore, the signal-to-multipath noise ratio at the output of the combiner with L branches can be approximated as

$$SNR_L = \frac{P_s L^2 |\beta|^2}{L P_{rm}} = \frac{L^2 G}{L(L-1)} \cong G \qquad L \gg 1 \tag{1.13}$$

If only one signal component is demodulated we will have

$$SNR_1 = P_s \beta^2 / P_{rm} = G/(L-1) \tag{1.14}$$

Of course, this should be considered only as a rough indication of the line of reasoning in using multipath diversity combining in spread spectrum receiver. A detailed discussion of this topic will be presented later.

1.1.3 Code Division Multiple Access

In order to use bandwidth efficiently, a number of different spread spectrum signals should coexist in the same frequency band. The case when each receiver receives a sum of the signals can be represented as

$$\sum_k s_{wk} = \sum_k \epsilon_k(s_{nk}) \tag{1.15}$$

where index k corresponds to the kth user in the same frequency band and $\epsilon_k(\ )$ defines the spreading operation of user k. In order to be able to coexist, the $\epsilon(\ )$

operation should meet the subsequently defined additional requirement. Let us consider the despreading of signal s_{wj} in receiver i.

$$\epsilon_i^{-1}(s_{wj}) = \epsilon_i(\epsilon_j(s_{nj})) = \epsilon_{ij}(s_{nj}) = \begin{cases} s_{nj} & i = j \\ s_{wij} & i \neq j \end{cases} \quad (1.16)$$

In other words, the despreading operation will produce the narrowband signal as long as $i = j$ and the wideband signal s_{wij} as long as $i \neq j$. After bandpass filtering we have

$$F(\epsilon_{ij}(s_{nj})) = \begin{cases} s_{nj} & i = j \\ s_{rij} & i \neq j \end{cases} \quad (1.17)$$

which means that the $F(\epsilon_i^{-1}(\ ))$ operation will reproduce the original signal for $i = j$ and will produce only low-level interference s_{wij} for $i \neq j$. As in the previous cases, the power of s_{rij} is less than the power of s_{wij}. So one despreading of signal i in the presence of $K - 1$ other signals belonging to different users results in

$$\epsilon_i^{-1}\left(\sum_k s_{wk}\right) = s_{ni} + \sum_{\substack{k=1 \\ k \neq i}}^{K} s_{wik} \quad (1.18)$$

or

$$F\left(\epsilon_i^{-1}\left(\sum_k s_{wk}\right)\right) = s_{ni} + \sum_{\substack{k=1 \\ k \neq i}}^{K} s_{rik} \quad (1.19)$$

Hence, mutual separation of the signals is based on a low correlation between operations ϵ_i and ϵ_k $\forall$ i, k. As will be shown later, $\epsilon(\ )$ operations are controlled with different codes, hence the name code division multiple access.

1.1.4 Capacity of a CDMA Network

Now suppose that we have K signals of the same power P_s present in the same frequency band. At the input of any receiver signal to noise ratio is

$$\gamma = P_s/(K - 1)P_s = 1/(K - 1) \quad (1.20)$$

where the presence of Gaussian thermal noise has been ignored for the moment. After despreading each interfering component coming from another user, which will be suppressed by factor G, the signal-to-noise ratio becomes

$$y_b = Gy = \frac{G}{K-1} \qquad (1.21)$$

If the thermal noise power is α, then

$$y_b = \frac{P_s G}{P_s(K-1) + \alpha} \qquad (1.22)$$

and the system capacity becomes

$$K = 1 + G/y_b - \alpha/P_s \cong G/y_b \qquad (1.23)$$

where y_b is the signal-to-noise ratio required for the given bit error rate in the system. The system capacity depends on the level of interference that can be tolerated, hence the name *interference-limited system*. In order to reduce y_b as much as possible, powerful error correcting coding is used.

In the limiting case (Shannon limit), if enough coding redundancy is used the required signal-to-noise ratio for error-free transmission is less than one (−1.6 dB, see Figure 1.4), in which case K becomes larger than G. In practice, a compromise between coding gain and hardware complexity must be used so that convolutional coding with code rates $r = 1/2$ or $1/3$ and constraint length 7 or 9 are the most often suggested options. A more detailed discussion on capacity will be presented later in the book.

1.2 EXAMPLES OF SPREAD SPECTRUM SYSTEMS

1.2.1 Direct Sequence Spread Spectrum System (DSSS)

The most often used form of spreading is obtained if a narrowband PSK signal is directly multiplied by a pseudorandom (*pseudonoise* or PN) sequence, hence the name. Very often the PN sequence is referred to as code, so the notation c (code) for the sequence will be used. A PN sequence is an ordered stream of binary ones (+1) and zeros (−1) referred to as chips rather than bits. Bit is the name reserved for the elements in an information-carrying stream. For an efficient spreading the chip interval T_c is much smaller than the bit interval T_b. So, if the primary modulated signal is represented in the form

$$PSK \Rightarrow s_n(t) = b(t)\cos \omega_0 t \qquad (1.24)$$

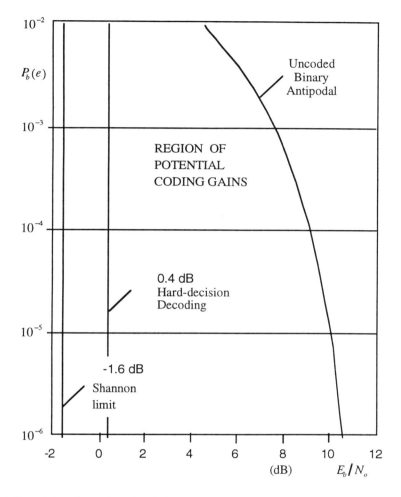

Figure 1.4 Potential coding gains of coded transmission with respect to binary uncoded antipodal transmission. (*From: Digital Transmission Theory* by S. Benedetto, E. Bigieri, and V. Castellani (Prentice Hall, 1987). Reprinted by permission of the authors.)

then $\epsilon(\ )$ represents multiplication of the PSK signal with code $c(t)$. It can be represented as

$$\epsilon(\ ) \Rightarrow s_w(t) = c(t) \cdot s_n(t) = c(t)b(t)\cos \omega_0 t \qquad (1.25)$$

The bandwidth of $s_w(t)$ is determined by $c(t)$ and is much higher than the bandwidth of $s_n(t)$ because $1/T_c \gg 1/T_b$. $\epsilon^{-1}(\ ) = \epsilon(\ )$, so again the despreading operation is multiplication by code $c(t)$. This time $s_w(t)$ is multiplied by the code and yields

$$\epsilon^{-1}(\ ) \Rightarrow c(t)s_w(t) = c^2(t)b(t)\cos \omega_0 t = b(t)\cos \omega_0 t = s_n(t) \qquad (1.26)$$

From the definition of the $\epsilon(\ )$ operation, one can see that for multipath separation $c(t) \cdot c(t-\tau)$ should remain wideband for $\tau > T_s$ (separation interval) and for multiple access capability $c_i(t) \cdot c_j(t-\tau)$ should remain wideband for all i, j, and τ. A direct extension to *quadratype phase shift keying* (QPSK) format can be obtained as

$$s_w(t) = c_1(t)b_1(t)\cos \omega_0 t - c_2(t)b_2(t)\sin \omega_0 t$$

where $c_1(t)$ and $c_2(t)$ are two different PN sequences and $b_1(t)$ and $b_2(t)$ two independent data streams. The complex representation of this signal is

$$s_w(t) = \text{Re}\{d(t)\exp(j\omega_0 t)\} \qquad (1.26a)$$

where

$$d(t) = d_r(t) + jd_i(t) = c_1(t)b_1(t) + jc_2(t)b_2(t)$$

In Figures 1.5 and 1.6 *binary phase shift keying* (BPSK)/DSSS and QPSK/DSSS systems are presented as illustrations. In practice product $c(t)b(t)$ in (1.25) can have additional components. For example, in the IS-95 standard, in order to maximize the system capacity $K = G/y_b$ by minimizing the required y_b for a given quality, $b(t)$ is encoded using convolutional codes and additionally modulated using orthogonal modulation with Walsh functions $W(t)$. So the equivalent data has form $b(t)*g(t)W(t)$ where $b(t)*g(t)$ stands for the encoded signal. In addition to this PN sequence, IS-95 is composed of two components, long code $lc(t)$, to separate signals coming from different cells and improve randomness, and short code $sc(t)$, called pilot code, to improve synchronization. So in (1.2) the product $c(t)b(t)$ now becomes $lc(t)sc(t)[b(t)*g(t)]W(t)$.

The overall signal demodulation process in the BPSK/DS system can be implemented by multiplying the input signal with a corresponding sequence and coherent carrier and then integrating in the bit interval T_b. The equivalent representation of the process is shown in Figure 1.5. So if K coherent signals are simultaneously transmitted (e.g., downlink transmission in cellular network), then the decision variable for the kth receiver can be represented as

$$d_k = \int_0^{T_b} 2c_k\cos \omega_0 t \left[\sum_{i=1}^{K} c_i b_i \cos \omega_0 t\right] dt \qquad (1.27)$$

$$= b_k \int_0^{T_b} c_k c_k \, dt + \sum_{i \neq k} b_i \int_0^{T_b} c_i c_k \, dt = b_k T_b + \sum b_i \rho_{ik} T_b$$

Fundamentals of Spread Spectrum Systems 11

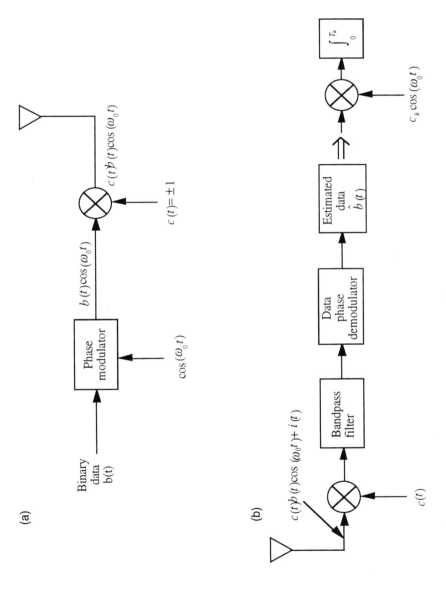

Figure 1.5 BPSK direct sequence spread spectrum (a) transmitter and (b) receivers.

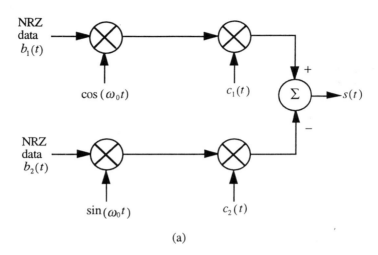

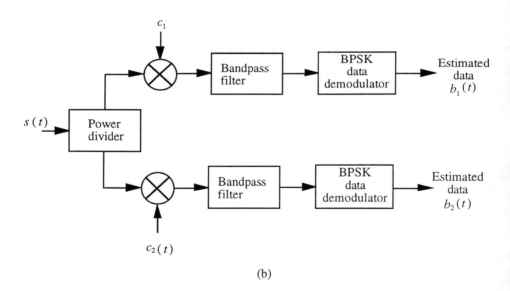

Figure 1.6 QPSK direct sequence spread spectrum (a) transmitter and (b) receiver.

where ρ_{ik} is the correlation function defined as

$$\rho_{ik} = \frac{1}{T_b}\int_0^{T_b} c_i c_k \, dt \qquad (1.28)$$

The second component in (1.27) is called *multiple access interference* (MAI) and should be kept as low as possible. One way to take care of it is to design a set of sequences with as low crosscorrelations as possible. We will address this problem within this section. Unfortunately there is a limit to what can be done in reducing the crosscorrelations within a given family of sequences. For this reason, additional methods for reducing MAI are discussed in Chapter 8.

1.2.2 Frequency Hopping Spread Spectrum System (FHSS)

In these systems for primary modulation *frequency shift keying* (FSK) (or *M-ary frequency shift keying* (MFSK)), a signal is used that can be represented as

$$\text{FSK} \Rightarrow s_n(t) = \cos(\omega_0 + b(t)\Delta\omega)t \qquad b(t) = \pm 1 \qquad (1.29)$$
$$= \cos(\omega_n t)$$

For the $\epsilon(\ )$ operation a frequency synthesizer is used that generates a continuous wave signal that changes its frequency in regular "hopping" intervals T_h that can be represented as

$$s_\epsilon(t) = \cos(\omega_1 + M(t, T_h)\Delta\omega)t = \cos \omega_\epsilon t \qquad (1.30)$$

where $M(t, T_h)$ is a random or pseudorandom number that can take different values from a set of size M. The $\epsilon(\ )$ operation consists of multiplying $s_n(t)$ and $s_\epsilon(t)$ and filtering out the signal $s_w(t)$ with frequency $\omega_n + \omega_\epsilon = \omega_t = \omega_0 + \omega_1 + [b(t) + M(t, T_h)]\Delta\omega$. Sometimes we refer to this process as the signal mixing up procedure. The frequency ω_t has a fixed component $\omega_0 + \omega_1 = \omega_c$ and variable component $[b(t) + M(t, T_h)]\Delta\omega$. So if $b \in [-1, 1]$ and $M(t, T_h) \in [0, 1, \ldots, M]$, $s_w(t)$ occupies bandwidth $B_w \rightarrow$ *from* $(\omega_c - \Delta\omega)$ *to* $\omega_c + (M + 1)\Delta\omega = (M + 2)\Delta\omega \cong M\Delta\omega$ for large M. If the bit interval $T_b > T_h$, we have fast hopping; if $T_h > T_b$, the system is referred to as slow hopping. The despreading operation is again the same. Multiplying signals $s_w(t)$ and locally generated signal $s_\epsilon(t)$ and filtering out the signal with frequency $\omega_t - \omega_\epsilon = \omega_n$ results into the original narrowband FSK signal.

If spreading and despreading operations are not synchronized, the random numbers at the transmitter $M_t(t, T_h)$ and the receiver $M_t(t - \tau, T_h) = M_r(t, T_h)$ will be different if $\tau > T_h$. So, the despreading operation will produce a signal of the frequency

$$\omega_t(t) - \omega_\epsilon(t - \tau) = \omega_o + [M_t(t, T_h) - M_r(t, T_h) + b(t)]\Delta\omega \qquad (1.31)$$
$$= \omega_o + [M_\epsilon(t, T_h) + b(t)]\Delta\omega$$

where $M_\epsilon(t, T_h) \in [-M, -(M-1), \ldots, 0, 1, \ldots, (M-1), M]$ and the resulting signal remains wideband.

For the multiple access application we have K signals and each spreading operation $\epsilon_k(\ )$, $k = 1, 2, \ldots, K$, uses a different number $M_k(t, T_h)$ to control its frequency synthesizer. The resulting signal is given by the sum

$$\sum_k \cos[\omega_c + (b_k + M_k)\Delta\omega]t \qquad (1.32)$$

where the time dependence of b_k and M_k is dropped for notation simplicity. When the signal of the user j is present in the despreading process in receiver i, the result is a signal of frequency $\omega_o + (M_j - M_i)\Delta\omega$, which remains wideband for as long as $M_i \neq M_j$ because $\Delta M = M_j - M_i \in [-M, (M-1), \ldots, 0, 1, \ldots, (M-1), M]$. Thus, for good multiple access performance in FH systems, the main challenge is to generate a set of random numbers M_i, in such a way that the number of time slots in which two or more of these numbers simultaneously have the same value is minimized. A block diagram of the FH transmitter and receiver is shown in Figure 1.7.

1.2.3 Hybrid DS/FH Schemes

In order to achieve spreading in as wide a frequency bandwidth as possible, sometimes, specially for antijamming applications, a combination of DS and FH systems is used in a so-called hybrid DS/FH configuration. A block diagram of this system is shown in Figure 1.8.

1.2.4 Chirp Signals for Pulse Compression

Although DS and FH signals are the most popular forms of spread spectrum signals, we should point out that some other forms of signal spreading are also used. One of these forms is the so-called chirp signal used in radar technology for pulse compression. The idea is to transmit a constant envelope signal with linearly sweeping (chirp) frequency, and at the receiver a matched filter will produce a signal of the same energy but in a much shorter time interval. This solves the problem of having limited output peak power in radar systems and having to receive weak signals from very

Fundamentals of Spread Spectrum Systems 15

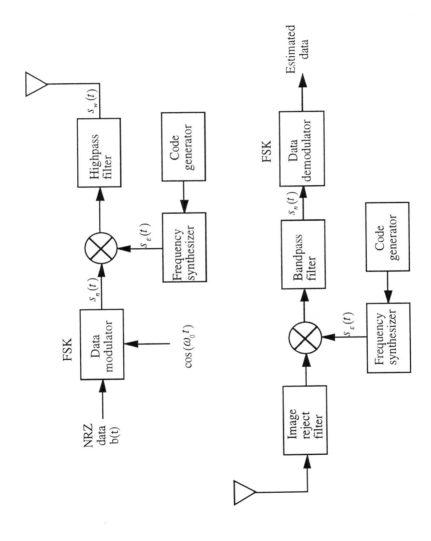

Figure 1.7 Frequency hop spread spectrum modem.

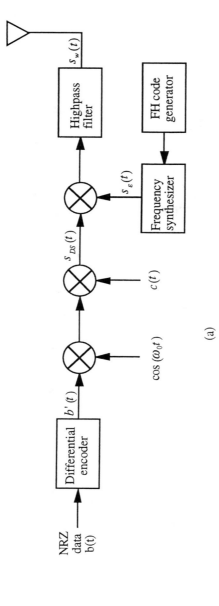

Figure 1.8 Hybrid direct sequence/frequency-hop spread spectrum (a) transmitter and (b) receiver.

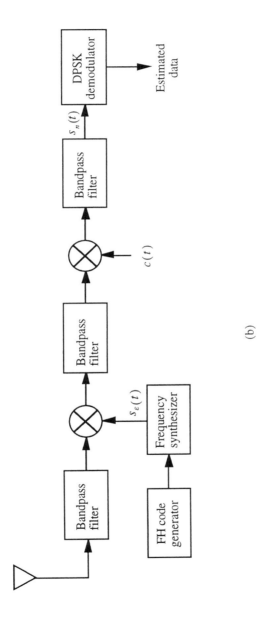

Figure 1.8 (continued).

close targets. More material covering this problem can be found in any textbook on radar technology. The reason we are mentioning this at all is that spread spectrum, first chirp signals, and later DS and FH (frequency agility) are also used in radar technology for pulse compression, jamming, and multireflection (clutter) reduction purposes. Spread spectrum communications owe a lot to the radar (ranging in general including deep space ranging) technology.

1.3 SEQUENCE GENERATION

1.3.1 Correlation Functions

As the first step we need to establish the relationship between a continuous and discrete correlation of a digital sequence $c(t)$. For this purpose let a code $c(t)$ be represented as

$$c(t) = \sum_{n=-\infty}^{\infty} c_n p(t - nT_c)$$

where $c_n = \pm 1$ is a bipolar sequence related to its unipolar version $a_n \in [0, 1]$ as $c_n = (-1)^{a_n}$ and $p(t)$ is a rectangular pulse shape in the chip interval $[0, T_c]$. So for two different codes $c(t)$ and $c'(t)$ the continuous crosscorrelation function is defined as

$$R_{cc'}(\tau) = \frac{1}{T} \int_0^T c'(t) c(t + \tau) \, dt$$

where $T = NT_c$ is the sequence period. For $c = c'$ it becomes the autocorrelation function $R_{cc}(\tau)$. When the two last equations are combined we have

$$R_{cc'}(\tau) = \frac{1}{T} \sum_m \sum_n c_n c'_m \int_0^T p(t - mT_c) p(t + \tau - nT_c) \, dt$$

This integral has nonzero value only when the two pulses overlap, which happens in the range $t - mT_c - (t + \tau - nT_c) < T_c$ because $p(t)$ is limited to the time interval $[0, T_c]$. So, if we use notation $\tau = kT_c + \epsilon$ for $(0 < \epsilon < T_c)$ and $T = NT_c$, the previous relation becomes

$$R_{cc'}(\tau) = R_{cc'}(k, \epsilon)$$

$$= \frac{1}{N} \sum_{m=0}^{N-1} c_m c'_{k+m} \frac{1}{T_c} \int_0^{T_c-\epsilon} p(\lambda) p(\lambda + \epsilon) \, d\lambda$$

$$+ \frac{1}{N} \sum_{m=0}^{N-1} c_m c'_{k+m+1} \frac{1}{T_c} \int_{T_c-\epsilon}^{T_c} p(\lambda) p(\lambda - T_c + \epsilon) \, d\lambda$$

The discrete correlation function is defined as

$$\theta_{cc'}(k) = \frac{1}{N} \sum_{n=0}^{N-1} c_n c'_{n+k}$$

Then $R_{cc'}(\tau)$ and $\theta_{cc'}(k)$ are related as

$$R_{cc'}(\tau) = R_{cc'}(k, \epsilon) = \left(1 - \frac{\epsilon}{T_c}\right) \theta_{cc'}(k) + \frac{\epsilon}{T_c} \theta_{cc'}(k+1)$$

and

$$R_c(\tau) = R_{cc}(\tau) = \left(1 - \frac{\epsilon}{T_c}\right) \theta_c(k) + \frac{\epsilon}{T_c} \theta_c(k+1)$$

In what follows we will present different classes of sequences and discuss their performance from the point of view of θ_c and $\theta_{cc'}$ as defined in the previous text.

1.3.2 Linear Recursions

If each element a_n of sequence $\{a_n\}$ can be calculated from its previous L elements as

$$a_n = -\sum_{i=1}^{L} g_i a_{n-i} \quad \text{for all } n \tag{1.33}$$

then it is said that the sequence satisfies linear recursion. Elements a_n and coefficients g_i are from the same field $\mathfrak{F}$. Most of the time field $\mathfrak{F}$ will be the Galois field of two elements (0, 1) and called the binary field, with standard notation $GF(2)$.

The generator structure containing L memory elements is given in Figure 1.9 and is often called the Fibonacci configuration, after the mathematician who studied linear recursions. The degree of the minimum-degree recursion that generates $\{a_n\}$ is called the linear span of the sequence. Using the formal power series representation of sequence $\{a_n\}$,

$$A(z) = \sum_{n=1}^{\infty} a_n z^{-n} \qquad (1.34)$$

By multiplying Eq. (1.33) by z^{-n}, introducing $g_0 = 1$, and summing over i for all n and after some manipulation we have

$$0 = \sum_{i=0}^{L} g_i a_{n-i} \Rightarrow 0 = \sum_{n=L+1}^{\infty} z^{-n} \sum_{i=0}^{L} g_i a_{n-i} = \sum_{i=0}^{L} g_i z^{-i} \sum_{n=L+1}^{\infty} a_{n-i} z^{-(n-i)} \qquad (1.35)$$
$$= G(z)A(z) - D(z)$$

where

$$G(z) = \sum_{i=1}^{L} g_i z^{L-i} \qquad (1.35a)$$

$$D(z) = \sum_{n=1}^{L} d_n z^{L-n} \qquad (1.35b)$$

$$d_n(z) = \sum_{i=0}^{n-1} g_i a_{n-i} \qquad n = 1, 2, \cdots, L \qquad (1.35c)$$

So, (1.35) can be written in the form

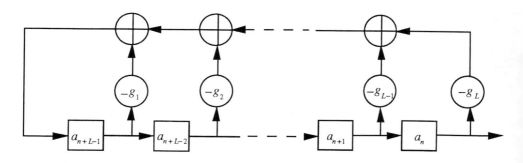

Figure 1.9 An L-stage linear feedback shift register in Fibonacci form.

$$A(z) = D(z)/G(z) \tag{1.36}$$

where coefficients of $G(z)$ are the weights used in the linear recursion and coefficients of $D(z)$ are related to the initial elements $a_1, a_2, \ldots, a_L$ by L independent equations. If N is the sequence period we have

$$a_n = a_{n+mN} \Rightarrow A(z) = \sum_{m=0}^{\infty} \sum_{n=1}^{N} a_n z^{-(mN+n)} = \sum_{m=0}^{\infty} (z^{-N})^m \sum_{n=1}^{N} a_n z^{-n} \tag{1.37}$$

$$= \frac{1}{z^N - 1} \sum_{n=1}^{N} a_n z^{N-n}$$

Dividing both the numerator and the denominator polynomial by their greatest common divisor yields

$$\gcd(z) = \gcd\left[\sum_{n=1}^{N} a_n z^{N-n}, z^N - 1\right]$$

Equation (1.37) can be expressed as

$$A(z) = \frac{P(z)}{Q(z)} \tag{1.38}$$

where $\gcd[P(z), Q(z)] = 1$.

With the proper normalization to have the coefficient of the highest power of z in $Q(z)$ unity, a "monic" form of $Q(z)$ is obtained. So, in this way a rational representation of $A(z)$ is obtained with the lowest possible degree of $Q(z)$; hence, the linear span L of $\{a_n\}$ is given by

$$L = \deg Q(z) \tag{1.39}$$

and $\{a_n\}$ satisfies the linear recursion equations (1.33) and (1.35a) with $g_i = q_i$ and $G(z) = Q(z)$. Such $Q(z)$ is called the *characteristic polynomial* of the sequence $\{a_n\}$ and the corresponding numerator polynomial $P(z)$ is called the initial condition polynomial of $\{a_n\}$.

From the previous discussion we can see that the period of the sequence $\{a_n\}$ generated with the characteristic polynomial $Q(z)$ is the smallest integer N such that $Q(z)$ divides $z^N - 1$. From this discussion we could see that the generator structure is defined by the polynomial $Q(z)$. If $Q(z)$ can be factored in the form

$$Q(z) = \prod_{j=1}^{J} Q_j^{p_j}(z) \tag{1.40}$$

where p_j is degree of polynomial Q_j, then relation (1.38) becomes

$$A(z) = \frac{P(z)}{\prod_{j=1}^{J} Q_j^{p_j}(z)} = \sum_{j=1}^{J} \frac{P_j(z)}{Q_j^{p_j}(z)} = \sum_{j=1}^{J} A_j(z) \qquad (1.41)$$

This means that sequence $\{a_n\}$ can be realized as a sum of J sequences $\{a_n\}$ generated with simpler characteristic polynomials Q^{p_j} and corresponding initialization $P_j(z)$. The overall linear span of such a sequence is

$$L = \sum_{j=1}^{J} p_j \deg Q_j(z) \qquad (1.42)$$

1.3.3 Sequence Generation by Using LFSR Long Division

If we go back to (1.38) and perform the first step in this formal long division we get

$$
\begin{array}{r|ll}
Q(z) & P(z) & = p_1 z^{-1} + \cdots \\
(z^L + q_1 z^{L-1} + \cdots + q_L) & p_1 z^{L-1} + p_2 z^{L-2} + \cdots + p_L & \\
p_1 z^{-1} Q(z) & = p_1 z^{L-1} + q_1 p_1 z^{L-2} + \cdots + q_{L-1} p_1 + q_L p_1 z^{-1} & \\
P(z) - p_1 z^{-1} Q(z) & = \quad r_1^{(1)} z^{L-2} + \cdots + r_{L-1}^{(1)} + r_L^{(1)} z^{-1} = z^{-1} R^{(1)}(z) &
\end{array}
\qquad (1.43)
$$

and the *linear feedback shift register* (LFSR) circuit implementation is given in Figure 1.10. In each next step of the long division process, the previous step's remainder is divided by $Q(t)$ and the quotient representing the next element in $\{a_n\}$

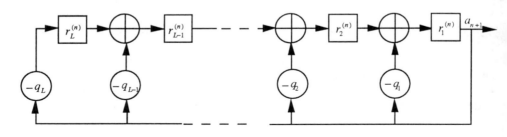

Figure 1.10 A minimum-memory LFSR in the Galois configuration for generating $\{a_n\}$.

and a new remainder are produced. By carrying out operation (1.43) n times we will end up with

$$z^{-n}R^{(n)}(z) = \sum_{i=1}^{L} r_i^{(n)} z^{L-i-n} \qquad n = 1, 2, \ldots \qquad (1.44)$$

on the right-hand side of (1.43) or with the result of the form

$$z^{-n}[R^{(n)}(z) - r_1^{(n)} z^{-1} Q(z)] = z^{-n-1} R^{(n+1)}(z) \qquad (1.45)$$

which can be transformed into

$$zR^{(n)}(z) - r_1^{(n)} Q(z) = R^{(n+1)}(z) \qquad (1.46)$$

which starts with

$$R^{(0)}(z) = P(z) \qquad (1.47)$$

producing

$$a_n = r_1^{(n-1)} \qquad (1.48)$$

1.3.4 Sequence Representation by Using Elements of an Extension Galois Field

Let $m_\alpha(z)$ be an irreducible polynomial of degree less than L over $GF(q)$ with root α from an extension Galois field $GF(q^L)$. $m_\alpha(z)$ is called the minimum polynomial of α over $GF(q)$. For given L and q we can have a set S_L of such polynomials and the size of the set is q^L

$$S_L = \left\{ x(z) : x(z) = \sum_{i=0}^{L-1} x_i z^i, \; x_i \in GF(q) \right\} \qquad (1.49)$$

It is well known that the elements of $GF(q^L)$ can be represented by the value of the polynomials from S_L evaluated at a root of an irreducible polynomial of degree L over $GF(q)$. So, the field element x from $GF(q^L)$ corresponding to the polynomial $x(z)$ in S_L can be represented as

$$x = x(\alpha) = \sum_{i=0}^{L-1} x_i \alpha^i \qquad (1.50)$$

In the previous section the contents of a shift register after n shifts were considered to be coefficients of a polynomial $R^{(n)}(z)$ of degree L, and successive register contests were related by the polynomial recursion defined by (1.46) to (1.48).

Suppose that $Q(z) = m_\alpha(z)$, that is, the characteristic polynomial of the sequence $\{c_n\}$ produced by the circuitry from Figure 1.10 is irreducible over $GF(q)$ and has root α in $GF(q^L)$. Bearing in mind that $m_\alpha(\alpha) = 0$, (1.46) for $z = \alpha$ becomes

$$\alpha R^{(n)}(\alpha) = R^{(n+1)}(\alpha) \tag{1.51}$$

Solving the recursion in the finite field gives

$$R^{(n)}(\alpha) = \alpha^n P(\alpha) \tag{1.52}$$

The smallest value of $n = N$ for which $\alpha^N = 1$ is called the exponent (or order) of α end is equal to $q^L - 1$ for as long as the elements of $GF(q^L)$ are generated by the primitive polynomials.

1.3.5 *m*-Sequences

From (1.52) one can see that if $Q(z) = m_\alpha(z)$ is a primitive polynomial after $n = N = q^L - 1$ iterations, $R^{(n)}(\alpha) = P(\alpha)$, which is the initial state of the sequence. This means that the sequence period is N. On the other hand, the maximum possible number of different states of the LFSR is q^L. One should be aware that ones in the all-zeros state LFSR could not change the state, so that this state is not allowed. This means that the maximum number of different states of LFSR generating a sequence is $q^L - 1$. Such a sequence of maximum possible length, called *m*-sequence, is generated if the generating (characteristic) polynomial is a primitive polynomial.

The number of cyclically distinct *m*-sequences over $GF(q)$ with linear span L is equal to the number $N_p(L)$ of primitive polynomial of degree L over $GF(q)$ and is given by

$$N_p(L) = \frac{q^L - 1}{L} \prod_{i=1}^{I} \frac{p_i - 1}{p_i} \tag{1.53}$$

where the I prime numbers p_i, $i = 1, 2, \ldots, I$, are determined from the prime decomposition of $q^L - 1$ that can be represented as

$$q^L - 1 = \prod_{i=1}^{I} p_i^{e_i} \tag{1.54}$$

and exponents e_i, $i = 1, \ldots, I$, are positive integers. For example, for $L = 5, 6$ and $q = 2$ we have

$$L = 5 \Rightarrow 2^5 - 1 = 31 \Rightarrow N_p(5) = \frac{31}{5} \cdot \frac{30}{31} = 6$$

$$L = 6 \Rightarrow 2^6 - 1 = 3^2 \cdot 7 \Rightarrow N_p(6) = \frac{63}{6} \cdot \frac{2}{3} \cdot \frac{6}{7} = 6$$

Statistical properties of *m*-sequences are discussed in many books [2], so to avoid repetition we only list some of them for binary *m*-sequences.

Balance property: For a binary sequence, the number N_1 of times that the symbol "1" occurs in one period of an *m*-sequence exceeds the number N_0 of times that "0" occurs in one period by one, that is, $N_1 = N_0 + 1$. This is a direct consequence of the fact that the all-zeros state of the LFSR is not allowed.

The shift and add property: If an *m*-sequence is added to its time-shifted replica, the result is the same *m*-sequence with a new shift. Using (1.38) an *m*-sequence represented by $A(z)$ and its delayed replica represented by $A'(z)$ after mod 2 summation will produce a sequence represented by

$$A''(z) = A(z) + A'(z) = \frac{1}{Q(z)}(P(z) + P'(z)) = \frac{P''(z)}{Q(z)} \quad (1.55)$$

where $P''(z)$ will represent again one of $2^L - 1$ possible initial states of the generator that will again produce the same sequence with a different shift.

Correlation properties: In general, when the field size q is a prime, the following mapping from a_n in $GF(q)$ to a complex qth root of unity can be constructed

$$c_n = \rho^{a_n} \quad (1.56)$$

where ρ is a primitive qth root of unity in the field of complex numbers

$$\rho = \exp(j2\pi/q) \quad (1.57)$$

For $q = 2 \Rightarrow \rho = -1$, a binary unipolar (0, 1) is mapped into a binary bipolar (±1) sequence. The autocorrelation function $\theta_{cc}(k)$ is defined in the general case as

$$\theta_{cc}(k) = \sum_{n=1}^{N} c_{n+k} c_n \quad (1.58)$$

It is obvious that for $k = 0$, this expression has value N; for $k \neq 0$ it becomes

$$\theta_{cc}(k) = \sum_{n=1}^{N} c_{n+k}c_n$$
$$= \sum_{n=1}^{N} (-1)^{a_{n+k}+a_n} = \sum_{n=1}^{N} (-1)^{a_{n+k'}} \leftarrow \text{due to shift and add property} \quad (1.59)$$
$$= N_0(-1)^0 + N_1(-1)^1 = N_0 - N_1 = -1 \leftarrow \text{due to balance property}$$

So, we can write

$$\theta_{cc}(k) = \begin{cases} N & k = 0 \\ -1 & k \neq 0 \end{cases} \quad (1.60)$$

1.3.6 The Trace Representation of m-Sequences

For any finite field $GF(q)$ within a larger field $GF(Q)$, where $Q = q^L$, the trace polynomial

$$\text{Tr}_q^Q(z) \triangleq \sum_{i=0}^{L-1} z^{q^i} \quad (1.61)$$

defines the mapping (called trace function) from the larger field $GF(Q)$ into the subfield $GF(q)$. For an element α in $GF(Q)$, the trace in $GF(q)$ is the trace polynomial evaluated at α, that is, $\text{Tr}_q^Q(\alpha)$. Using (1.52) we have

$$\alpha^n = P^{-1}(\alpha)R^{(n)}(\alpha) = \beta \sum_{i=1}^{L} r_i^{(n)} \alpha^{L-i} \quad (1.62)$$

where $\beta = P^{-1}(\alpha)$ with α and β in $GF(Q)$. Applying the trace function to both sides of (1.62) and simplifying notation $\text{Tr}_q^Q(\alpha) \Rightarrow \text{Tr}$ we have

$$\text{Tr}(\alpha^n) = \sum_{i=1}^{L} \text{Tr}(\beta \alpha^{L-i}) r_i^{(n)} \quad (1.63)$$

Bearing in mind that $\text{Tr}(\ ) \in GF(q)$ and $r_i^{(n)}$ can be considered as a delayed version of an m-sequence, by the shift and add property (1.63) represents an m-sequence with an arbitrary shift m; so, in general, we can write

$$\{\text{Tr}(\alpha^{m+n})\} = \{a_n\} \quad (1.64)$$

1.3.7 Decimation of Sequences

If for an integer I we create a new sequence $\{d_n\}$ from sequence $\{a_n\}$ as

$$d_n = a_{In} \quad \text{for all } n \tag{1.65}$$

then the sequence $\{d_n\}$ is said to be the decimation by I of the sequence $\{a_n\}$. In other words, the nth bit of the sequence $\{d_n\}$ is the Inth bit of the sequence $\{a_n\}$. If period N of $\{a_n\}$ and I are relatively prime, then the period of $\{d_n\}$ is also N. Let us look at the correlation function of $\{d_n\}$. By definition we have

$$\theta_{dd}(\tau) = \sum_{n=1}^{N} a_{I(n+\tau)} a_{In} = \sum_{k} a_{k+I\tau} a_k = \theta_{aa}(I\tau) \tag{1.66}$$

where k is defined from $In = k \mod N$. Using the trace representation of these sequences we have

$$\{a_n\} = \text{Tr}(\alpha^n) \tag{1.67}$$

and the decimation by I of $\{a_n\}$ results in the sequence

$$\{d_n\} = \text{Tr}\{\alpha^{In}\} \tag{1.68}$$

where α^I is another primitive element of $GF(Q)$ and thus, $\{d_n\}$ is an m-sequence with the minimum polynomial $m_{\alpha^I}(z)$ for its characteristic polynomial.

1.3.8 Crosscorrelation of Binary m-Sequences

Let $\{a_n\} = \{\text{Tr}_2^Q(\alpha^n)\}$ be an m-sequence with $Q = 2^L$ and $\{c_n\}$ its bipolar version $c_n = (-1)^{a_n}$. The periodic crosscorrelation between $\{c_n\}$ and its rth decimation $\{c_n'\} = \{c_{rn}\}$ is given as

$$\theta_{cc'}(\tau) = \sum_{n=1}^{2^L-1} c_{n+\tau} c_{rn} = \sum_{n=1}^{2^L-1} (-1)^{a_{n+\tau} + a_{rn}} = \sum_{n=1}^{2^L-1} (-1)^{\text{Tr}(\alpha^{n+\tau}) + \text{Tr}(\alpha^{rn})} \tag{1.69}$$

since α is primitive, α^n scans through the nonzero elements of $GF(Q)$ as n varies. If we add and substract the term obtained for the zero element of the field we have

$$\theta_{cc'}(\tau) = -1 + \sum_{x \in GF(Q)} (-1)^{\text{Tr}(xy+x^r)} = -1 + \Delta_r(y) \tag{1.70}$$

where $y = \alpha^\tau$. In what follows we will be interested in a special case where

$$r = 2^k + 1 \qquad e = \gcd(2k, L) \qquad (1.71)$$

where e divides k. In order to have r and $N = 2^L - 1$ relatively prime we need $L/\gcd(k, L)$ to be odd. Let us start with evaluating the expression

$$[\Delta_r(y)]^2 = \sum_{x,z \in GF(Q)} \sum (-1)^{\text{Tr}[y(x+z)+x^r+z^r]} \qquad (1.72)$$

By introducing $z = w + x$ and bearing in mind the form of r we have

$$[\Delta_r(y)]^2 = \sum_{x,w \in GF(Q)} \sum (-1)^{\text{Tr}[yw+x^r+(w+x)(w+x)^{2^k}]} \qquad (1.73)$$

$$= \sum_{x,w \in GF(Q)} \sum (-1)^{\text{Tr}[yw+x^r+(w+x)(w^{2^k}+x^{2^k})]}$$

Due to the trace linearity and the fact that wx^{2^k} and $xw^{2^{L-k}}$ have the same trace since they are conjugates, we have

$$[\Delta_r(y)]^2 = \sum_{w \in GF(Q)} (-1)^{\text{Tr}(yw+w^r)} \sum_{x \in GF(Q)} (-1)^{\text{Tr}[x(w^{2^k}+w^{2^{L-k}})]} \qquad (1.74)$$

Now when x varies over $GF(Q)$, the trace in the inner sum is zero and one equally often as long as $\beta = w^{2^k} + w^{2^{L-k}} \neq 0$. So the sum over x equals zero if $\beta \neq 0$ or 2^L if $\beta = 0$. Condition $\beta = 0$ can be further modified to

$$w^{2^k} + w^{2^{L-k}} = 0$$

$$w^{2^k} = w^{2^L} \cdot w^{2^{-k}} = w^{2^L-1+1} \cdot w^{2^{-k}} = w \cdot w^{2^{-k}} \qquad (1.75)$$

$$w^{2^{2k}-1} = 1$$

which implies $w \in GF(2^{2k})$ together with the initial condition $w \in GF(2^L)$. So (1.74) can be written as

$$[\Delta r(y)]^2 = 2^L \sum_{w \in \Omega} (-1)^{\text{Tr}(yw+w^r)} \qquad (1.76)$$

where

$$\Omega = \{w : w \in GF(2^{2k}), w \in GF(2^L)\} \tag{1.77}$$

The intersection of $GF(2^{2k})$ and $GF(2^L)$ is $GF(2^e)$ where $e = \gcd(2k, L)$. The order of every w in $GF(2^e)$ divides $2^e - 1$ and, bearing in mind the initial assumption that e divides k, we have

$$w^r = w^{(2^e)^c} \cdot w = w^2 \tag{1.78}$$

where c is an integer. Since the trace of w and w^2 are identical, (1.76) becomes

$$[\Delta_r(y)]^2 = 2^L \sum_{w \in GF(2^e)} (-1)^{\operatorname{Tr}_2^{2^e}\{[\operatorname{Tr}_{2^e}^{2^L}(y) + \operatorname{Tr}_{2^e}^{2^L}(1)]w\}} \tag{1.79}$$

due to trace property TP.5 (see Appendix A1). On the other hand, due to TP.4, $\operatorname{Tr}_{2^e}^{2^L}(y + 1)$ is zero for 2^{L-e} values of y, so

$$[\Delta_r(y)]^2 = 2^{L+e} \quad \text{for } 2^{L-e} \text{ values of } y \tag{1.80}$$

When $\operatorname{Tr}_{2^e}^{2^L}(y + 1)$ is nonzero, the outer trace in (1.79) is zero and one equally often as w scans through $GF(2^{2e})$, so

$$[\Delta_r(y)]^2 = 0 \quad \text{for } 2^L - 2^{L-e} \text{ values of } y \tag{1.81}$$

Combining (1.70) with (1.80) and (1.81) yields

$$\theta_{cc'}(\tau) = \begin{cases} -1 & \text{for } 2^L - 2^{L-e} - 1 \text{ values of } \tau \\ -1 + 2^{(L+e)/2} & \text{for } 2^{L-e-1} + 2^{(L-e-2)/2} \text{ values of } \tau \\ -1 - 2^{(L+e)/2} & \text{for } 2^{L-e-1} - 2^{(L-e-2)/2} \text{ values of } \tau \end{cases} \tag{1.82}$$

1.3.9 Composite Binary Sequences

For two binary sequences $\{c_n\}$ and $\{c'_n\}$ obtained from unipolar sequences $\{a_n\}$ and $\{a'_n\}$ having linear span L and L', respectively, the correlation function is given by

$$\theta_{cc'}(\tau) = \sum_{n=1}^{N} (-1)^{a_{n+\tau} + a'_n} \tag{1.83}$$

where $N = \operatorname{lcm}(2^L - 1, 2^{L'} - 1)$. If we now create a new family of sequences $\{a_n^{(j)}\}$ where

$$a_n^{(j)} = a_{n+j} + a'_n \tag{1.84}$$

the crosscorrelation function between two sequences $\{a_n^{(j)}\}$ and $\{a_n^{(k)}\}$ will be

$$\begin{aligned}\theta_{jk}(\tau) &= \sum_{n=1}^{N} (-1)^{a_{n+\tau}^{(j)} + a_n^{(k)}} \\ &= \sum_{n=1}^{N} (-1)^{a_{n+j+\tau} + a'_{n+\tau} + a_{n+k} + a'_n} \\ &= \sum_{n=1}^{N} (-1)^{a_{n+k+\tau'(j+\tau-k)} + a'_{n+\tau''(\tau)}}\end{aligned} \tag{1.85}$$

If $\tau + j - k \neq 0 \bmod(2^L - 1)$ & $\tau \neq 0 \bmod(2^{L'} - 1)$, we have

$$\theta_{jk}(\tau) = \theta_{cc'}(k + \tau'(\tau + j - k) - \tau''(\tau)) \tag{1.86}$$

where $\theta_{cc'}(\tau)$ is given by (1.83).

If $\tau + j - k \neq 0 \bmod(2^L - 1)$ & $\tau = 0 \bmod(2^{L'} - 1)$, a'_n cancels in (1.85) and we have

$$\theta_{jk}(\tau) = \sum_{n=1}^{N} (-1)^{a_{n+k+\tau'(\tau+j-k)}} = -N/(2^L - 1) \tag{1.87}$$

If $\tau + j - k = 0 \bmod(2^L - 1)$ & $\tau \neq 0 \bmod(2^{L'} - 1)$, we have

$$\theta_{jk}(\tau) = \sum_{n=1}^{N} (-1)^{a'_{n+\tau''(\tau)}} = -N/(2^{L'} - 1) \tag{1.88}$$

If $\tau + j - k = 0 \bmod(2^L - 1)$ & $\tau = 0 \bmod(2^{L'} - 1)$,

$$\theta_{jk}(\tau) = N \tag{1.89}$$

This occurs only when $j = k$ & $\tau = 0$. The size of the set of different sequences $\{a_n^{(j)}\}$ is $N_d = \gcd(2^L - 1, 2^{L'} - 1)$.

1.3.10 Gold Sequences

We will now use results obtained in Sections 1.3.8 and 1.3.9 to discuss a special case where two binary m-sequences $\{a_n\}$ and $\{a'_n\}$ with the same linear span L, generated by characteristic polynomials $m_\alpha(z)$ and $m_{\alpha'}(z)$ (decimation by r), form a new composite sequence in accordance with definition (1.84). For the special case where

$$r = 2^{[(L+2)/2]} + 1 \tag{1.90}$$

we have Gold sequences, where $[x]$ stands for integer part of x. This choice of r satisfies condition (1.71) with

$$e = \begin{cases} 1 & L \text{ odd} \\ 2 & L \text{ even} \end{cases} \tag{1.91}$$

Under these conditions, (1.82) becomes

$$\theta_{cc'} = \begin{cases} -1 \\ -1 + 2^{[(L+2)/2]} \\ -1 - 2^{[(L+2)/2]} \end{cases} \tag{1.92}$$

Gold's design provides the set of sequences of size $2^L + 1$, including the component sequences.

1.3.11 The Small Set of Kasami Sequences

Composite sequences obtained by combining an m-sequence of period $2^L - 1$, L even, and characteristic polynomial $m_\alpha(z)$, $\alpha \in GF(2^L)$, and another m-sequence with period $2^{L/2} - 1$ and characteristic polynomial $m_{\alpha'}(z)$ where decimation parameter r is given as $r = 2^{L/2} + 1$. α' being an element of $GF(2^{L/2})$ are referred to as the small set of Kasami sequences. Using the same procedure as in Section 1.3.8, for this new set of parameters, and results from Section 1.3.9 one can show that the crosscorrelation over $2^L - 1$ elements is

$$\theta_{cc'} = \begin{cases} -1 \\ -1 + 2^{L/2} \\ -1 - 2^{L/2} \end{cases}$$

Adding the long period m-sequence to this set gives $2^{L/2}$ sequences.

References

[1] Glisic, S., and P. Leppänen (Eds.), *Code Division Multiple Access Communications*, Boston, MA: Kluwer, 1995.
[2] Simon, M., et al., *Spread Spectrum Communications Handbook*, New York: McGraw-Hill, 1994.

Selected Bibliography

Alltop, W. O., "Complex Sequences with Low Periodic Correlations," *IEEE Trans. on Information Theory*, Vol. IT-26, May 1980, pp. 350–354.

Antweiler, M., and L. Bömer, "Complex Sequences Over $GF(p^m)$ with a Two-Level Autocorrelation Function and Large Linear Span," *IEEE Trans. on Information Theory*, Vol. IT-38, Jan. 1992, pp. 120–130.

Blake, I. F., and J. Mark, "A Note on Complex Sequences with Low Correlations," *IEEE Trans. on Information Theory*, Vol. IT-28, Sept. 1982, pp. 814–816.

Boztas, S., R. Hammons, and P. V. Kumar, "4-Phase Sequences with Near-Optimum Correlation Properties," *IEEE Trans. on Information Theory*, Vol. IT-38, May 1992, pp. 1101–1113.

Glisic, G. S., "Power Density Spectrum of the Product of Two Time Displaced Versions of a Maximum Length Binary Pseudonoise Signal," *IEEE Trans. on Communications*, Vol. COM-31, Feb. 1983, pp. 281–286.

Gold, R., "Maximal Recursive Sequences for Spread Spectrum Multiplexing," *IEEE Trans. on Information Theory*, Vol. IT-14, Jan. 1968, pp. 154–156.

Gold, R., "Optimal Binary Sequences for Spread Spectrum Multiplexing," *IEEE Trans. Information Theory*, Vol. IT-13, Oct. 1967, pp. 619–621.

Golomb, S. W., *Shift Register Sequences*, San Francisco: Holden-Day, 1967.

Krone, S. M., and D. V. Sarwate, "Quadriphase Sequences for Spread-Spectrum Multiple Access Communications," *IEEE Trans. on Information Theory*, Vol. IT-30, May 1984, pp. 520–529.

Kumar, P. V., "Frequency Hopping Code Designs Having Large Linear Span," *IEEE Trans. on Information Theory*, Vol. IT-34, Jan. 1988, pp. 146–151.

Kumar, P. V., and C. M. Liu, "Lower Bounds on the Maximum Correlation of Complex Roots-of-Unity Sequences," *IEEE Trans. on Information Theory*, Vol. IT-36, May 1990, pp. 633–640.

Kumar, P. V., and O. Moreno, "Prime-Phase Sequences with Periodic Correlation Properties Better than Binary Sequences," *IEEE Trans. on Information Theory*, Vol. IT-37, May 1991, pp. 606–616.

Kumar, P. V., and R. A. Scholtz, "Bounds on the Linear Span of Bent Sequences," *IEEE Trans. on Information Theory*, Vol. IT-29, Nov. 1983, pp. 854–862.

Kumar, P. V., and R. A. Scholtz, "Generalized GMW Sequences and Applications to Frequency Hopping," *Cryptologic Quart.*, Vol. 3, Spring/Summer 1984.

Kumar, P. V., R. A. Scholtz, and L. R. Welch, "Generalized Bent Functions and Their Properties," *J. Combin. Theory Ser. A*, Vol. 40, No.1, Sept. 1985, pp. 90–107.

Lempel, A., and H. Greenberger, "Families of Sequences with Optimal Hamming Correlation Properties," *IEEE Trans. on Information Theory*, Vol. IT-20, Jan. 1974, pp. 90–94.

Lempel, A., and M. Cohn, "Maximal Families of Bent Sequences," *IEEE Trans. on Information Theory*, Vol. IT-28, Nov. 1982, pp. 865–868.

Maric, S. V., and E. L. Titlebaum, "Frequency Hop Multiple Access Codes Based Upon the Theory of Cubic Congruences," *IEEE Trans. on Aerospace and Electronic Systems*, Vol. 25, Nov. 1990, pp. 1035–1039.

McEliece, R. J., *Finite Fields for Computer Scientists and Engineers*, Boston, MA: Kluwer, 1989.

Mersereau, R. M., and T. S. Seay, "Multiple Access Frequency Hopping Patterns with Low Ambiguity," *IEEE Trans. on Aerospace and Electronic Systems*, Vol. AES-17, 1981, pp. 571–578.

No, J. S., and P. V. Kumar, "A New Family of Binary Pseudorandom Sequences Having Optimal Correlation Properties and Large Linear Span," *IEEE Trans. on Information Theory*, Vol. IT-35, March 1989, pp. 371–379.

Olsen, J. D., R. A. Scholtz, and L. R. Welch, "Bent-Function Sequences," *IEEE Trans. on Information Theory*, Vol. IT-28, Nov. 1982, pp. 858–864.

Popovic, B., "Comment on 'Code Acquisition for a Frequency Hopping System'," *IEEE Trans. on Communications*, May 1989.

Popovic, B., "New Sequences for Frequency Hopping Multiplex," *Electronics Letters*, Vol. 24, June 1986.

Pursley, M. B., "Frequency-Hop Transmission for Satellite Packet Switching and Terrestrial Packet Radio Networks," *IEEE Trans. on Information Theory*, Vol. IT-32, Sept. 1986, pp. 652–667.

Pursley, M. B., "Reed-Solomon Codes for Frequency-Hop Communications," *Reed-Solomon Codes and Their Applications*, S. B. Wicker and V. K. Bhargava (eds.), Piscataway, NJ: IEEE Press, 1994.

Pursley, M. B., and D. V. Sarwate, "Evalution of Correlation Parameters for Periodic Sequences," *IEEE Trans. on Information Theory*, Vol. IT-23, July 1977, pp. 508–513.

Reed, I. S., "K-Th Order Near-Orthogonal Codes," *IEEE Trans. on Information Theory*, Vol. IT-15, Jan. 1971, pp. 116–117.

Roth, G. M., and G. Seroussi, "On Cyclic MDS Codes of Length Over $GF(q)$," *IEEE Trans. on Information Theory*, Vol. IT-32, March 1986, pp. 284–285.

Sarwate, D. V., "Bounds on Crosscorrelation and Autocorrelation of Sequences," *IEEE Trans. on Information Theory*, Vol. IT-25, Nov. 1979, pp. 720–724.

Sarwate, D. V., and M. B. Pursley, "Crosscorrelation Properties of Pseudorandom and Related Sequences," *Proc. IEEE*, Vol. 68, May 1980, pp. 593–619.

Scholtz, R. A., and L. R. Welch, "GMW Sequences," *IEEE Trans. on Information Theory*, Vol. IT-30, May 1984, pp. 548–553.

Scholtz, R. A., and L. R. Welch, "Group Characters: Sequences with Good Correlation Properties," *IEEE Trans. on Information Theory*, Vol. IT-24, Sept. 1968, pp. 537–545.

Shaar, A. A., C. F. Woodstock, and P. A. Davies, "Number of One-Coincidence Sequence Sets for Frequency-Hopping Multiple Access Communication Systems," *IEEE Proc.—Part F: Communications, Radar, and Signal Processing*, Vol. 131, Dec. 1984, pp. 725–728.

Shaar, A. A., and P. A. Davies, "A Survey of One Coincidence Sequences for Frequency-Hopped Spread-Spectrum," *IEEE Proc.—Part F: Commununications, Radar, and Signal Processing*, Vol. 131, Dec. 1984, pp. 719–724.

Sidelnikov, V. M., "On Mutual Correlation of Sequences," *Soviet Math. Dokl.*, Vol. 12(1), 1971, pp. 197–201.

Solomon, G., "Optimal Frequency Hopping for Multiple Access," *Proc. 1977 Symp. Spread Spectrum Communications*, Naval Electronics Laboratory Center, San Diego, CA, March 13–16, 1973, pp. 33–35.

Song, H. Y., I. S. Reed, and S. W. Golomb, "On the Nonperiodic Cyclic Equivalence Classes of Reed-Solomon Codes," *IEEE Trans. on Information Theory*, Vol. IT-39, July 1993, pp. 1431–1434.

Suehiro, N., and M. Hatori, "Modulatable Orthogonal Sequences and Their Application to SSMA Systems," *IEEE Trans. on Information Theory*, Vol. IT-34, Jan. 1988, pp. 93–100.

Tietäväinen, A., "On the Cardinality of Sets of Sequences with a Given Maximum Correlation," *Discrete Math.*, Vol. 106/107, 1992, pp. 471–477.

Titlebaum, E. L., "Time Frequency Hop Signals, Part I: Coding Based Upon the Theory of Linear Congruences," *IEEE Trans. on Aerospace and Electronic Systems*, Vol. AES-17, July 1981, pp. 490–494.

Titlebaum, E. L., "Time Frequency Hop Signals, Part II: Coding Based Upon the Theory of Quadratic Congruences," *IEEE Trans. on Aerospace and Electronic Systems*, Vol. AES-17, July 1981, pp. 494–501.

Welch, L. R., "Lower Bounds on the Maximum Crosscorrelation of Signals," *IEEE Trans. on Information Theory*, Vol. IT-20, May 1974, pp. 397–399.

Ziemer, R. E., and R. L. Peterson, *Digital Communications and Spread Spectrum Systems*, New York: Macmillan, 1985.

Appendix A1 The Trace Function

As already given by (1.61), the trace polynomial from the larger field $GF(q^L)$ to $GF(q)$ is defined as

$$\mathrm{Tr}_q^Q(z) = \sum_{j=0}^{L-1} z^{q^j} \qquad (A1.1)$$

where $Q = 2^L$. The trace in $GF(q)$ of an element α in $GF(Q)$ is then defined as the trace polynomial evaluated at α, that is, $\mathrm{Tr}_q^Q(\alpha)$. In what follows we will discuss some basic properties of the trace function.

Property 1. For $\alpha \in GF(Q)$, $\mathrm{Tr}_q^Q(\alpha)$ has values in $GF(q)$.

To prove this we start with polynomial $R(z)$ defined over $GF(q)$ and prove

$$(R(z))^q = R(z^q) \qquad (A1.2)$$

for any $q = p^n$, p being a prime number.

$$R(z) = \sum_{i=0}^{L} a_i z^i \qquad a_i \in GF(q) \qquad (A1.3)$$

We will represent this polynomial as

$$R(z) = R_1(z) + R_2(z) = R \qquad (A1.4)$$

with

$$R_1(z) = a_L z^L = R_1 \qquad R_2(z) = R(z) - R_1(z) = R_2 \qquad (A1.5)$$

where z is dropped for simplicity. Let us start with

$$(R(z))^p = (R_1(z) + R_2(z))^p = \sum_{j=0}^{p} \binom{p}{j} R_1^j R_2^{p-j} \qquad (A1.6)$$

where $\binom{p}{j}$ counts the number of times a term appears and is the element of $GF(q)$ corresponding to the sum of $\binom{p}{j}$ unit elements. In $GF(q)$, integer elements are added mod p. Since p is prime, $\binom{p}{j}$ is divisible by p unless $j = p$ or $j = 0$; so we have

$$\binom{p}{j} \bmod p = \begin{cases} 1 & \text{if } j = p \text{ or } j = 0 \\ 0 & \text{otherwise} \end{cases} \qquad (A1.7)$$

and (A1.6) results in

$$R^p = R_1^p + R_2^p \tag{A1.8}$$

If the procedure is repeated n times we have

$$(R(z))^{p^n} = \sum_{i=0}^{L} (a_i)^{p^n}(z^{p^n})^i \tag{A1.9}$$

Since a_i is in $GF(q)$ where $q = p^n$, the order of a_i divides $q - 1$ so that $a_i^q = a_i$ and we have

$$(R(z))^q = \sum_{i=0}^{L} a_i z^{q^i} = R(z^q) \tag{A1.10}$$

Using (A1.1) and (A1.10) we have

$$[\text{Tr}_q^Q(\alpha)]^q = \left(\sum_{j=0}^{L-1} \alpha^{q^j}\right)^q = \sum_{j=0}^{L-1} \alpha^{q^{j+1}} \tag{A1.11}$$
$$= \sum_{j=1}^{L-1} \alpha^{q^j} + \alpha^{q^L} = \sum_{j=0}^{L-1} \alpha^{q^j} = \text{Tr}_q^L(\alpha)$$

since $\alpha \in GF(q^L) \Rightarrow \alpha^{q^L} = \alpha$. From (A1.11) we can see that $\text{Tr}_q^Q(\alpha)$ is a field element whose qth power equals itself. Only elements of $GF(q)$ have this property (see Appendix A2).

Property 2. Conjugate field elements have the same trace, that is,

$$\text{Tr}_q^Q(\alpha^q) = \text{Tr}_q^Q(\alpha) \quad \text{for } \alpha \in GF(Q) \tag{A1.12}$$

The proof of this property follows directly from (A1.12) and (A1.10).

Property 3. The trace is linear. For $a,b \in GF(q)$ and $\alpha,\beta \in GF(Q)$

$$\text{Tr}_q^Q(a\alpha + b\beta) = a\text{Tr}_q^Q(\alpha) + b\text{Tr}_q^Q(\beta) \tag{A1.13}$$

To prove this property we start with

$$\mathrm{Tr}_q^Q(a\alpha + b\beta) = \sum_{j=0}^{L-1}(a\alpha + b\beta)^{q^j}$$

$$= \sum_{j=0}^{L-1}(a^{q^j}\alpha^{q^j} + b^{q^j}\beta^{q^j}) \quad \text{using (A1.10)} \tag{A1.14}$$

$$= \sum_{j=0}^{L-1} a\alpha^{q^j} + b\beta^{q^j} \leftarrow (a \in GF(q) \Rightarrow a^q = a)$$

$$= a\mathrm{Tr}_q^Q(\alpha) + b\mathrm{Tr}_q^Q(\beta)$$

Property 4. There are q^{L-1} elements in $GF(Q)$ that have trace value a for each a in $GF(q)$.

An element in $GF(Q)$ has

$$\mathrm{Tr}_q^Q(\alpha) = a \tag{A1.15}$$

if and only if α is a root of the trace equation

$$z + z^q + z^{q^2} + \cdots + z^{q^{L-1}} - a = 0 \quad a \in GF(q) \tag{A1.16}$$

Every element of $GF(Q)$ must be the root of exactly one such equation. Each of the q equations, for q different values of a, has exactly q^{L-1} roots. Since all the roots of the trace equations are accounted for by elements of $GF(Q)$, there must be exactly q^{L-1} elements of $GF(Q)$ with trace a in $GF(q)$.

Property 5. If $GF(q) \subset GF(q^k) \subset GF(q^L)$, then

$$\mathrm{Tr}_q^{q^L}(\alpha) = \mathrm{Tr}_q^{q^k}(\mathrm{Tr}_{q^k}^{q^L}(\alpha)) \tag{A1.17}$$

The right-hand side can be written as

$$\mathrm{Tr}_q^{q^k}(\mathrm{Tr}_{q^k}^{q^L}(z)) = \sum_{j=0}^{k-1}\left(\sum_{i=0}^{L/k-1} z^{q^{ki}}\right)^{q^j}$$

$$= \sum_{j=0}^{k-1}\sum_{i=0}^{L/k-1} z^{q^{ki+j}} \leftarrow \text{using (A1.10)}$$

$$= \sum_{n=0}^{L-1} z^{q^n} \leftarrow \text{using } (n = ki + j)$$

$$= \mathrm{Tr}_q^{q^L}(z)$$

which for $z = \alpha$, $\alpha \in GF(q^L)$, gives (A1.17).

Appendix B1 Binary Field and Vector Space

In this section, we briefly review basic terminology and definitions from discrete algebra that have been already used in Chapter 1. Due to limited space, many details will be omitted.

B1.1 Binary Arithmetic and Field

Consider the binary set {0, 1} and define two binary operations, called addition "+" and multiplication "·", on {0, 1} as

$$\begin{array}{ll} 0 + 0 = 0 & 0 \cdot 0 = 0 \\ 0 + 1 = 1 & 0 \cdot 1 = 0 \\ 1 + 0 = 1 & 1 \cdot 0 = 0 \\ 1 + 1 = 0 & 1 \cdot 1 = 1 \end{array} \tag{B1.1}$$

These two operations are commonly called modulo-2 addition and multiplication, respectively. The modulo-2 addition can be implemented with an X-OR gate, and the modulo-2 multiplication can be implemented with an AND gate. The set {0, 1} together with modulo-2 addition and multiplication is a binary field, denoted $GF(2)$. For a general definition of a field, a concept of group is needed with a set of properties that will be discussed later.

B1.2 Vector Space Over $GF(2)$

A binary n-tuple is an ordered sequence,

$$(a_1, a_2, \ldots, a_n) \tag{B1.2}$$

with components from $GF(2)$, that is, $a_i = 0$ or 1 for $1 \leq i \leq n$. There are 2^n distinct binary n-tuples. Let us define an addition operation for any two binary n-tuples as

$$(a_1, a_2, \ldots, a_n) + (b_1, b_2, \ldots, b_n) = (a_1 + b_1, a_2 + b_2, \ldots, a_n + b_n) \tag{B1.3}$$

where $a_i + b_i$, is carried out in modulo-2 addition. The addition of two binary n-tuples results in a third binary n-tuple.

Let us define a scalar multiplication between an element c in $GF(2)$ and a binary n-tuple $(a_1, a_2, \ldots, a_n)$ as

$$c \cdot (a_1, a_2, \ldots, a_n) = (c \cdot a_1, c \cdot a_2, \ldots, c \cdot a_n) \tag{B1.4}$$

where $c \cdot a_i$ is carried out in modulo-2 multiplication. The scalar multiplication also results in a binary n-tuple. Let V_n denote the set of all 2^n binary n-tuples. The set V_n together with the addition defined for any two binary n-tuples in V_n and the scalar multiplication defined between an element in $GF(2)$ and a binary n-tuple in V_n is called a *vector space* over $GF(2)$. The elements in V_n are called *vectors* and will be designated as v. Note that V_n contains the all-zero n-tuple, $(0, 0, \ldots, 0)$, and

$$(a_1, a_2, \ldots, a_n) + (a_1, a_2, \ldots, a_n) = (0, 0, \ldots, 0) \tag{B1.5}$$

Example 1. Let $n = 4$. The vector space V_n consists of the following 16 vectors:

$$\begin{aligned}
v_1 &= (\,0\,0\,0\,0\,) & v_9 &= (\,0\,0\,0\,1\,) \\
v_2 &= (\,0\,0\,1\,0\,) & v_{10} &= (\,0\,0\,1\,1\,) \\
v_3 &= (\,0\,1\,0\,0\,) & v_{11} &= (\,0\,1\,0\,1\,) \\
v_4 &= (\,0\,1\,1\,0\,) & v_{12} &= (\,0\,1\,1\,1\,) \\
v_5 &= (\,1\,0\,0\,0\,) & v_{13} &= (\,1\,0\,0\,1\,) \\
v_6 &= (\,1\,0\,1\,0\,) & v_{14} &= (\,1\,0\,1\,1\,) \\
v_7 &= (\,1\,1\,0\,0\,) & v_{15} &= (\,1\,1\,0\,1\,) \\
v_8 &= (\,1\,1\,1\,0\,) & v_{16} &= (\,1\,1\,1\,1\,)
\end{aligned}$$

According to the rule for vector addition,

$$(\,0\,1\,0\,1\,) + (\,1\,1\,1\,0\,) = (\,0{+}1,\,1{+}1,\,0{+}1,\,1{+}0\,) = (\,1\,0\,1\,1\,)$$

According to the rule for scalar multiplication,

$$1 \cdot (\,1\,0\,1\,1\,) = (\,1{\cdot}1,\,1{\cdot}0,\,1{\cdot}1,\,1{\cdot}1\,) = (\,1\,0\,1\,1\,)$$
$$0 \cdot (\,1\,0\,1\,1\,) = (\,0{\cdot}1,\,0{\cdot}0,\,0{\cdot}1,\,0{\cdot}1\,) = (\,0\,0\,0\,0\,)$$

A subset S of V_n is called a subspace of V_n if (1) the all-zero vector is in S and (2) the sum of two vectors in S is also a vector in S.

Example 2. The following set of vectors

$$\begin{array}{ll}
(\,0\,0\,0\,0\,) & (\,0\,1\,0\,1\,) \\
(\,1\,0\,1\,0\,) & (\,1\,1\,1\,1\,)
\end{array}$$

forms a subspace of the vector space V_n.

B1.3 Linear Combination

A linear combination of k vectors, $v_1, v_2, \ldots, v_k$ in V_n, is a vector of the form

$$u = c_1 v_1 + c_2 v_2 + \cdots + c_k v_k \quad (B1.6)$$

where $c_i \in GF(2)$ are called the coefficients of v_i. There are 2^k such linear combinations of $v_1, v_2, \ldots, v_k$. These 2^k linear combinations give 2^k vectors in V_n, which form a subspace of V_n. A set of vectors, $v_1, v_2, \ldots, v_k$ in V_n, is said to be linearly independent if

$$c_1 v_1 + c_2 v_2 + \cdots + c_k v_k \neq 0 \quad (B1.7)$$

unless all $c_1, c_2, \ldots, c_k$ are the zero elements in $GF(2)$. The subspace formed by the 2^k linear combinations of k linearly independent vectors $v_1, v_2, \ldots, v_k$ in V_n is called a k-dimensional subspace of V_n.

B1.4 Dual Space

The inner product of two vectors, $a = (a_1, a_2, \ldots, a_n)$ and $b = (b_1, b_2, \ldots, b_n)$, is defined as

$$a \cdot b = a_1 \cdot b_1 + a_2 \cdot b_2 + \cdots a_n \cdot b_n \quad (B1.8)$$

where $a_i \cdot b_i$ and $a_i \cdot b_i + a_{i+1} \cdot b_{i+1}$ are carried out in modulo-2 multiplication and addition.

Example 3.

$$(1\ 1\ 0\ 1\ 1) \cdot (1\ 0\ 1\ 1\ 1)$$
$$= 1 \cdot 1 + 1 \cdot 0 + 0 \cdot 1 + 1 \cdot 1 + 1 \cdot 1$$
$$= 1 + 0 + 0 + 1 + 1$$
$$= 1$$

Two vectors, a and b, are said to be orthogonal if

$$a \cdot b = 0 \quad (B1.9)$$

Example 4.

$$(1\ 0\ 1\ 1\ 0) \cdot (1\ 1\ 0\ 1\ 1)$$
$$= 1 \cdot 1 + 0 \cdot 1 + 1 \cdot 0 + 1 \cdot 1 + 0 \cdot 1$$
$$= 1 + 0 + 0 + 1 + 0$$
$$= 0$$

Let S be a k-dimensional subspace of V_n. Let S_d be the subset of vectors in V_n such that, for any a in S and any b in S_d,

$$a \cdot b = 0 \qquad (B1.10)$$

then S_d is called the *dual space* of S. The dimension of S_d is $n - k$.

Example 5. Consider V_5, the vector space of all 5-tuples over $GF(2)$; then a possible pair of S and S_d is

S	S_d
(0 0 0 0 0)	(0 0 0 0 0)
(1 1 1 0 0)	(1 0 1 0 1)
(0 1 0 1 0)	(0 1 1 1 0)
(1 0 0 0 1)	(1 1 0 1 1)
(1 0 1 1 0)	
(0 1 1 0 1)	
(1 1 0 1 1)	
(0 0 1 1 1)	

B1.5 Binary Irreducible Polynomials

A polynomial with coefficients from the binary field $GF(2)$ is called a *binary* polynomial. For example, $1 + X^2$, $1 + X + X^3$, $1 + X^3 + X^5$ are binary polynomials. A binary polynomial $p(X)$ of degree m is said to be *irreducible* if it is *not divisible* by any binary polynomial of degree less than m and greater than zero. For example, $1 + X + X^2$, $1 + X + X^3$, $1 + X + X^5$, and $1 + X^2 + X^5$ are irreducible polynomials. For any positive integer $m \geq 1$, there exists at least one irreducible polynomial of degree m. An irreducible polynomial $\bar{p}(X)$ of degree m is said to be *primitive* if the *smallest* positive integer n for which $\bar{p}(X)$ divides $X^n + 1$ is $n = 2^m - 1$. For example, $1 + X + X^4$ is a primitive polynomial. The smallest positive integer n for which $1 + X + X^4$ divides $X^n + 1$ is $n = 2^4 - 1 = 15$. For any positive integer m, there exists a primitive polynomial of degree m. Table B1.1 gives a list of primitive polynomials.

Table B1.1
A List of Primitive Polynomials

m	Primitive Polynomials
3	$1 + X + X^3$
4	$1 + X + X^4$
5	$1 + X^2 + X^5$
6	$1 + X + X^6$
7	$1 + X^3 + X^7$
8	$1 + X^2 + X^3 + X^4 + X^8$
9	$1 + X^4 + X^9$
10	$1 + X^3 + X^{10}$
11	$1 + X^2 + X^{11}$
12	$1 + X + X^4 + X^6 + X^{12}$
13	$1 + X + X^3 + X^4 + X^{13}$
14	$1 + X + X^6 + X^{10} + X^{14}$
15	$1 + X + X^{15}$
16	$1 + X + X^3 + X^{12} + X^{16}$
17	$1 + X^3 + X^{17}$
18	$1 + X^7 + X^{18}$
19	$1 + X + X^2 + X^5 + X^{19}$
20	$1 + X^3 + X^{20}$
21	$1 + X^2 + X^{21}$
22	$1 + X + X^{22}$
23	$1 + X^5 + X^{23}$
24	$1 + X + X^2 + X^7 + X^{24}$

B1.6 Galois Fields

A field is a set of elements (or symbols) in which we can do addition, subtraction, multiplication, and division without leaving the set. Addition and multiplication satisfy the commutative, associative, and distributive laws. The system of real numbers is a field, called the *real-number field*. The system of complex numbers is also a field, known as the *complex-number field*. The complex-number field is actually constructed from the real-number field by requiring the symbol

$$i = \sqrt{-1}, \quad (B1.11)$$

as a *root* of the irreducible (over the real-number field) polynomial $X^2 + 1$, that is,

$$(\sqrt{-1})^2 + 1 = 0 \quad (B1.12)$$

Every complex number is of the form, $a + bi$ where a and b are real numbers. The complex-number field contains the real-number field as a *subfield*. The complex-number field is an *extension* field of the real-number field. The complex-number and real-number fields have infinite elements.

B1.7 Finite Fields

It is possible to construct fields with a finite number of elements. Such fields are called *finite fields*. Finite fields are also known as *Galois* fields after their discoverer. For any positive integer $m \geq 1$, there exists a Galois field of 2^m elements, denoted $GF(2^m)$. The construction of $GF(2^m)$ is very much the same as the construction of the complex-number field from the real-number field. We begin with a primitive (irreducible) polynomial $p(X)$ of degree m with coefficients from the binary field $GF(2)$. Since $p(X)$ has degree m, it must have roots somewhere. Let α be the root of $p(X)$, that is,

$$p(\alpha) = 0 \tag{B1.13}$$

(just as we let the symbol $i = \sqrt{-1}$ be the root of the irreducible polynomial $X^2 + 1$ over the real-number field). Then the set

$$\{0, 1, \alpha, \alpha^2, \cdots, \alpha^i, \cdots, \alpha^{2^m-2}\} \tag{B1.14}$$

forms a Galois field of 2^m elements, where α^i is the ith power of α. Since $p(X)$ divides X^{2^m-1}, the element α must also be a root of X^{2^m-1}. Hence,

$$\alpha^{2^m-1} + 1 = 0 \tag{B1.15}$$

This implies that

$$\alpha^{2^m-1} = 1 \tag{B1.16}$$

since $1 + 1 = 0$ (modulo-2 addition). We also use the notation that $\alpha^0 = 1$.
Multiplication is carried out as follows. For $0 \leq i, j < 2^m - 1$,

$$\alpha^i \cdot \alpha^j = \alpha^{i+j} = \alpha^r \tag{B1.17}$$

where r is the remainder resulting from dividing (real number division) $i + j$ by $2^m - 1$. Note that

$$\alpha^i \cdot \alpha^{2^m-1-i} = \alpha^{2^m-1} = 1 \tag{B1.18}$$

Hence α^i is called the multiplicatibve inverse of α^{2^m-1-i} and vice versa. Note also that

$$\alpha^{2^m-1-i} = \alpha^{2^m-1} \cdot \alpha^{-i} = \alpha^{-i} \tag{B1.19}$$

Division is carried out as

$$\alpha^i \div \alpha^j = \frac{\alpha^i}{\alpha^j} = \alpha^{i-j} \tag{B1.20}$$

Now we consider *addition*. First we define that

$$\begin{aligned} 0 + 0 &= 0 \\ 0 + 1 &= 1 + 0 = 1 \\ 1 + 1 &= 0 \\ 0 + \alpha^i &= \alpha^i + 0 = \alpha^i \end{aligned} \tag{B1.21}$$

Then,

$$\begin{aligned} \alpha^i + \alpha^i &= \alpha^0 \cdot \alpha^i + \alpha^0 \cdot \alpha^i \\ &= (\alpha^0 + \alpha^0) \cdot \alpha^i \\ &= (1 + 1) \cdot \alpha^i \\ &= 0 \cdot \alpha^i \\ &= 0 \end{aligned} \tag{B1.22}$$

For $i \neq j$, $\alpha^i + \alpha^j \neq 0$, this is best explained by an example.

Example 6. Let $m = 4$. The polynomial $p(X) = X^4 + X + 1$ is a binary primitive polynomial of degree 4. Let α be a root of $p(X)$. Then, $p(\alpha) = \alpha^4 + \alpha + 1 = 0$. Using the fact that $\alpha^4 + \alpha^4 = 0$ and $\alpha^4 + 0 = \alpha^4$, we have

$$\alpha^4 = \alpha + 1. \tag{B1.23}$$

Now we consider the set

$$\{0, 1, \alpha, \alpha^2, \alpha^3, \alpha^4, \alpha^5, \alpha^6, \alpha^7, \alpha^8, \alpha^9, \alpha^{10}, \alpha^{11}, \alpha^{12}, \alpha^{13}, \alpha^{14}\} \tag{B1.24}$$

Note that $\alpha^{15} = 1$. Using the identity $\alpha^4 = \alpha + 1$, every power α^i can be expressed as a polynomial of α with degree 3 or less as shown in Table B1.2.
For example,

$$\begin{aligned}
\alpha^5 &= \alpha \cdot \alpha^4 = \alpha \cdot (\alpha + 1) = \alpha^2 + \alpha \\
\alpha^6 &= \alpha \cdot \alpha^5 = \alpha \cdot (\alpha^2 + \alpha) = \alpha^3 + \alpha^2 \\
\alpha^7 &= \alpha \cdot \alpha^6 = \alpha \cdot (\alpha^3 + \alpha) = \alpha^4 + \alpha^3 \\
&= \alpha + 1 + \alpha^3 = \alpha^3 + \alpha + 1 \\
&\vdots
\end{aligned} \qquad (B1.25)$$

Addition is done in *polynomial* form. Let

$$\begin{aligned}
\alpha^i &= a_0 + a_1 \alpha + a_2 \alpha^2 + a_3 \alpha^3 \\
\alpha^j &= b_0 + b_1 \alpha + b_2 \alpha^2 + b_3 \alpha^3
\end{aligned} \qquad (B1.26)$$

where $a_i, b_i \in GF(2)$. Then,

Table B1.2
The Galois Field $GF(2^4)$ Generated by $p(X) = X^4 + X + 1$

Power Representation	Polynomial Representation	4-Tuple Representation
0	0	(0 0 0 0)
1	1	(1 0 0 0)
α	α	(0 1 0 0)
α^2	α^2	(0 0 1 0)
α^3	α^3	(0 0 0 1)
α^4	$1 + \alpha$	(1 1 0 0)
α^5	$\alpha + \alpha^2$	(0 1 1 0)
α^6	$\alpha^2 + \alpha^3$	(0 0 1 1)
α^7	$1 + \alpha \quad + \alpha^3$	(1 1 0 1)
α^8	$1 \quad + \alpha^2$	(1 0 1 0)
α^9	$\alpha \quad + \alpha^3$	(0 1 0 1)
α^{10}	$1 + \alpha + \alpha^2$	(1 1 1 0)
α^{11}	$\alpha + \alpha^2 + \alpha^3$	(0 1 1 1)
α^{12}	$1 + \alpha + \alpha^2 + \alpha^3$	(1 1 1 1)
α^{13}	$1 \quad + \alpha^2 + \alpha^3$	(1 0 1 1)
α^{14}	$1 \quad\quad + \alpha^3$	(1 0 0 1)

$$\alpha^i + \alpha^j = (a_0 + a_1\alpha + a_2\alpha^2 + a_3\alpha^3)$$
$$+ (b_0 + b_1\alpha + b_2\alpha^2 + b_3 + \alpha^3) \qquad \text{(B1.27)}$$
$$= (a_0 + b_0) + (a_1 + b_1)\alpha + (a_2 + b_2)\alpha^2 + (a_3 + b_3)\alpha^3$$
$$= \alpha^k \quad \text{(from Table B1.2)}$$

where $a_i + b_i$ is carried out with modulo-2 addition. For example,

$$\alpha^5 + \alpha^{13} = (\alpha + \alpha^2) + (1 + \alpha^2 + \alpha^3)$$
$$= 1 + \alpha + \alpha^3 = \alpha^7$$
$$\alpha^{11} + \alpha^3 = (\alpha + \alpha^2 + \alpha^3) + \alpha^3 \qquad \text{(B1.28)}$$
$$= \alpha + \alpha^2 = \alpha^5$$
$$\alpha^7 + \alpha^7 = (1 + \alpha + \alpha^3) + (1 + \alpha + \alpha^3)$$
$$= 0$$

Since $\alpha^i + \alpha^i = 0$, α^i is its own additive inverse, that is,

$$\alpha^i = -\alpha^i \qquad \text{(B1.29)}$$

Hence,

$$\alpha^i - \alpha^j = \alpha^i + (-\alpha^j) = \alpha^i + \alpha^j \qquad \text{(B1.30)}$$

Subtraction is identical to addition. This completes our construction of Galois field $GF(2^4)$. We say that $GF(2^4)$ is generated by the primitive polynomial $p(X) = X^4 + X + 1$. Note that there is a one-to-one correspondence between the polynomial $a_0 + a_1\alpha + a_2\alpha^2 + a_3\alpha^3$ and the 4-tuple (a_0, a_1, a_2, a_3).

Hence every element in $GF(2^4)$ has three forms: the power form, the polynomial form, and the vector form, as shown in Table B1.2.

The primitive polynomial $p(X) = X^4 + X + 1$ has four roots that are all in $GF(2^4)$. They are

$$\alpha, \alpha^2 \qquad \alpha^{2^2} = \alpha^4 \qquad \alpha^{2^3} = \alpha^8 \qquad \text{(B1.31)}$$

For example,

$$p(\alpha^4) = (\alpha^4)^4 + (\alpha^4) + 1 = \alpha^{16} + \alpha^4 + 1$$
$$= \alpha \cdot \alpha^{15} + \alpha^4 + 1$$
$$= \alpha + \alpha^4 + 1 \qquad \text{(B1.32)}$$
$$= \alpha^4 + \alpha + 1 = 0$$

α^2, α^4, and α^8 are called the conjugate roots of α. We can easily show that

$$p(X) = (X + \alpha)(X + \alpha^2)(X + \alpha^4)(X + \alpha^8) = X^4 + X + 1 \tag{B1.33}$$

B1.8 Primitive Elements

Concider the Galois field $GF(2^m)$ generated by the primitive polynomial

$$p(X) = p_0 + p_1 X + \cdots + p_{m-1} X^{m-1} + X^m \tag{B1.34}$$

The element α (a root of $p(X)$) whose powers generate all the nonzero elements of $GF(2^m)$ is called a *primitive* element of $GF(2^m)$. In fact, any element β in $GF(2^m)$ whose powers generate all the nonzero elements of $GF(2^m)$ is a primitive element.

Example 7. Consider the Galois field $GF(2^4)$ given in Table B1.2. The powers of α^4 are

$(\alpha^4)^0 = 1$ $\qquad (\alpha^4)^1 = \alpha^4$ $\qquad (\alpha^4)^2 = \alpha^8$
$(\alpha^4)^3 = \alpha^{12}$ $\qquad (\alpha^4)^4 = \alpha^{16} = \alpha$ $\qquad (\alpha^4)^5 = \alpha^{20} = \alpha^5$
$(\alpha^4)^6 = \alpha^{24} = \alpha^9$ $\qquad (\alpha^4)^7 = \alpha^{28} = \alpha^{13}$ $\qquad (\alpha^4)^8 = \alpha^{32} = \alpha^2$
$(\alpha^4)^9 = \alpha^{36} = \alpha^6$ $\qquad (\alpha 4)^{10} = \alpha^{40} = \alpha^{10}$ $\qquad (\alpha^4)^{11} = \alpha^{44} = \alpha^{14}$
$(\alpha^4)^{12} = \alpha^{48} = \alpha^3$ $\qquad (\alpha^4)^{13} = \alpha^{52} = \alpha^7$ $\qquad (\alpha^4)^{14} = \alpha^{56} = \alpha^{11}$

which generate all the 15 nonzero elements of $GF(2^4)$. Thus α^4 is a primitive element. α^7 is also a primitive element.

B1.9 Minimum Polynomials

Consider the Galois field $GF(2^m)$ generated by a primitive polynomial $p(X)$ of degree m. Let β be a nonzero element of $GF(2^m)$. Consider the powers,

$$\beta, \beta^2, \beta^{2^2}, \cdots, \beta^{2^i}, \cdots \tag{B1.35}$$

Let e be the smallest non-negative integer for which

$$\beta^{2^e} = \beta \tag{B1.36}$$

The integer e is called the exponent of β. The powers

$$\beta, \beta^2, \beta^{2^2}, \cdots, \beta^{2^{e-1}} \qquad (B1.37)$$

are distinct and called conjugates of β. The product

$$\phi(X) = (X + \beta)(X + \beta^2) \cdots (X + \beta^{2^{e-1}}) \qquad (B1.38)$$
$$= a + a_1 X + \cdots + a_{e-1} X^{e-1} + X^e$$

is a polynomial of degree e.

1. $\phi(X)$ is binary and irreducible over $GF(2)$.
2. $\phi(X)$ is called the minimal polynomial of the element β.
3. $\phi(X)$ is the binary irreducible polynomial of minimum degree that has β as a root.
4. $\phi(X)$ has $\beta, \beta^2, \cdots, \beta^{2^{e-1}}$ as all its roots.

Example 8. Consider the field $GF(2^4)$ given in Table B1.2. Let $\beta = \alpha^3$. We form the following power sequence:

$$\begin{aligned}
\beta &= \alpha^3 \\
\beta^2 &= \alpha^6 \\
\beta^{2^2} &= \beta^4 = \alpha^{12} \\
\beta^{2^3} &= \beta^8 = \alpha^{24} = \alpha^9 \\
\beta^{2^4} &= \beta^{16} = \alpha^{48} = \alpha^3 = \beta
\end{aligned} \qquad (B1.39)$$

Since $\beta^{2^4} = \beta$, the exponent of β is 4. We see that $\beta = \alpha^3$, $\beta^2 = \alpha^6$, $\beta^{2^2} = \alpha^{12}$, and $\beta^{2^3} = \alpha^9$ are all distinct. The minimum polynomial of $\beta = \alpha^3$ is

$$\begin{aligned}
\phi(X) &= (X + \beta)(X + \beta^2)(X + \beta^{2^2})(X + \beta^{2^3}) \\
&= (X + \alpha^3)(X + \alpha^6)(X + \alpha^{12})(X + \alpha^9) \\
&= X^4 + (\alpha^3 + \alpha^6 + \alpha^9 + \alpha^{12})X^3 \\
&\quad + (\alpha^9 + \alpha^{12} + \alpha^{15} + \alpha^{15} + \alpha^{18} + \alpha^{21})X^2 \\
&\quad + (\alpha^{18} + \alpha^{21} + \alpha^{24} + \alpha^{27})X + \alpha^{30} \\
&= X^4 + X^3 + X^2 + X + 1
\end{aligned} \qquad (B1.40)$$

which is irreducible.

CHAPTER 2

Code Acquisition

Code synchronization consists of two steps, acquisition and tracking. In Chapter 3 we will analyze code tracking, which is part of the code synchronization problem in which the relative delay τ, between the input and locally generated code, is within the tracking range. Most of the time, this range is defined as $\tau \leq T_c$ where T_c is the code chip interval. However, the initial delay τ between the two codes is much larger and a separate, most-of-the-time completely different algorithm should be used to reduce this delay to the tracking range. These algorithms are the subject of this chapter.

We start with the maximum likelihood approach in order to get an insight into what is the best possible action that we can take. Unfortunately the best from the performance point of view means the highest complexity. So, to find a compromise between the two, a series of approximations is presented. The most widely used algorithm is the so-called *serial search strategy*, where the phase of the local code is changed step by step, in equal increments, resulting in a serial search of the code delay uncertainty region until the synchro position is found. This approach is very much the same for both direct sequence (DS) and frequency hopping (FH) systems.

For each value of the phase of the local sequence, a correlation between the input signal and the local signal replica is formed and compared to a threshold. A high value of the correlation (above the threshold) indicates the synchro position. A proper threshold set-up procedure is a crucial part of the receiver initialization. This will be also discussed within this chapter.

If the SNR is high, the sequence chips in the DS signal can be detected and loaded in the sequence generator as an initial state. If n chips of the sequence are correctly detected, where n is the length of the generator shift register, the synchronization is acquired. In FH systems, an equivalent procedure would require detection of the instantaneous value of FH signal frequency, its conversion into a corresponding code initial value, and consequent initialization of the sequence generator. Due to limited space this type of code acquisition will not be discussed within this book. A section of this chapter will deal with the mathematical tools available for the analysis of the code acquisition algorithms. The performance measure for those algorithms will be the average acquisition time and its variance, although for some applications these criteria will be modified.

For the case when a certain prior knowledge (probability density function) about the synchro position is available, a number of so-called modified search strategies will be described and discussed.

2.1 MAXIMUM LIKELIHOOD PARAMETER ESTIMATION

We start with a simple problem where for a received signal $r(t) = s(t, \theta) + n(t)$ we have to estimate a generalized time-invariant parameter θ (frequency, phase, or delay) of a signal $s(t, \theta)$ in the presence of Gaussian noise $n(t)$. The best that we can do is to find an estimate $\hat{\theta}$ of the parameter θ for which the aposterior probability $p(\hat{\theta}/r)$ is maximum, hence the name *maximum aposterior probability* (MAP) estimate. In other words, the chosen estimate based on the received signal r is correct with the highest probability. Practical implementation requires us to locally generate a number of trial values $\tilde{\theta}$, to evaluate $p(\tilde{\theta}/r)$ for each such value, and then to choose $\hat{\theta} = \tilde{\theta}$ for which $p(\tilde{\theta}/r)$ is maximum. Analytically this can be expressed as

$$\text{MAP} \Rightarrow \hat{\theta} = \arg \max_{\tilde{\theta}} p(\tilde{\theta}/r) \qquad (2.1)$$

Very often in practice, evaluation of $p(\tilde{\theta}/r)$ in closed form is not possible. Using the Bayesian rule for the joint probability distribution function

$$p(r, \tilde{\theta}) = p(r)p(\tilde{\theta}/r) = p(\tilde{\theta})p(r/\tilde{\theta}) \qquad (2.2)$$

and assuming a uniform prior distribution of θ, maximizing $p(\tilde{\theta}/r)$ becomes equivalent to maximizing $p(r/\tilde{\theta})$, a function that can be determined much more easily. This algorithm is known as the *maximum likelihood* (ML) estimation and can be defined analytically as

$$\text{ML} \Rightarrow \hat{\theta} = \arg \max_{\tilde{\theta}} p(r/\tilde{\theta}) \qquad (2.3)$$

It is straightforward to show that in the case of the Gaussian noise the ML principle requires searching for such a value of θ that would maximize the likelihood function defined as

$$p(r/\tilde{\theta}) = C_1 \mathrm{Exp}\{-C_2 \int [r(t) - s(t, \tilde{\theta})]^2 \, dt \quad (2.4)$$

where C_1 and C_2 are constants independent on $\tilde{\theta}$. This is equivalent to maximizing

$$\lambda(\tilde{\theta}) = 2\int r(t)s(t, \tilde{\theta})\, dt - \int s^2(t, \tilde{\theta})\, dt \quad (2.5)$$

where $s(t, \tilde{\theta})$ is the locally generated replica of the signal with a trial value $\tilde{\theta}$. For the given signal power, the second term in (2.5) is a constant, so the maximization is equivalent to the maximization of the first term only. This can be expressed as

$$\lambda(\tilde{\theta}) = \int r(t)s(t, \tilde{\theta})\, dt \quad (2.6)$$

The ML structure based on (2.6) is shown in Figure 2.1(a). An additional insight into the meaning of this principle can be obtained by replacing $r(t)$ in the previous equation, which gives

$$\lambda(\tilde{\theta}) = \int s(t, \theta)s(t, \tilde{\theta})\, dt + \int n(t)s(t, \tilde{\theta})\, dt \quad (2.7)$$

The second term (crosscorrelation) has zero value because $n(t)$ and $s(t, \theta)$ are independent and uncorrelated signals. The first crosscorrelation has its maximum for $\theta = \tilde{\theta}$ when this term becomes the autocorrelation function at zero delay. Instead of a single parameter, $\theta \rightarrow \boldsymbol{\theta}$ might simultaneously represent a set of parameters (vector) and (2.3) will then represent a joint maximization of a ML function with respect to all parameters. More detail about these standard principles, widely used in the field of synchronization theory, can be found in the textbooks [1].

Instead of searching for the maximum of $\lambda(\tilde{\theta})$ in a so-called open loop configuration, an equivalent procedure would be to find zero of the first derivative of $\lambda(\tilde{\theta})$

$$\mathrm{MLT} \Rightarrow \hat{\theta} = \arg_{\tilde{\theta}} \mathrm{zero}\, \frac{\partial \lambda(\tilde{\theta})}{\partial \tilde{\theta}} = \arg_{\tilde{\theta}} \mathrm{zero} \int r(t) \frac{\partial s(t, \tilde{\theta})}{\partial \tilde{\theta}}\, dt \quad (2.8)$$

This structure is known as the *maximum likelihood tracker* (MLT), and its mechanization is shown in Figure 2.1(b). In a practical realization of the tracker, the signal derivative is often approximated by the signal difference

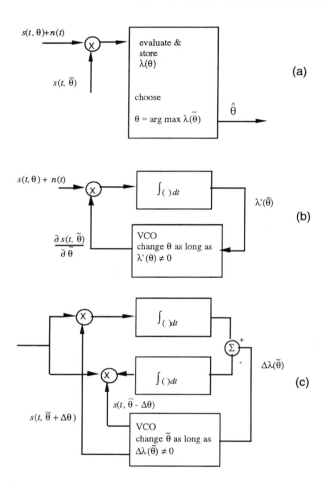

Figure 2.1 (a) ML structure, (b) MLT structure, and (c) ELT structure.

$$\frac{\partial s(t, \tilde{\theta})}{\partial \tilde{\theta}} = \frac{1}{2\Delta\theta}\{s(t, \tilde{\theta} + \Delta\theta) - s(t, \tilde{\theta} - \Delta\theta)\} \qquad (2.A)$$

where $s(t, \tilde{\theta} + \Delta\theta)$ and $s(t, \tilde{\theta} - \Delta\theta)$ are so-called early and late versions of the local signal with respect to the generalized parameter θ to be estimated. This results in

$$ELT \Rightarrow \hat{\theta} = \arg \text{zero}\{E(t, \tilde{\theta}) - L(t, \tilde{\theta})\} \qquad (2.9)$$

where

$$E(t, \tilde{\theta}) = \frac{1}{2\Delta\theta}\int r(t)s(t, \tilde{\theta} + \Delta\theta)\,dt$$

$$L(t, \tilde{\theta}) = \frac{1}{2\Delta\theta}\int r(t)s(t, \tilde{\theta} - \Delta\theta)\,dt \qquad (2.10)$$

The structure is known as *early late tracker* (ELT), and its mechanization is shown in Figure 2.1(c). The previous relations will be the basis for deriving receiver structures for code acquisition and tracking.

2.2 ML-MOTIVATED CODE SYNCHRONIZATION

2.2.1 Problem Definition

In order to get an initial insight into the problem, let us first assume a biphase modulated DSSS receiver, whose input waveform $r(t)$ is given by

$$r(t) = \sqrt{2S}b(t)c(t + \zeta T_c)\cos(\omega_c t + \omega_D t + \theta_c) + n(t) \qquad (2.11)$$

where the parameters S stands for the waveform's average power; ω_c and θ_c denote the carrier radian frequency and phase, respectively; ω_D is a (possibly present, in mobile environments) radian frequency shift (Doppler offset) between transmitter and receiver; $c(t)$ is a ±1-valued spreading code with chip rate $R_c = T_c^{-1}$ chips/s, delayed by ζT_c sec with respect to an arbitrary time reference; $b(t)$ is a binary data sequence (which might or might not be present during the acquisition mode); and $n(t)$ is additive noise or different type of interference or both. *The synchronizer's task consists of providing the receiver with reliable estimates $\hat{\zeta}$, $\hat{\omega}_D$, and $\hat{\theta}_c$ of the corresponding unknown quantities so that despreading and demodulation can follow.* For the code synchronization part, in particular, the equivalent problem is to align the unknown phase ζT_c of the incoming code as closely as possible with the known (and controllable) phase $\hat{\zeta} T_c$ of a locally generated (at the receiver) identical despreading code $c(t + \hat{\tau} T_c)$.

As mentioned, (2.11) refers to DS spreading. The corresponding received waveform in the ith hopping interval $((i - 1)T_h + \zeta T_h, iT_h + \zeta T_h)$ for a typical FH system with, say, MFSK modulation would be similar to (2.11):

$$r(t) = \sqrt{2S}\cos(\omega_{c;i}t + \omega_{b;i}t + \omega_D t + \theta_{c;i}) + n(t) \qquad (2.12)$$

where $\omega_{c;i}$ is the hopping radian frequency generated by a PN code sequence during the ith hop; $\omega_{b;i}$ is the data symbol radian frequency (we assume for simplicity an one-hop-per-symbol system); ω_D is the Doppler offset, assumed independent of the hop index; and ζT_h is the unknown timing offset of the received waveform with respect to the receiver time reference. Again, the job of the synchronizer would be to align the incoming and local hopping patterns (sequences) and provide an estimate $\hat{\zeta}$ for the timing offset, along with the other parameters. Thus, from a conceptual standpoint, the FH synchronization problem is identical to the previously stated DS problem. Furthermore, it is typically true that FH code acquisition is faster and easier to perform than DS in benign noise environments due to higher SNR and smaller code uncertainty regions encountered (this might not be true in the presence of jamming or other strong interference).

2.2.2 Uncertainty Range of the Signal Parameters

The uncertainty about the values of the parameters to be synchronized can be attributed to a number of factors whose importance varies with the application. For example, in asynchronous communication, there is no timing information provided initially to the transmitter/receiver pair. Even in the synchronous case, however (i.e., when the users are connected to some sort of universal timing system or so-called synchro channel), uncertainties arise due to the relative range and dynamic movement (velocity, acceleration) of the transmitter with respect to the receiver—parameters that may only be vaguely known. This gives rise to the delay tracking problem, which will be discussed in Chapter 3. Moreover, it may be necessary to account for time differences arising from clock instability and the relative drift. Even for very accurate clocks, relative phase shifts are inevitable in the long run. The uncertainty in range, denoted by ΔR, implies an unknown propagation delay for the spread signal, which in turn translates into a temporal uncertainty region. The corresponding number N_{unc} of code chips in that region, for an one-way propagation link, will be given by

$$N_{unc} = \Delta R \frac{R_c}{V_{EM}} \qquad (2.13)$$

where R_c is the code rate and V_{EM} is the velocity of the electromagnetic wave. The value of N_{unc} in (2.13) could vary from a few chips to a whole code-length or many thousands of chips, depending on the application. The other important aspect is the relative frequency shift (Doppler) ω_D, caused by relative radial velocity between transmitter and receiver. If $\Delta \dot{R}$ denotes the difference in range rate of the pair, then the corresponding relative Doppler shift is approximately given by

$$\omega_D = \frac{(\Delta \dot{R})\omega_c}{V_{EM}} \qquad (2.14)$$

For predictably moving objects, such as satellites, ω_D can be precomputed and thus compensated for; whereas for fast and unpredictably moving objects (such as terrestrial high-speed vehicles or fighter aircraft), Doppler offset might be a significant consideration. In either case, bounds on $\Delta \dot{R}$ are usually specified, from which bounds on ω_D can be derived. From now on, we let Δt and Δf denote the total time and frequency uncertainties, respectively. Reasons similar to this explain the ignorance of the receiver about the carrier phase θ_c. Typically, an estimate $\hat{\theta}_c$ is provided to the receiver (if needed for coherent demodulation) via one of the familiar carrier tracking loops. In most applications, code synchronization and despreading are performed prior to carrier tracking due to the fact that the spread signal does not provide enough SNR for the carrier tracking loop to operate successfully [2]. Code synchronization in the absence of carrier phase information is designated as *noncoherent* code synchronization. In a few circumstances, enough SNR may exist before despreading for the carrier loop to track; in that case, code synchronization could be performed with θ_c (and ω_D) assumed known (*coherent* code synchronization), resulting in a slightly different detector structure.

2.2.3 Detector Structures: DS Signal

To motivate the detector structure, let us first assume for simplicity a coherently received waveform with no data modulation (i.e., let $b(t) = 1$ and assume ω_D, θ_c known in (2.11)). The question then is: How should the receiver go about estimating ζ? Given a T-sec observation $r(t)$, $0 \le t \le T$, with T assumed fixed, a desirable answer might be the ML criterion, as described in the previous section. The corresponding estimate, $\hat{\zeta}_{ML}$, is the argument that maximizes the *likelihood function* (LF). In this case, the general expression (2.6) becomes $\lambda(\tilde{\zeta}) = \Lambda\{r(t)|c(t; \tilde{\zeta}); 0 \le t \le T\}$, where the notation $c(t; \tilde{\zeta})$ is used to denote $c(t + \tilde{\zeta} T_c)$. In accordance with (2.6) we have

$$\lambda(\tilde{\zeta}) = \int_0^T r(t) c(t; \tilde{\zeta}) dt \qquad (2.15)$$

which is the time-domain (coherent) correlation operation between the observed waveform and the local code, positioned at the candidate offset $\tilde{\zeta} T_c$.

To understand the motivation for the receiver structure in the noncoherent environment let us use only the first term of (2.7). (We ignore the noise term for the moment.) Let us also use the normalized signal $r(t) = c(t, \zeta)\cos(\omega_c t + \omega_D t + \theta_c)$ and assume that ω_D is known to the receiver and θ_c is not. So the local signal is $c(t, \tilde{\zeta})\cos(\omega_c t + \omega_D t)$. The first term of (2.7) becomes

$$Y_c(\tilde{\zeta}) = \int c(t,\zeta)c(t,\tilde{\zeta})\{\cos(\omega_c t + \omega_D t)\cos\theta_c - \sin(\omega_c t + \omega_D t)\sin\theta_c\}\cos(\omega_c t + \omega_D t)\,dt$$

$$= \frac{1}{2}\cos\theta_c \int c(t,\zeta)c(t,\tilde{\zeta})\,dt \qquad (2.15a)$$

which is based on the fact that all terms with frequency $2\omega_c$ will be averaged out to zero. In order to eliminate of the unknown parameter θ_c, a similar term obtained by correlating the input signal with $c(t,\tilde{\zeta})\sin(\omega_c t + \omega_D t)$ can be expressed as

$$Y_s(\tilde{\zeta}) = \frac{1}{2}\sin\theta_c \int c(t,\zeta)c(t,\tilde{\zeta})\,dt \qquad (2.15b)$$

Using these two parameters we have

$$\lambda_{nc}(\tilde{\zeta}) = \sqrt{Y_c^2(\tilde{\zeta}) + Y_s^2(\tilde{\zeta})}$$

$$Y_c(\tilde{\zeta}) = \sqrt{2}\int_0^T r(t)c(t;\tilde{\zeta})\cos(\omega_c t + \omega_D t)\,dt$$

$$Y_s(\tilde{\zeta}) = \sqrt{2}\int_0^T r(t)c(t;\tilde{\zeta})\sin(\omega_c t + \omega_D t)\,dt$$

$$(2.15c)$$

Once again we note in (2.15a) that the Doppler offset is assumed known since it is used in the carrier-mixing operations in the in-phase and quadrature correlations. If it is not, then it becomes one additional parameter to be estimated (or averaged out), meaning that the LF is now a function of the two parameters ζ and ω_D.

2.2.4 Detector Structures: FH Signal

The prior formulation regarded DS systems, whose modeling tends to be a bit easier. For FH systems, the designer of the code synchronization circuits is faced with a number of modeling details regarding the presence of *fast FH* (FFH) or *slow FH* (FFH), a distinction that depends on factors such as the number of data symbols per hop, the exact type of data modulation in the SFH case, and the presence or not of a known data preamble.[1]

---

[1] Note that, in the DS case, the presence of unknown phase-modulated data can be absorbed in the uniformly random carrier phase θ_c, leading to the same noncoherent formulation as before, provided that the integration time is one bit (or symbol) time; otherwise, modeling considerations arise as in the FH case.

As is explained in Section 1, in a FH system with a total of G carrier hopping frequencies, the hopping pattern is dictated by a PN code–derived, random number $M_h(t)$ whose purpose is to determine the carrier hopping frequency at each hop; thus, a timing uncertainty manifests itself in the shifted argument $M_h(t + \zeta T_c) \equiv M_h(t; \zeta)$. As an example of the likelihood formulation in the SFH case, assume that a total of N_h hops is observed, each hop carrying $N_{s/h}$ symbols, where each symbol is drawn randomly from an MFSK alphabet and is unknown to the synchronizing unit (no preamble). Furthermore, the carrier phase is assumed to shift randomly from symbol to symbol through the channel (taking into account any deliberate phase continuity is possible but would just escalate the complexity of the formulation). Then, by starting with (2.4), the LF conditioned on an assumed ζ (with Doppler ambiguity neglected) would average over all possible data-filter envelopes in each symbol, for all symbols in a hop as well as for all hops, resulting in

$$\Lambda\{r(t)|c_h(t; \zeta); 0 \le t \le T = N_h T_h = N_h N_{s/h} T_s\}$$

$$= \prod_{i=1}^{N_h} \prod_{l=1}^{N_{s/h}} \left\{ \frac{1}{M} \sum_{m=1}^{M} I_0\left(\frac{2\sqrt{S}}{N_0} Y^{(i,l,m)}(\zeta)\right) \right\} \quad (2.16)$$

where $c_h(t; \zeta)$ is the PN-derived hopping code dictating the values of $M_h(t; \zeta)$; $I_0(\cdot)$ is the familiar zeroth-order modified Bessel function; and the envelope $Y^{(i,l,m)}(\zeta)$ is defined via the following two quadrature components, similar to (2.15c):

$$Y_c^{(i,l,m)}(\zeta) = \sqrt{2} \int_{(i-1)T_h+(l-1)T_s}^{(i-1)T_h+lT_s} r(t)\cos(\omega_{I_h(t;\zeta)}t + \omega_m t)\,dt$$

$$Y_s^{(i,l,m)}(\zeta) = \sqrt{2} \int_{(i-1)T_h+(l-1)T_s}^{(i-1)T_h+lT_s} r(t)\sin(\omega_{I_h(t;\zeta)}t + \omega_m t)\,dt \quad (2.16a)$$

Since (2.16) appears complicated, simpler implementations would just energy-detect over the total M/T_s-Hz data band around each candidate carrier hopping frequency (assuming an orthogonal T_s^{-1} symbol-frequency spacing) and then create quasi-likelihood variants such as choosing the maximum over each hop as the ML estimate of the hopping frequency and then accumulating such maxim over the observed hops. Still, this should be done, in principle, for all candidate offsets ζ.

2.2.5 Parallel Versus Serial Implementation

The ML synchronizing receiver implied by (2.15) and (2.16) should, in principle, create all possible time-offset versions of the known code waveform, correlate all of them with the received chunk of data, and choose the ζ corresponding to the largest correlation as its estimate, $\hat{\zeta}_{ML}$. In view of the continuous range of values of ζ (and ω_D, if also required), this is an impossibility and some type of range quantization is necessary, which is indeed what is done in practice. The resulting candidate values are called *bins* or *cells*. Instead of simultaneous generating $\lambda(\tilde{\zeta})$ for all cells and choosing $\hat{\zeta} = \tilde{\zeta}$ that gives the largest $\lambda(\tilde{\zeta})$ this can be implemented in a serial maner by creating $\lambda(\tilde{\zeta})$ for each cell, one in time, memorizing all of them, and then choosing the cell with the largest $\lambda(\tilde{\zeta})$. For q cells this means replacing q correlators with one correlator and a memory of size q. Further reduction is obtained if each new generated $\lambda(\tilde{\zeta})$ is compared to a given threshold and then cell rejected or accepted as synchro cell depending on whether or not $\lambda(\tilde{\zeta})$ is larger than the threshold. One should be aware that reduction in complexity as previously mentioned will increase the average acquisition time. This is an issue that will be discussed later in this chapter. This is exactly the *coarse code synchronization* or *code acquisition* problem, the result of which is to resolve the code phase (or epoch) ambiguity within the size of the cell. Since this remaining error is typically larger than desired, further operations are required to reduce it to acceptable levels. This remaining part of the synchronization task, namely, that of *fine synchronization* or *code tracking*, is performed by one of the available code-tracking loops, which we discuss in Chapter 3.

2.2.6 Two-Dimensional Search

For the two-dimensional problem mentioned previously (time plus frequency uncertainty), the corresponding cells will also be two-dimensional. Their total number will equal $q = (\Delta t/\delta t)(\Delta f/\delta f)$, where δt and δf stand for the cell dimensions in the respective directions. For accurate estimation, their sizes should be as small as possible; however, they cannot be too small, as that would either overburden the receiver complexity or extend the search time, since the total search space is obviously proportional to q. Typical choices for δt are of the order of the code chip time T_c or a large fraction thereof (such as 1/2 or 1/4) and of the order of the data bandwidth for δf. Similarly for FH, δt is chosen as a fraction of T_h. This ensures that at least one cell is close to the peak of the (noiseless) correlation output of (2.15) or (2.16).

Once the nature and size of these cells have been determined, the next question is how to go about performing the search most successfully. Clearly, the strategy will depend on a variety of factors such as criteria of performance, degree of complexity, computational power available (directly related to cost), and prior available information about the location of the correct cell. As mentioned earlier, a brute-force approach would try to create a bank of parallel correlation branches, each matched to a possible quantized value of the timing offset; it would then process the received waveform through all of them simultaneously, pick the largest, and declare a candidate solution. Unless the uncertainty region (number of cells) is small, corresponding to either a small code period or a small initial uncertainty, such a solution (which we may call the "totally parallel solution") becomes obviously unwieldy in complexity very quickly. We note, however, that small uncertainty regions may be encountered in a nested design, whereby a multitude of different-period codes are combined for precisely the purpose of aiding acquisition. Furthermore, neural network structures are currently being explored for this purpose, where the neural network is trained for all possible such values. Such a scheme would emulate the spirit (if not the exact statistical processing) of these solutions.

2.2.7 Serial Search

In current practice, total parallelism is out of the question when the number of cells is very large (although it appears feasible for smaller uncertainty regions [3–5]) and simpler solutions are necessary. One of the most familiar such approaches is the simple technique of serial search, namely, one where the search starts from a specific cell and serially examines the remaining cells in some direction and in a pre specified order until the correct cell is found [6,7]. Hence, serial search techniques do not account for any additional information gathered during the past search time, which could conceivably be used to alter the direction of search toward cells that show increased posterior likelihood of being the correct ones [8]. (See also [9,10] for related concepts in the radar target-search problem.) A serial search starts from a cell that could be chosen totally arbitrarily (no prior information), or by some prior knowledge about a likely cell, and proceeds in a simple and easily implementable predirected manner. When the uncertainty space is two-dimensional as discussed and searching all possible cells serially appears very time consuming, then a speed-up may be achieved by employing a bank of filters, each matched to a possible Doppler offset [11]. The same idea can be applied to the one-dimensional case (no frequency uncertainty), where now a bank of correlators may be employed, each starting from a different point of the uncertainty region, which effectively amounts to dividing the search in many parallel subsearches and therefore reducing the total search time by a proportional amount.

One should be aware that although it holds true that only one cell contains the exact delay and Doppler offsets of the incoming code, the set of desirable cells acceptable to the receiver includes a number of cells adjacent to the exact one. Indeed, the receiver will terminate acquisition and initiate tracking the first time a cell is reached (and correctly identified), which is close enough to the true synchronization so that the tracking loop can pull in and perform the remaining synchronization operation successfully. All these desirable cells are collectively called *hypothesis* H_1, and the remaining nondesirable "out-of-sync" cells comprise *hypothesis* H_0. As an example, consider the case where the receiver examines the code delay uncertainty in steps of half a chip time ($\delta t = T_c/2$) and there is no frequency uncertainty. Then all four cells located in the interval $(-T_c, T_c)$ around the true delay of the incoming code are included in hypothesis H_1—since some amount of code correlation exists for each one of these cells, an amount that can initiate the code tracking loop.

This definition of cells and hypotheses implies that each test does not pertain to a single value of the unknown parameter ζ but rather to a range of values. It is straightforward to show that, under mild conditions and approximations pertaining to the pseudorandom nature of the code, this reformulated hypothesis testing results in a statistic (correlation) and threshold setting that do not depend on the given (tested) value of the unknown parameter (a *uniformly most powerful test* (UMPT)). This is because the threshold value is set by the desirable probability of a false alarm per cell, which is independent of ζ under H_0.

The generic code acquisition model of Figure 2.2 recapitulates the basic ideas discussed so far. The two-dimensional time/frequency code offset uncertainty within the noisy received waveform is quantized into a number of cells, which are typically searched in a serial fashion by a correlation receiver, although parallel multiple branches are also possible. Motivated by an ML argument, the receiver creates a crosscorrelation between the incoming waveform and the local code at a specific offset, whose output is used to decide whether the currently examined cell is desirable (H_1). The process continues until one such cell is correctly identified. At that point, acquisition is terminated and tracking is initiated.

2.3 PERFORMANCE MEASURES

A number of sources contribute to the randomness of the acquisition process, including:

- Initial uncertainty about the code-phase offset;
- Channel distortion (e.g., fading channels);

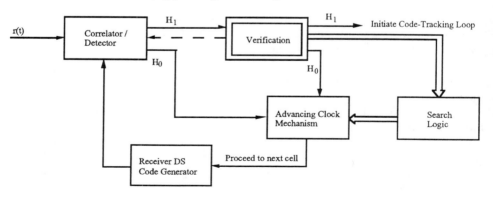

Figure 2.2 Generic model for the serial-search likelihood-ratio code acquisition receiver.

- Noise plus interference (e.g., narrowband and multiuser);
- Random data;
- Unknown carrier phase (noncoherent receivers) and Doppler offset;
- Partial code correlation.

The receiver's set of adjustable factors includes threshold settings, correlation times, number of tests per code chip, search strategy, and cell logic (i.e., verification strategy). For a well-optimized design, implicit is the knowledge of important parameters such as the SNR, code rate, code length, code-uncertainty region, and reset penalty-time. Regarding performance measures for code acquisition, the dominant parameter of interest is the random time that elapses prior to acquisition, T_{acq}, and its statistics. We may distinguish two basic scenarios.

In case I, although the fastest possible acquisition is desired, no absolute time limit (stop time, T_{stop}) exists, which arises when the transmitted waveform is always present as, for instance, in satellite-based position location systems.

In case II, for "push-to-talk" and other packetized or frame-based systems, data transmission starts after a certain time interval T_{stop}, which signifies the end of a preamble.

Then, it is imperative that acquisition be performed in that specific interval with very high probability, for otherwise communication is impossible. To characterize both scenarios completely, knowledge of the *cumulative probability distribution function* (CPDF) $F_{T_{acq}}$ of T_{acq} would be needed, which might be more than required in order to identify a good design (not to mention that $F_{T_{acq}}(t)$ is difficult to obtain except for simple cases). Hence, one typically settles for either the mean acquisition time $E\{T_{acq}\}$ and, possibly, the variance var$\{T_{acq}\}$ (case I). For case II, however, the corresponding performance measure is the probability of prompt acquisition $\Pr\{T_{acq} \leq T_{stop}\} = F_{T_{acq}}(T_{stop})$.

Under what circumstances will the code not be acquired? One possibility is the existence of the stop time as described, and the inability of the search process to locate the correct cell in that interval of time due to noise, interference, and fading. Another is the case of a *catastrophic false alarm* (FA), that is, when the acquisition mechanism erroneously decides that code synchronization has occurred and a very prolonged tracking effort via a code-tracking loop is initiated. The *overall probability of missing the code*, P_M^{ov}, can be computed from total probability as

$$P_M^{ov} = \Pr\{\text{FA before } T_{stop}\} + \Pr\{\text{neither FA nor ACQ before } T_{stop}\} \quad (2.17)$$

where ACQ denotes the correct acquisition event. The complementary *overall probability of detection*, P_D^{ov}, is many times cited as the key performance measure in such circumstances. It is not necessarily true that all false alarms result in catastrophic delays. In fact, certain correlation-based logic can be employed (sometimes known as "lock detector logic") that detects the out-of-sync situation quickly and proceeds to discard the examined cell. In reality, the time required for indicating false lock is a random variable. In this case, we refer to the false-alarm state as a *returning state* associated with a random penalty time T_p, as opposed to the previous catastrophic case with an *absorbing* FA state [6]. For computational ease, T_p can also be modeled as a fixed time [12–14]. The type of performance criterion comprises a first partition of acquisition receivers. Further distinctions concern the basic unit of any design, which is the per-cell decision-making device (*detector*). This can be classified as either *coherent* or *noncoherent* (see the introduction) and can be categorized according to the specific statistical testing philosophy employed for each cell test, that is, Bayes, Neyman–Pearson, or other. In a Neyman–Pearson type of test, the detector thresholds are set in a fixed position for all cells, which for stationary interference results in a fixed *false alarm probability per cell*, $P_{FA,c}$. That is essentially equivalent to disregarding any available prior information about the correct cell position and treating all cells as equally likely candidates. This is simpler to implement than a receiver that attempts to make good use of any cell-position prior information; in the latter case, a Bayes test is more appropriate.

An acquisition detector will always consist of an initial testing level (called the *first dwell*), which may or may not be followed by a *verification mode* (logic), whose purpose is to secure the initial positive (sync) indication and prevent eventual false alarm. The verification mode is typically called a "multiple-dwell detector" [15], and consists of a series of successive tests of the same cell. Since a cell can potentially be rejected at any such stage, and since there is always the possibility (however remote) of a false alarm, it follows that the time spent by the detector in each cell is a random variable. This is true regardless of the nature of the individual dwell times. Those can be either *fixed*, in which case decisions are made at prearranged time intervals, or *random*, when a variable-time test is employed (the so-called Wald sequential test [16]), arising from the random crossing time of appropriate thresholds [17–19]. We note that the advent of digitized receivers and clocks tends to increasingly favor fixed-dwell tests, which can be designed to be very competitive performance-wise with sequential tests. Detectors can employ either *full-period* or *partial-period* correlation, the latter being favored at the early stages of multiple-dwell tests. Verification modes can employ *immediate-rejection logic* or *non-immediate-rejection logic*, the terminology pertaining to the action taken by the supervising logic once a cell fails to pass an intermediate-dwell test [20]. Another esoteric distinction comes about because successive multiple-dwell decisions can be made either statistically *independent* (integrate-and-dump kind of operation) or *dependent* (continuous type of integration, with sampling but no dumping).

Finally, detectors are classified according to whether they employ *passive* or *active* correlation [6,21]. The former term refers to a passive filter matched to a segment of the known spreading code. Ideally, this creates no correlation with the received waveform until the identical segment in the incoming code matches perfectly in time with the receiver filter impulse-response, at which moment maximum correlation is created. The alternative of active correlation is when the two codes (local and incoming) stay at a fixed phase offset, corresponding to the cell under examination, for the whole cell correlation period. This can be done, for example, by running the receiver code at the same rate with the incoming code while correlating. If the decision at the end of that time interval is negative, a clock pulse advances the local code forward to the next cell position and the process is repeated. We note that passive correlation inherently enjoys a *fast decision rate* (equal to or a multiple of the code rate), while active correlation is confined to slow decision rates. *Hybrids* of the above are also possible, whereby the first-stage decision is based on passive and the verification on active correlations.

The remaining important aspect of the acquisition process is the search strategy, that is, the procedure adopted by the receiver in its search through the uncertainty region. A host of serial search strategies was proposed and analyzed in [6,22,23]; see also [24]. The main serial search types are the *straight search*, the *Z-search*, and the *expanding-window search*. More details will be given in the following sections. The reader may also consult the aforementioned references as well as [5,11,13,14,19,20,25–42]. The problem of analyzing performance under nonwhite noise (narrowband interference and jamming, mostly) has been addressed in [43–46]. The impact of code acquisition on packet radio networks has been addressed in [47], whereas adaptive-array techniques have been studied in [48]. A fundamentally different class of search is *sequential state estimation*, which refers to those techniques that try to estimate directly the state of the pseudonoise code dictating the spreading code and to load that estimate onto the receiver shift register [49–51]. Since these techniques are based on hard-decision chip estimates made in a spread-SNR environment (i.e., estimating the ±1 value of each chip separately), and since a single chip error makes the whole state estimate useless, their utility is constrained to relatively benign, high chip-SNR environments where they can be very fast (the time it takes to estimate n chips for an n-stage shift register).

2.4 PERFORMANCE ANALYSIS TOOLS

A significant amount of literature and effort has been devoted to analyzing the performance of code acquisition, the goal of the analysis being the maximization of some performance criteria discussed in previous sections (such as minimum acquisition time) by means of appropriate parameter selection such as threshold values in acquisition detectors or search lock strategy. The analysis is necessarily specific to the structure under consideration, but there exist few key tools and general concepts that recur in all analytical efforts.

As we identified previously, the two building blocks of any acquisition receiver is the detector and the search strategy. Accordingly, the purpose of analysis is to combine these two elements in order to derive the statistics of the acquisition time or the overall probability of detection. In fact, the impact of the specific detector structure on the performance measure can typically be summarized by few important parameters per cell, as we will illustrate with a simple example. This implies that optimization of the detector structure and the choice of the search strategy can proceed independently.

There exist two general analytical directions for this type of analysis: the time domain combinatorial technique and the transform domain (or circular-state diagram) technique. The former proceeds along total-probability arguments and is occasionally faster and more insightful, whereas the latter seems more systematic and able to handle complicated search techniques.

2.4.1 Transform Domain Analysis

Signal Flow Graph Method

The code acquisition decision process can be represented using signal flow graph techniques. Each cell is represented by a node of a graph, and transitions between nodes depend on the outcome of the decision in a given cell. These transitions are characterized by branches connecting the nodes called transmission functions. To motivate operation in a transform domain let us consider a simple model of a process represented by graph in Figure 2.3.

As the first step let us evaluate the probability $p_{ac}(t)$ that the process will move from a to c in exactly t seconds. To do it we will introduce an additional variable τ to designate time needed for the process to move from a to b characterized with probability $p_{ab}(\tau)$. The parameter $p_{ac}(t, \tau)$ represents the probability that the process moves from a to c in t sec given the fact that it moves from a to b in τ sec. This probability can be represented as

$$p_{ac}(t, \tau) = p_{ab}(\tau)p_{bc}(t - \tau) \qquad (2.18)$$

resulting in

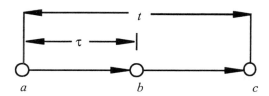

Figure 2.3 Signal flow graph.

$$p_{ac}(t) = \int p_{ac}(t, \tau) d\tau = \int p_{ab}(\tau) p_{bc}(t - \tau) d\tau$$
$$= p_{ac}(t) * p_{bc}(t) \tag{2.19}$$

In other words, the overall probability $p_{ac}(t)$ is a convolution of the two internode transition probabilities p_{ab} and p_{bc}. It is clear that for a graph with a large number of nodes we will have to deal with multiple convolutions giving rise to computational complexity. In this case, people involved in electrical engineering prefer to move to either a Laplace (s) transform domain for continuous variables or a z domain for discrete variables. So, instead of the previous equation we would be deal with

$$P_{ac}\left(\begin{smallmatrix}s\\\text{or}\\z\end{smallmatrix}\right) = P_{ab}\left(\begin{smallmatrix}s\\\text{or}\\z\end{smallmatrix}\right) \cdot P_{bc}\left(\begin{smallmatrix}s\\\text{or}\\z\end{smallmatrix}\right) \tag{2.20}$$

where multiple convolutions become multiple products and make life much easier. For most cases in which we will be interested, there will be constant integration time to create the decision variable in each cell. This leads to using the z-transform for the decision process flow graph representation. If $p_{ij}(n)$ is the probability for the process to move from node i to node j in exactly n steps, then its z-transform

$$P_{ij}(z) = \sum_{n=0}^{\infty} z^n p_{ij}(n) \tag{2.21}$$

is called the "generating function" or the geometric transform. For the analysis that follows we are going to need a few relations derived from this definition. First of all, the first and the second derivatives of this function can be represented as

$$\frac{\partial}{\partial z} P_{ij}(z) = \sum_{n=0}^{\infty} n p_{ij}(n) z^{n-1} \tag{2.22}$$

$$\frac{\partial^2}{\partial z^2} P_{ij}(z) = \sum_{n=0}^{\infty} n(n-1) p_{ij}(n) z^{n-2} \tag{2.23}$$

The average number of steps required to move from node i to node j is

$$\bar{n} = \sum_{n=0}^{\infty} n p_{ij}(n) = \left.\frac{\partial}{\partial z} P_{ij}(z)\right|_{z=1} \tag{2.24}$$

and the average time to do it can be represented as

$$\bar{t}_{ij} = T_{ij} = \bar{n}T = \left(\frac{\partial}{\partial z}P_{ij}(z)\bigg|_{z=1}\right) \cdot T \qquad (2.25)$$

where T is the *cell observation time* time needed to create the decision variable. For the variance we start with the definition

$$\sigma_T^2 = (\overline{n^2} - \bar{n}^2)T^2 \qquad (2.26)$$

Bearing in mind that the second derivative of the generating function can be represented as

$$\frac{\partial^2}{\partial z^2}P_{ij}(z)\bigg|_{z=1} = \sum_{n=0}^{\infty} n^2 p_{ij}(n) - \sum_{n=0}^{\infty} np_{ij}(n) = \overline{n^2} - \bar{n} \qquad (2.27)$$

the variance of time t_{ij} can be expressed in the form

$$\sigma_T^2 = \left[\frac{\partial^2 P_{ij}(z)}{\partial z^2} + \frac{\partial P_{ij}(z)}{\partial z} - \left(\frac{\partial P_{ij}(z)}{\partial z}\right)^2\right]_{z=1} T^2 \qquad (2.28)$$

In what follows we will use these few relations to analyze serial search code acquisition. In order to get an initial insight into this method, we assume that there are q cells to be searched. Now q may be equal to the length of the PN code to be searched or some multiple of it (e.g., if the update size is one-half chip, q will be twice the code length to be searched). Further assume that if a "hit" (output is above threshold) is detected by the threshold detector, the system goes into a verification mode that may include both an extended duration dwell time and an entry into a code loop tracking mode. In any event, we model the "penalty" of obtaining a FA as $K\tau_D$ sec, and the dwell time itself as τ_D sec. If a true hit is observed, the system has acquired the signal, and the search is completed. Assume the FA probability P_{FA} and the probability of detection P_D are given. We will also assume that only one cell represents the synchro position. Let each cell be numbered from left to right so that the kth cell has an a priori probability of having the signal present, given that it was not present in cells 1 through $k-1$, of

$$P_k = \frac{1}{q + 1 - k} \qquad (2.29)$$

The generating function flow diagram is given in Figure 2.4 using the rule that at each node the sum of the probability emanating from the node equals unity.

The unit time representations τ_D sec and $K\tau_D$ sec are represented in z-transform by z^K. Consider node 1. The a priori probability of having the signal present is

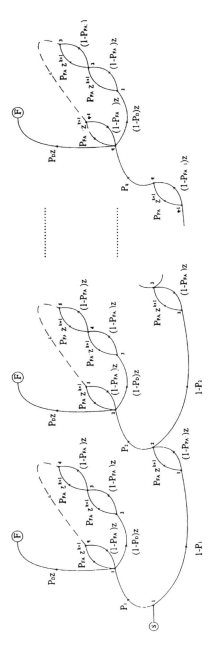

Figure 2.4 Decision process flow diagram.

$P_1 = 1/q$, and the probability of it not being present in the cell is $1 - P_1$. Suppose the signals were not present. Then we advance to the next node (node 1a); since it corresponds to a probabilistic decision and not a unit time delay, no z multiplies the branch going to it. At node 1a a FA occurs, with probability $P_{FA} = \alpha$, which requires one unit of time to determine (τ_D sec), and then K units of time ($K\tau_D$ sec) are needed for verification. At the same node there will be no FA with probability $(1 - \alpha)$, which takes on dwell time to determine, and is represented by $(1 - \alpha)z$ branch going to node 2.

Now consider the situation at node 1 when the signal does occur. If it happens, then acquisition, as we have defined it, occurs and the process is terminated, hence the node F denoting "finish." If there was no hit at node 1 (the integrator output was below the threshold), which occurs with probability $1 - P_D$, one unit of time would be consumed. This is represented by the branch $(1 - P_D)z$, leading to node 2. At node 2 in the upper left part of the diagram either a FA occurs with probability α and delay $(K + 1)$ or a FA does not occur with a delay of 1 unit. The remaining portion of the generating function flow graph is a repetition of the portion just discussed with the appropriate node changes. By using standard signal flow graph reduction techniques one can show that the overall transfer function between nodes S and F can be represented as

$$U(z) = \frac{(1 - \beta)}{1 - \beta z H^{q-1}} \frac{1}{q}\left[\sum_{l=0}^{q-1} H^l(z)\right] \tag{2.30}$$

a moment generating function where

$$H(z) = \alpha z^{K+1} + (1 - \alpha)z \text{ and } \beta = 1 - P_D \tag{2.31}$$

Using (2.25) the mean acquisition time is given (after some algebra) by

$$\overline{T} = \frac{2 + (2 - P_D)(q - 1)(1 + KP_{FA})}{2P_D}\tau_d \tag{2.32}$$

with τ_D included in the formula to translate from our unit time scale.

For the usual case when $q \gg 1$, $\overline{T}$ is given by

$$\overline{T} = \frac{(2 - P_D)(1 + KP_{FA})}{2P_D}(q\tau_D) \tag{2.33}$$

The variance of the acquisition time is given by (2.28). It can be shown that the expression for σ_T^2 is

$$\sigma_T^2 = \tau_D^2 \Bigg\{ (1 + KP_{FA})^2 q^2 \left(\frac{1}{12} - \frac{1}{P_D} + \frac{1}{P_D^2} \right) + 6q[K(K+1)P_{FA}(2P_D - P_D^2)$$
$$+ (1 + P_{FA}K)(4 - 2P_D - P_D^2)] + \frac{1 - P_D}{P_D^2} \Bigg\} \qquad (2.34)$$

In addition, when $K(1 + KP_{FA}) \ll q$, then

$$\sigma_T^2 = \tau_D^2 (1 + KP_{FA})^2 q^2 \left(\frac{1}{12} - \frac{1}{P_D} + \frac{1}{P_D^2} \right) \qquad (2.35)$$

As a partial check on the variance result, let $P_{FA} \to 0$ and $P_D \to 1$. Then we have

$$\sigma_T^2 = \frac{(q\tau_D)^2}{12} \qquad (2.36)$$

which is the variance of a uniformly distributed random variable, as one would expect for the limiting case. These results provide a useful estimate of acquisition time for an idealized PN type system. Two basic modifications should be made to make the estimates reflect actual hardware or software systems. First, Doppler effects should be taken into account. The result of code Doppler is to smear the relative code phase during the acquisition dwell time, which increases or reduces the probability of detection depending on the code phase and the algebraic sign of the code Doppler rate. The Doppler also affects the effective code sweep rate, which in the extreme case can reduce it to zero to cause the search time to increase greatly. This topic will be discussed later. The second refinement to the model concerns the handover process between acquisition and tracking. Typically after a "hit" the code tracking loop is turned on to attempt to pull the code into tight lock. Further, often in low-SNR systems where both acquisition (pull-in) bandwidth and tracking bandwidth are used, multiple code loop bandwidths will be employed in order to soften the transition between acquisition and tracking modes. Consequently, the probability of going from the acquisition mode to the final code loop bandwidth in the tracking mode occurs with some probability less than 1. The estimation of this probability is at best a very difficult problem (although, some approximate results have been developed). At high SNRs this probability quickly approaches 1, so it is not a problem. At low SNRs the formula for acquisition time should replace P_D with P_D'

$$P_D' = P_D P_{HO} \qquad (2.37)$$

with P_{HO} being the probability of handover.

2.4.2 Direct Method

The mean synchronization time can be calculated in a straightforward manner by considering all possible sequences of events leading to a correct code acquisition and by direct evaluation of the probabilities that these sequences will occur, hence the name *direct method*. In a serial search of the code delay uncertainty region represented by q cells with synchro cell having index n, the acquisition will be completed after a given number of missed detections j of the correct cell and a given number of FAs k in all incorrect cells. Parameters in this analysis are as follows:

- Probability of detection when the correct cell is tested, P_d;
- Probability of FA when an incorrect cell is tested, P_{fa};
- Integration (or dwell) time in cell, T_i;
- Time required to reject an incorrect cell when FA occurs.

The total synchronization time for a particular event defined by a set of parameters (n, j, k) is

$$T(n, j, k) = nT_i + jqT_i + kT_{fa} \tag{2.38}$$

The FA penalty T_{fa} may be many times larger than T_i, so FAs are undesirable events. For low P_{fa}, the threshold should be high. At the same time if the threshold is too high, P_d will decrease, causing the possibility of missing the synchro cell. This leads to a need to repeat visit all the cells in order to get a new chance to detect a signal in the correct cell. The total number of cells tested for this event is $(n + jq)$, the total number of correct cells is $(j + 1)$, and the total number of incorrect cells is $(n + jq - j - 1)$. The probability of the correct cell being the nth cell is $1/q$, and the probability of j missed detections followed by a correct detection is $P_d(1 - P_d)^j$. The k FAs can occur in any order within the $(n + jq - j - 1) \triangleq K$ incorrect cells. The probability of a particular ordering is $P_{fa}^k(1 - P_{fa})^{K-k}$, and there are $\binom{K}{k}$ orderings of k FAs in K cells.

Combining all of these, the probability of the event (n, j, k) is

$$\Pr(n, j, k) = \frac{1}{q} P_d (1 - P_d)^j \binom{K}{k} P_{fa}^k (1 - P_{fa})^{K-k} \tag{2.39}$$

The mean synchronization time is

$$\overline{T}_s = \sum_{n,j,k} T(n,j,k) \Pr(n,j,k) \tag{2.40}$$

The correct cell number can range over $(1, q)$, the number of missed detections can range from zero to infinity, and the number of FAs ranges over $[0, K]$. Thus the mean synchronization time is

$$\overline{T}_s = \frac{1}{q} \sum_{n=1}^{q} \sum_{j=0}^{\infty} \sum_{k=0}^{K} [(n + jq)T_i + kT_{fa}] \binom{K}{k} P_{fa}^k (1 - P_{fa})^{K-k} P_d (1 - P_d)^j$$

$$= \frac{1}{q} \sum_{n=1}^{q} \sum_{j=0}^{\infty} (n + jq)T_i \left[\sum_{k=0}^{K} \binom{K}{k} P_{fa}^k (1 - P_{fa})^{K-k} \right] P_d (1 - P_d)^j$$

$$+ \frac{1}{q} \sum_{n=1}^{q} \sum_{j=0}^{\infty} T_{fa} \left[\sum_{k=0}^{K} \binom{K}{k} k P_{fa}^k (1 - P_{fa})^{K-k} \right] P_d (1 - P_d)^j \quad (2.41)$$

The first summation over k is evaluated using the identity

$$\sum_{k=0}^{K} \binom{K}{k} a^k b^{K-k} = \binom{K}{0} b^K + \binom{K}{1} a b^{K-1} + \cdots + \binom{K}{k} b^K = (b + a)^K \quad (2.42)$$

with $a = P_{fa}$ and $b = 1 - P_{fa}$, $b + a = 1.0$ and the first summation over k equal to unity. The second sum over k is the mean of a discrete random variable that has a binomial distribution. The mean value is KP_{fa}. Thus (2.41) simplifies to

$$\overline{T}_s = \frac{1}{q} \sum_{n=1}^{q} \sum_{j=0}^{\infty} [(n + jq)T_i + KT_{fa}P_{fa}] P_d (1 - P_d)^j \quad (2.43)$$

Expanding this expression using $K = n + jq - j - 1$ and reordering yields

$$\overline{T}_s = \frac{1}{q} \sum_{n=1}^{q} \sum_{j=0}^{\infty} [(n - 1)T_{da} + (j + 1)T_i + j(q - 1)T_{da}] P_d (1 - P_d)^j \quad (2.44)$$

where

$$T_{da} = T_i + T_{fa} P_{fa} \quad (2.45)$$

is the average dwell time at an incorrect phase cell. Equation (2.44) can be evaluated completely using the identities

$$\sum_{i=1}^{L} i = L \left(\frac{L + 1}{2} \right)$$

$$\sum_{i=0}^{\infty} m^i = \frac{1}{1 - m} \quad m < 1$$

$$\sum_{i=0}^{\infty} i m^i = \frac{m}{(1 - m)^2} \quad m < 1 \quad (2.46)$$

After some straightforward algebraic manipulation, the result is

$$\overline{T}_s = (q - 1)T_{da}\left(\frac{2 - P_d}{2P_d}\right) + \frac{T_i}{P_d} \quad (2.47)$$

This expression for mean synchronization time is a function of P_{fa} through the definition of T_{da}. This result was derived in Section 2.4.1 using signal flow graph techniques. In Appendix A2 it is shown that the variance of T_s can be expressed as

$$\sigma_{T_s}^2 = \overline{T_s^2} - (\overline{T}_s)^2$$

$$= \left[\frac{q^2 - 1}{12} - \frac{(q-1)^2}{P_d} + \frac{(q-1)^2}{P_d^2}\right]T_{da}^2$$

$$+ (2q - 1)\frac{1 - P_d}{P_d^2}T_i^2 + (2q - 1)\frac{1 - P_d}{P_d^2}T_i T_{fa} P_{fa} \quad (2.48)$$

$$- (q - 1)\frac{2 - P_d}{2P_d}T_{fa}^2 P_{fa}^2 + (q - 1)\frac{2 - P_d}{2P_d}T_{fa}^2 P_{fa}$$

This result is similar to the result derived in Section 2.4.1 using signal flow graph techniques. For $q \gg 1$, $1 - P_d \ll 1$, and $P_{fa} \ll 1.0$, the variance is approximated by

$$\sigma_{T_s}^2 \approx T_{da}^2 q^2 \left(\frac{1}{12} - \frac{1}{P_d} + \frac{1}{P_d^2}\right) \quad (2.49)$$

which agrees exactly with (2.35). For $P_d = 1.0$, $P_{fa} = 0.0$, and $q \gg 1$, the variance is

$$\sigma_{T_s}^2 = T_{da}^2 q^2 \left(\frac{1}{12}\right) \quad (2.50)$$

This is equal to the variance of a random variable uniformly distributed over the range $(0, qT_{da})$. The variance increases as P_d falls below unity. One of the goals of the spread spectrum system designer will be to design a synchronization system that minimizes mean synchronization time. Equation (2.44) indicates that mean synchronization time is a function of P_d, P_{fa}, T_i, T_{fa}, and q. The designer has some degree of control over all these variables, including q. Even though the phase uncertainty region in seconds is fixed by system requirements, the region can be subdivided into any number of cells q by the system designer. The remaining four variables are not independent of one another. It will be seen quantitatively later that high P_d together with low P_{fa} implies large T_i. Thus there will be an optimum set of P_d, P_{fa}, T_i, T_{fa} that minimizes mean synchronization time. It is not correct to assume that minimum average synchronization time will always be achieved with $1 - P_d \ll 1.0$, so the correct phase cell is detected on the first sweep. In some cases the selection of a moderate P_d will result in a much lower T_i than a high P_d and, thus, will result in reduced average T_s, even though several sweeps of the uncertainty region may be required.

2.4.3 Modified Search Strategies

If a probability distribution function of the synchro cell position is known, then some other-than-uniform search strategies have proven to be more efficient. One example of a practical situation where the *probability distribution function* (pdf) of the synchro cell can be approximated as triangular or Gaussian is resynchronization, where after some active period the input signal is not available for some time (e.g., push-to-talk communications and slotted packetized communications), or acquisition in the network with universal time, where the code delay uncertainty region is defined by the clock instability. At the beginning of a new active period in all these cases the initial synchronization error will be smaller with higher probability and vice versa. The variance of this error prior pdf is higher if the inactive time is longer. For such a case we will shown that a nonuniform search of the delay uncertainty region is more efficient, resulting in shorter code acquisition time.

The focus here will be on the two classes of nonuniform strategies depicted in Figure 2.5, namely, the z-search and the expanding-window search. These two classes can be further subdivided into broken or continuous searches (depending on whether the receiver employs rewinding in order to skip certain cells) and edge or center searches (depending on where the search and each subsequent sweep are initiated). For a straight serial search, the process can be modeled by a circular state diagram shown in Figure 2.6 with $v + 2$ states, where $v - 1$ of these correspond to out-of-sync cells (hypothesis H_0), one to the collecting state (hypothesis H_1), one to the absorbing correct state (ACQ), and one to the possibly absorbing FA state. The FA state is absorbing if there is nonzero probability of an erroneous decision in the verification mode. Along the branches between these various states are found generalized gains $H(z)$ that represent the generating functions of the individual discrete-time detection processes associated with the corresponding paths. Applying standard flow graph reduction techniques to the circular state diagrams then allows the evaluation of the moment-generating function $U(z)$ of the underlying acquisition process. To gain an initial insight into the computational power offered by this method we will first define some general relations that will be modified for specific search strategies later.

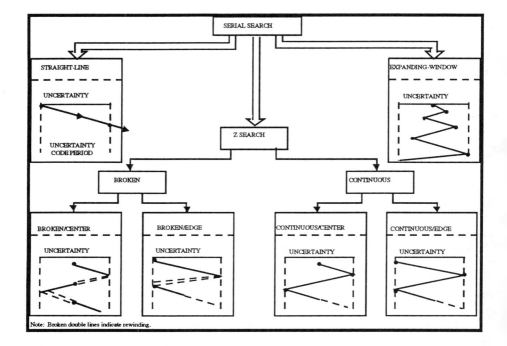

Figure 2.5 Structure of serial-search strategies. (© 1984 IEEE. *From*: [6].)

2.4.3.1 Circular State Diagram

In general, let us assume a serial search system represented in Figure 2.6 where H_1 denotes the hypothesis that the received and local codes are misaligned by less than a code chip (some times designated as a "hit") and H_0 the alternate hypothesis; that is, the relative alignment is greater than or equal to a chip. This can be represented by a $(v + 2)$-state flow graph, where $v - 1$ of the states belong to the cells corresponding to H_0, and one state (the vth) corresponds to H_1. These v states are indexed in a circular arrangement (see Figure 2.6) with the ith state, $i = 1, 2, \ldots, v - 1$ corresponding to the ith offset code phase position to the right of the "true" sync position (H1). The remaining two states are the FA state and the ACQ state. These are also indicated in Figure 2.6, where it can be seen that the acquisition state can be directly reached only from the vth (H_1) state, whereas the FA state can be directly reached from any of the $v - 1$ states corresponding to the offset cells (H_0). The first way in which this flow graph is more general than the one from the previous section is that the a priori probability distribution $\{\pi_j; j = 1, 2, \ldots, v\}$ assigned to the v states at which the search process can be entered is arbitrary. In the absence of any a priori

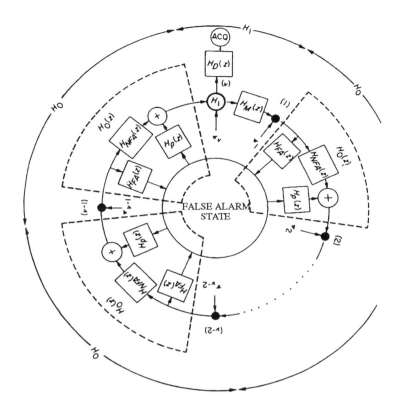

Figure 2.6 A flow graph representation of a generalized serial search acquisition system—the circular state diagram.(© 1984 IEEE. *From*: [6].)

information concerning the initial relative position of the codes, the system designer would assign a uniform distribution to the model, that is, $\{\pi_j = 1/v; j = 1, 2, \ldots, v\}$. This is the case that we have discussed thus far. The second way in which Figure 2.6 is a generalization of the previous notions is that each branch of the flow graph is assigned a generalized gain $H(z)$ that characterizes all the possible ways by which the process can move along that branch. The significance of the subscripts on these branches is as follows:

$H_D(z)$ = gain for verification of detection;
$H_M(z)$ = gain for missed verification of detection;
$H_{FA}(z)$ = gain for false alarm occurrence;
$H_{NFA}(z)$ = gain for no false alarm occurrence;
$H_P(z)$ = gain for penalty after false alarm occurrence.

Properly combining these gains then facilitates computing the gain associated with the "generalized" branch between any pair of nodes, for example,

$$H_0(z) = H_{\text{NFA}}(z) + H_{\text{FA}}(z)H_P(z) \qquad (2.51)$$

representing the gain in going from node i to node $i + 1$ for $i = 1, 2, \ldots, v - 1$. As follows, the flow graph representation of the system is used to compute the moment-generating function $U(z)$ of the underlying acquisition process. Using standard flow graph reduction techniques, it can be shown that

$$U(z) = \frac{H_D(z)}{1 - H_M(z)H_0^{v-1}(z)} \sum_{i=1}^{v} \pi_i H_0^{v-1}(z) \qquad (2.52)$$

Using (2.25) and (2.28) one can get mean acquisition time and its variance for any particular $H_0(z)$ and $H_D(z)$ and specified search strategy. For simplicity in what follows we will use the notation $H(z) \to H$ so that at all times one should be aware that these transfer functions are functions of z.

2.4.3.2 Nonuniform Serial Search

Continuous/Center z Serial Search

To apply this approach to the nonuniform search case, one merely translates the motion of the specific search strategy under consideration into a circle motion along an equivalent circular state diagram analogous to Figure 2.6. To demonstrate how this is accomplished, let us consider first the continuous/center z serial search illustrated in Figure 2.7. Here the search is initiated at the center of the code phase uncertainty region and proceeds following the arrows in the manner shown; that is, it reverses direction every time the boundaries are reached. Assuming that the location of the true sync state (H_1) is at the shaded cell, the search process will meet it once during each sweep at the dotted positions. We indicate the starting cell by $v - k$ where for the H_1 cell to be in the indicated side k must satisfy $1 \leq k \leq (v - 1)/2$ (for convenience, we assume v to be odd). Similar diagrams can be drawn for $k = 0$ or $(v + 1)/2 \leq k \leq v - 1$; however, that will not be necessary due to the symmetry of the problem. Furthermore, we note that since the search is always initiated at the center, π_j should be interpreted as the probability that the central (entrance) cell is not the vth (H_1) but the jth; in other words, π_j stands for the probability that H_1 is actually $v - j$ positions to the right (if $j \geq (v + 1)/2$) or j positions to the left (if $j \leq (v - 1)/2$). Translating the z motion of the search into an equivalent motion along an equivalent circular path leads to the circular state diagram of Figure 2.8, which for purposes of deriving the transfer function $U_{v-k,\text{ACQ}}(z)$ from state $v - k$ to state

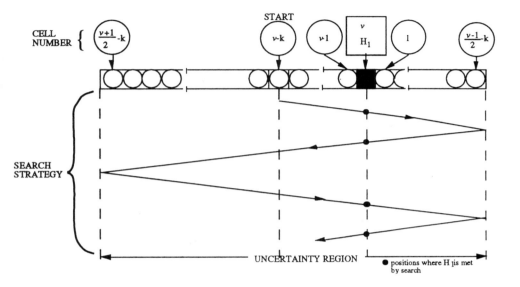

Figure 2.7 Cell numbering for the continuous/center z serial search with $1 \leq k \leq (v-1)/2$. (© 1984 IEEE. From: [23].)

ACQ can be consolidated into Figure 2.9, which contains two forward paths and one feedback loop. By summing up all possible paths from "enter" to "acq," we get for the transfer function (for $0 < k < (v-1)/2$)

$$U_{v-k,\text{ACQ}} = \frac{H_0^k H_D [1 + H_M H_0^{v-2-2k}]}{1 - H_M^2 H_0^{2(v-2)}} \qquad (2.53)$$

Finally, averaging $U_{v-k,\text{ACQ}}(z)$ over the a priori probability distribution of the code phase uncertainty and taking certain symmetries into account give the desired result for the acquisition process generating function $U(z)$, namely,

$$\begin{aligned}
U &= \sum_{k=0}^{v-1} \pi_k U_{v-k,\text{ACQ}} \\
&= H_D \Bigg\{ \frac{\sum_{j=1}^{(v-3)/2} H_j^0 [\pi_{v-j} + \pi_j H_0^{v-1}][1 + H_M H_0^{v-2-2j}]}{1 - H_M^2 H_0^{2(v-2)}} \\
&\quad + \frac{H_0^{(v-1)/2}[\pi_{(v+1)/2} + \pi_{(v-1)/2} H_0^{v-1}]}{1 - H_M H_0^{2v-3}} + \frac{\pi_v}{1 - H_M H_0^{v-2}} \Bigg\} \qquad (2.54)
\end{aligned}$$

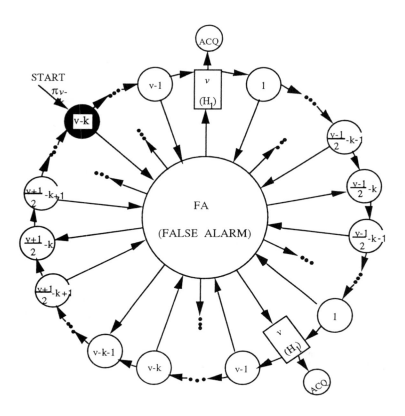

Figure 2.8 Equivalent circular-state diagram for the continuous/center z serial search, $1 \leq k \leq (v-1)/2$. (© 1984 IEEE. From: [23].)

Equation (2.54) can be combined with any a priori distribution to provide specific results. For example, for the symmetric triangular distribution

$$\pi_j = \begin{cases} \left(\dfrac{2}{v-1}\right)\left[1 - \left(\dfrac{2}{v-1}\right)j\right] & j = 1, \ldots, \dfrac{v-1}{2} \\ \pi_{v-j} & j = \dfrac{v+1}{2}, \ldots, v-1 \\ \left(\dfrac{2}{v-1}\right) & j = v \end{cases} \quad (2.55)$$

which can be also used as an approximation to a truncated Gaussian distribution, and U becomes

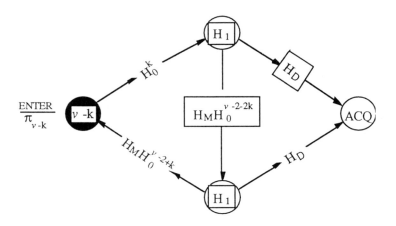

Figure 2.9 Flow graph and corresponding path gains for the continuous/center z serial search entering at node $v - k$, $0 < k < (v - 1)/2$. (© 1984 IEEE. From: [23].)

$$U = \left(\frac{2}{v-1}\right)\left(\frac{H_D}{1 - H_M^2 H_0^{2(v-2)}}\right)$$

$$\times \left\{\frac{H_0(1 + H_0^{v-1})}{1 - H_0}\left[1 - \frac{2}{v-1}\left(\frac{1 - H_0^{(v-1)/2}}{1 - H_0}\right)\right.\right.$$

$$\left.\left. - H_0^{v-3} H_M\left(1 - \frac{2}{v-1}\left(\frac{1 - H_0^{(v-1)/2}}{H_0^{(v-3)/2}(1 - H_0)}\right)\right)\right] + 1 + H_M H_0^{v-2}\right\} \quad (2.56)$$

From the generating function U we can obtain the mean acquisition time $\bar{T}$. For example, for a single dwell system described in the previous section, differentiation and evaluating the result at $z = 1$ gives

$$\bar{T} = \tau_d\left\{\frac{1}{P_D} + \frac{1 + KP_{FA}}{3P_D(2 - P_D)(v - 1)}[6v^2 - 18v + 12 \right.$$

$$\left. - P_D(6v^2 - 15v + 9) + P_D^2(2v^2 - 4v)]\right\} \quad (2.57)$$

which for large v reduces to

$$\bar{T} = \tau_d\left\{\frac{1}{P_D} + \frac{2(1 + KP_{FA})v(3 - 3P_D + P_D^2)}{3P_D(2 - P_D)}\right\} \quad (2.58)$$

This will be compared later with the same parameter for broken/center z search.

Broken/Center z Search

This kind of search is similar to the continuous/center z, with the exception that the same cells are not searched twice in a row; instead, when one of the two boundaries is reached, the local code is quickly rewound to the center and the search continues in the opposite direction. Clearly, for prior distributions that are peaked around the center, an improvement in acquisition performance should be expected with respect to the continuous/center z strategy. Using identical steps as before and denoting by T_r the time required to rewind the code, U is found to be

$$U = H_D \left\{ \sum_{j=1}^{(v-1)/2} [\pi_{v-j} + \pi_j z^{T_r/\tau_d} H_0^{(v+1)/2}] H_0^j + \frac{\pi_v}{1 - z^{T_r/\tau_d} H_M H_0^{(v-1)/2}} \right\} \quad (2.59)$$

which for the case of triangular symmetric prior and large v reduces to

$$U = \left(\frac{2}{v-1}\right) H_D \left\{ \left(\frac{H_0}{1-H_0}\right)\left(\frac{1 + z^{T_r/\tau_d} H_0^{(v+1)/2}}{1 - z^{2T_r/\tau_d} H_M H_0^v}\right) \cdot \left[1 - \left(\frac{2}{v-1}\right)\left(\frac{1 - H_0^{(v-1)/2}}{1 - H_0}\right)\right]$$
$$+ \frac{1}{1 - z^{T_r/\tau_d} H_M H_0^{(v-1)/2}} \right\} \quad (2.60)$$

Once again applying the necessary differentiation, the mean acquisition time for a single dwell system becomes

$$\overline{T} = \tau_d \left\{ \frac{1}{P_D} + \frac{T_r}{\tau_d} \frac{[[(v-3)/2]P_D + 2(v-2)(1-P_D)]}{(v-1)P_D} \right.$$
$$\left. + \frac{(1 + KP_{FA})}{(v-1)P_D} \left[\frac{5}{12}(v^2 - 2v - 3)P_D + (v^2 - 2v - 1)(1-P_D)\right] \right\} \quad (2.61)$$

which for large v reduces to

$$\overline{T} = \tau_d \left\{ \frac{1}{P_D}\left[1 + \left(\frac{4 - 3P_D}{2}\right)\frac{T_r}{\tau_d}\right] + \frac{(1 + KP_{FA})v}{P_D}\left(1 - \frac{7}{12}P_D\right) \right\} \quad (2.62)$$

To illustrate the improvement in mean acquisition time performance by using a broken, rather than a continuous, center z search, we can take the ratio of the latter terms in (2.58) and (2.62) since, for large enough v, the first term in these equations can be ignored. Thus, to a good approximation

$$\frac{\overline{T}_{\text{cont}}}{\overline{T}_{\text{broken}}} = \frac{2(3 - 3P_D + P_D^2)}{3(2 - P_D)(1 - P_D 7/12)} \quad (2.63)$$

Figure 2.10 is a plot of this mean acquisition time improvement factor versus P_D. We observe that the maximum improvement occurs for $P_D = 1$ (absolute probability of detecting the correct cell once it is reached), in which case (2.63) reduces to

$$\frac{\overline{T}_{\text{cont}}}{\overline{T}_{\text{broken}}} = \frac{8}{5} = 1.6 \quad (2.64)$$

that is, a 37.5% savings in acquisition time. In the more general case where the a priori probability distribution of the code phase uncertainty is arbitrary (but symmetric), for $P_D = 1$ and v large, it is simple to show that

$$\frac{\overline{T}_{\text{cont}}}{\overline{T}_{\text{broken}}} = \frac{2\sum_{j=1}^{v/2} j\pi_j + \frac{v}{2}}{\sum_{j=1}^{v/2} j\pi_j + \frac{v}{4}} \quad (2.65)$$

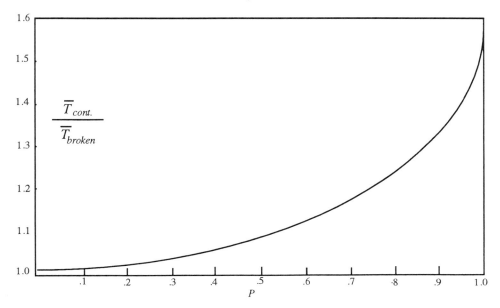

Figure 2.10 Mean acquisition time improvement factor versus detection probability for single dwell acquisition system with triangular a priori code phase uncertainty distribution. (© 1994. *From: Spread Spectrum Communication Handbook* by M. Simon, J. Omura, R. Shultz, and B. Levitt. Reproduced with permission of The McGraw-Hill Companies.)

which is lower and upper bounded by

$$\frac{6}{5} \leq \frac{\overline{T}_{\text{cont}}}{\overline{T}_{\text{broken}}} = 2 \qquad (2.66)$$

corresponding to the a priori distributions

$$\pi_1 = \pi_\nu = \frac{1}{2}$$
$$\pi_j = 0 \qquad j \neq 1, \nu \qquad (2.67)$$

and

$$\pi_{\nu/2} = \pi_{\nu/2+1} = \frac{1}{2}$$
$$\pi_j = 0 \qquad j \neq \nu/2, \nu/2 + 1 \qquad (2.68)$$

Thus, regardless of the a priori probability distribution of the code phase uncertainty, the broken/center z search potentially offers an improvement of at least 20% and at most 100% over the continuous/center z search. Of course, for $P_D < 1$, these improvements will decrease accordingly.

Expanding-Window Search Strategies

The previously discussed concept can be further elaborated to develop a search strategy where cells with high a priori probability of synchro state will be visited more often. These algorithms are known as expanding-window search strategies. Two representative cases of such algorithms (A and B) are shown in Figure 2.11.

The two cases differ in the way in which the search is continued once the entire uncertainty region has been covered without success. In particular, case A repeats the search starting from the window-covering range R_1, while case B continues by repeating the R_{NSW} window. The circular state diagrams for these algorithms will now have N_{SW} levels where N_{SW} is the number of steps (sweeps) in which the window reaches the maximum size. Without going into details of the analysis, some results are shown in Figures 2.12 to 2.14. In these figures the normalized acquisition time versus the number of partial windows is shown, P_D being a parameter. For each figure a different a priori distribution for code delay uncertainty is assumed.

We observe from these figures that except for $P_D = 1$, there always exists an optimum number of partial windows in the sense of minimizing mean acquisition time. For $P_D = 1$, one window, that is, a continuous/center z search, is optimum.

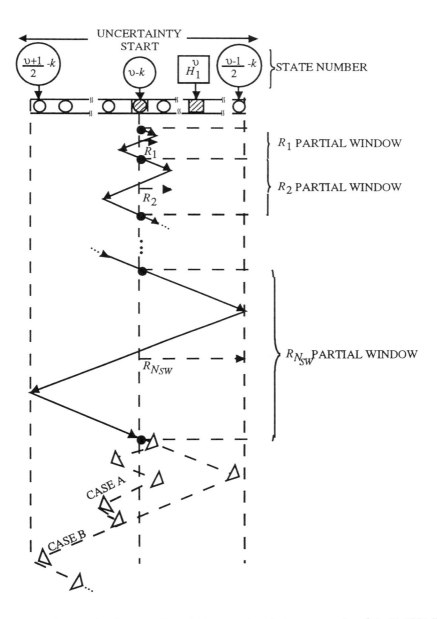

Figure 2.11 Definitions for the expanding window search technique—cases A and B. (© 1984 IEEE. From: [23].)

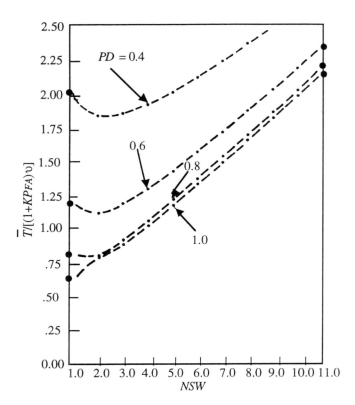

Figure 2.12 Normalized mean acquisition time versus number of partial windows for expanding window search strategy; single dwell system with triangular a priori distribution for code phase uncertainty. (© 1984 IEEE. *From*: [23].)

Furthermore, the more peaked the distribution—for example, Gaussian rather than triangular, or Gaussian with $v/2 = 5\sigma$ rather than Gaussian with $v/2 = 3\sigma$—the more there is to be gained by using an expanding-window rather than a z-type search. Also, the sensitivity of using more than the optimum number of partial windows decreases as the distribution becomes more peaked.

2.5 AUTOMATIC DECISION THRESHOLD LEVEL CONTROL (ADTLC)

For efficient operation of a DSSS receiver, an automatic level control for the decision threshold must be used to provide the decision conditions where the FA probability $P_a \to 0$ and the probability of signal correct detection $P_d \to 1$. A simplified version would be reduced to the problem of setting the threshold at the level where P_a is

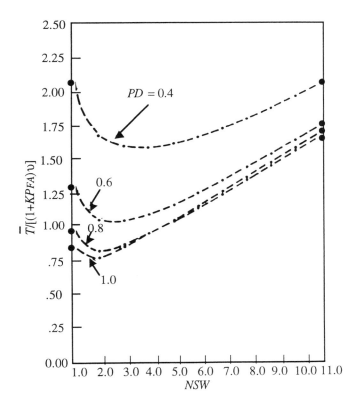

Figure 2.13 Normalized mean acquisition time versus number of partial windows for expanding window search strategy; single dwell system with Gaussian a priori distribution for code phase uncertainty, $v/2 = 3\sigma$. (© 1984 IEEE. *From*: [23].)

constant independent of the variation of the signal or noise level. This class of algorithms is referred to as *constant false alarm rate* (CFAR) algorithm. In this section, the model of the ADTLC scheme is described. As a first step we have to specify the structure of the DSSS receiver itself. We consider a simple DSSS receiver with a single or multiple dwell serial PN acquisition system, which employs a standard, noncoherent (square-law) detector. The received signal plus noise is actively correlated with a local replica of the PN code and then passed through a bandpass filter. The filter output is then square-law envelope detected with the detector output being integrated for a fixed time duration τ_d (the "dwell time") in an integrate-and-dump circuit and then compared to a preset threshold b to detect sync or out-of-sync conditions. A general block diagram for ADTLC circuitry is shown in Figure 2.15 where T_c is the chip period and $c(t - k_1 T_c)$ and $c(t - k_2 T_c)$ are time-displaced versions of the local PN sequence. In order to avoid the situation where

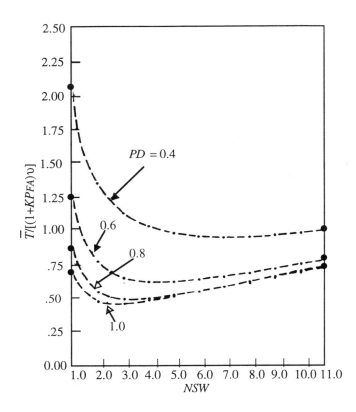

Figure 2.14 Normalized mean acquisition time versus number of partial windows for expanding window search strategy; single dwell system with Gaussian a priori distribution for code phase uncertainty $v/2 = 5\sigma$. (© 1984 IEEE. *From*: [23].)

the sync signal is partly present in both channels we must have $\mathrm{abs}(k_1 - k_2) > 1$. Hardware implementation of the circuitry may be obtained by combining components from Figure 2.15, with the existing components in the receiver. For example, if $\mathrm{mod}(k_1 - k_2) = 2$, then the first four elements in each branch from Figure 2.15 can be used from the DLL tracking loop of the receiver described in Section 3. If the time delay in the DLL loop is only T_c then the DLL cannot be used for this purpose. In this case, the serial synchronization (acquisition) circuitry can be used at the same time to realize the upper branch of ADTLC circuitry and the only additional hardware would be the lower branch from Figure 2.15. In this case, $k_1 = 0$ and $k_2 > 2$. Samples S_1 and S_2, taken every τ_d sec, are used in the "threshold setting" block for threshold updating. Two algorithms for the threshold updating are considered: an *instantaneous threshold setting* (ITS) algorithm and a CFAR algorithm.

Code Acquisition

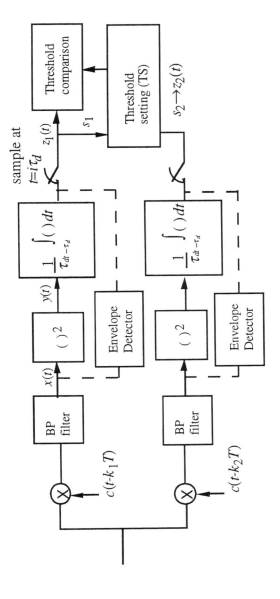

Figure 2.15 General block diagram of ADTLC circuitry. (© 1991 IEEE. *From:* [52].)

2.5.1 Instantaneous Threshold Setting Algorithm

This algorithm is defined by

$$b = k \cdot \min(S_1, S_2) \qquad (2.69)$$

where k is a parameter to be optimized. For the further analysis of this algorithm, the circuit from Figure 2.15 is modified so that the square-law plus integrate-and-dump circuit is replaced by an envelope detector. In this case, the threshold b is determined as a sum of the smallest samples taken in k successive sampling intervals

$$b = (1/k_1) \sum_{i=l}^{k} \min(S_1, S_2) \qquad (2.69a)$$

where k and k_1 are constants to be optimized.

2.5.2 Constant False Alarm Rate Algorithm

For an a priori probability of FA $p_a = r/R$ where r and R are integers, the algorithm operates as follows. First, each τ_d sec the parameter

$$y = \min(S_1, S_2) \qquad (2.70)$$

is formed and compared to the current value of the threshold to form the new variable $\bar{y}$ defined as

$$\bar{y} = \begin{cases} 1 & y \geq b \\ 0 & y < b \end{cases} \qquad (2.71)$$

After R successive decisions a new variable B is formed as the number of samples y having a value larger than b or

$$B = \sum_{i=1}^{R} \bar{y}_i \qquad (2.72)$$

Depending on B, the CFAR algorithm is defined as

$$b \Rightarrow b + \epsilon \Delta b \qquad (2.73)$$

where

$$\epsilon = \begin{cases} -1 & \text{if } B < r - \Delta r \\ 0 & \text{if } r - \Delta r \leq B \leq r + \Delta r \\ +1 & \text{if } B > r + \Delta r \end{cases} \quad (2.74)$$

It should be noted that the average value of $p_a = r/R$ will be better approximated if r and R are larger (keeping p_a constant). Unfortunately, for larger R, the system will be slower. Δr in (2.74) is a parameter to be optimized.

A performance analysis of these algorithm is based on Markov chain models and details can be found in [52]. Some of the results of such analyses are shown in Figures 2.16 to 2.20. There results are obtained given the assumption that the probability that a sample of signal plus noise (z_1) is larger than a sample of noise only (z_0) is much larger than probability that $z_0 > z_1$. Using this assumption, the pdf for the threshold in the ITS system versus the normalized threshold $b^* = bN_B/2\sigma^2$ is shown in Figure 2.16 where b is the threshold. The integrate-and-dump output

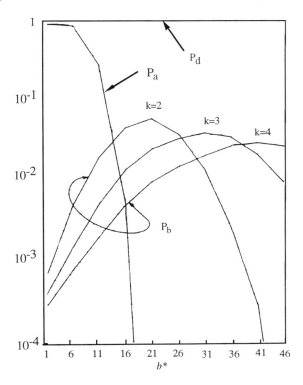

Figure 2.16 Probability density function for the threshold in ITS-system versus b^* with k being a parameter, $N_B = 10$ and signal-to-noise ratio $y = 10$ dB. (© 1991 IEEE. *From:* [52].)

signal is approximated by N_B discrete samples, and σ^2 is the variance of the noise. In the figure, k and N_B are parameters. We can see that the maximum of these curves is in the range where P_a is rather small and that P_d is large (close to 1) at the same time. By increasing k from 1 to 4 the maximum of the P_b curves is moved to the range where P_a is lower. We should be aware of the fact that for a certain value of b^* the probability P_d starts to decrease. This is shown in Figure 2.17 with SNR as a parameter. This means that there is an optimum value of k that simultaneously provides the minimum of P_a with P_d close to 1. From Figure 2.17 we can see that for $y = 6$ dB, k should be 3 or 4; if $y = 12$ dB, k should be somewhere between 5 and 8. The same set of curves from Figure 2.17 is shown in Figure 2.18 for $N_B = 6$. The mutual relationship between curves remains approximately the same, but the curves are now slightly moved to the left where b^* is smaller. This was excepted because the number of samples N_B is now decreased. In the case of the CFAR algorithm, the probability density function of the threshold is given in Figure 2.19. It can be seen that for a given $P_a = 10^{-2}$ the sharp maximum of the

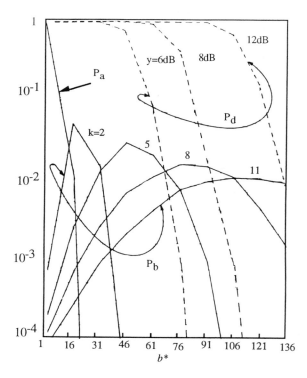

Figure 2.17 Probability density function for the threshold in ITS-system with k and y being parameters and $N_B = 10$. (© 1991 IEEE. From: [52].)

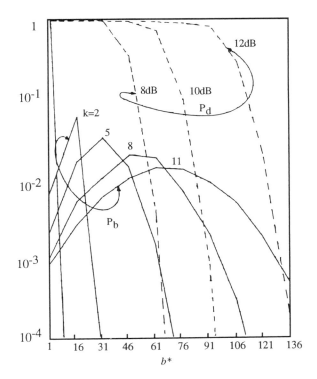

Figure 2.18 Probability density function for the threshold in ITS-system with k and y being parameters and $N_B = 6$. (© 1991 IEEE. From: [52].)

curve is indeed in the range of b^*, which provides $P_a = 10^{-2}$ (note that the b^* scale is now different compared to the abscissa on the previous figures). The good threshold pdf of the CFAR and the small time constant of the ITS algorithms are combined in the MITS algorithm. By a proper choice of the parameters k and k_1 we can control the position and the spreading of the curves representing the pdf for the normalized threshold b_0. This is illustrated in Figure 2.20 for three different pairs k, k_1.

2.6 SEQUENTIAL DETECTION

For a given threshold and fixed integration time the next step in the code acquisition process will depend on whether or not the correlator output is larger than the threshold. Besides that, any additional information about how much beyond or below the threshold the signal is can be used to make these decisions more reliable. In other words, if the signal is much beyond or below the threshold we can be rather

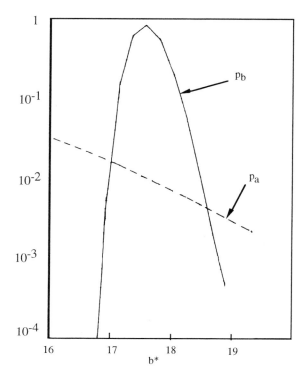

Figure 2.19 Approximated probability density function for the threshold in CFAR system with $R = 500$, $r = 5$, $\Delta r = 1$, and $p_a = 10^{-2}$. (© 1991 IEEE. From: [52].)

sure that our positive or negative decision, respectively, is correct. On the other hand, if the signal is close, either above or below the threshold, it might be better to continue integration in order to improve the likelihood of the correct decision. So, if $d(t)$ is the decision variable, the decision will be modified as follows

$$\begin{cases} \text{If } d(\tau_d) = \int_0^{\tau_d} x(t)\,dt > th + D & \text{Go to verification mode} \\ \text{If } d(\tau_d) < th & \text{Go to next cell} \\ \text{If } th < d(\tau_d) < th + D & \text{Continue the integration} \end{cases} \quad (2.75)$$

where D is decision uncertainty region, th is a threshold, and $x(t)$ is the output of square law detector. So, the decision is based on a sequence of subdecisions, hence the name *sequential detection*. The theory of sequential detection can be found in

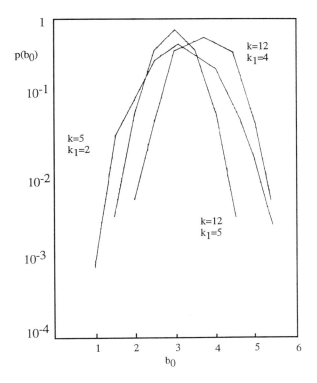

Figure 2.20 Probability density function for the threshold in MITS algorithm versus b_0, k_1, and k being parameters. (© 1991 IEEE. *From*: [52].)

different textbooks and will not be repeated here. Instead we will present a modification of this idea that is very often used in practice. Equation (2.75) can be rearranged as

$$\begin{cases} 1.\ \text{If}\ d(\tau_d) - th > D & \text{Go to verification mode} \\ 2.\ \text{If}\ d(\tau_d) - th < 0 & \text{Go to next cell} \\ 3.\ \text{If}\ 0 < d(\tau_d) - th < D & \text{Continue the integration} \end{cases} \quad (2.76)$$

and implemented as shown in Figure 2.21. Parameter th is created by integrating bias signal b

$$th = \int_0^{\tau_D} b\, dt \quad (2.77)$$

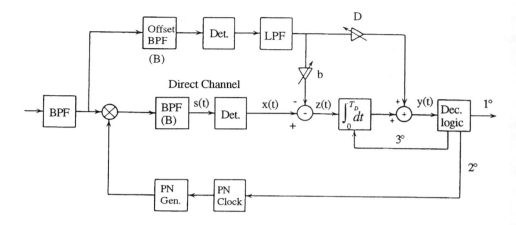

Figure 2.21 Sequential detection acquisition circuit (biphase signal). (© 1982 John Wiley & Sons, Inc. From: *Coherent Spread Spectrum Systems* by Jack Holmes. Reprinted by permission of John Wiley & Sons, Inc.)

In order to gain a better insight into the operation of this circuitry let us first assume that there is no bias b. In this case the square-law detector output alone would be integrated, resulting in

$$d(t) = \int_0^t (s(t) + n(t))^2 dt$$
$$= \int_0^t (s^2(t) + 2s(t)n(t) + n^2(t)) dt \qquad (2.78)$$

The average value of this signal can be represented as

$$\overline{d}(t) = \int_0^t (\overline{s}^2(t) + \overline{n}^2(t)) dt$$
$$= N(1 + \gamma)t = N_0 B(1 + \gamma)t \qquad (2.79)$$

where $N = N_0 B$ is the average noise power $\gamma = S/N$ is the SNR and $\overline{2s(t)n(t)} = 0$ because $n(t)$ and $s(t)$ are uncorrelated. If now bias b is added and $d(t)$ compared with D, the overall signal $y(t)$, shown in Figure 2.22, is obtained

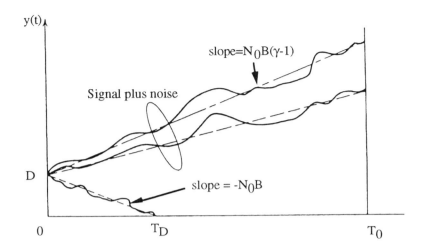

Figure 2.22 Integrator plus threshold output (sequential detector). (© 1982 John Wiley & Sons, Inc. From: *Coherent Spread Spectrum Systems* by Jack Holmes. Reprinted by permission of John Wiley & Sons, Inc.)

$$\bar{y}(t) = D + d(t) - \int_0^{\tau_D} b\,dt = D + N_0 B(1 + \gamma)t - bt \quad (2.80)$$

For $b = 2N_0 B$ we have

$$\bar{y}(t) = D + N_0 B(\gamma - 1)t \quad (2.81)$$

So, in the absence of the signal $y(t)$ on average will decrease from D to zero with slope $(N_0 B - b)$. In order to have negative slope bias b should be larger than $N_0 B$. The cell is dismissed as soon as $y(t)$ reaches zero or after the time-out $t = T_0$. Otherwise the cell is accepted as the synchro cell. For system optimization, both D and b should be carefully chosen.

2.7 LOCK DETECTORS

When code synchronization is acquired and tracking mode is entered, there is a need for constant monitoring of the system in order to be able to take action immediately when the synchronization for any reason has been lost. This function is performed by a so-called lock detector. This can be implemented by checking the level of correlation with a prolonged integration time or by creating the correlation in the same dwell interval t_D and then observing a given sequence of subsequent decisions. A set of possible sequences of these decisions is given in Figure 2.23.

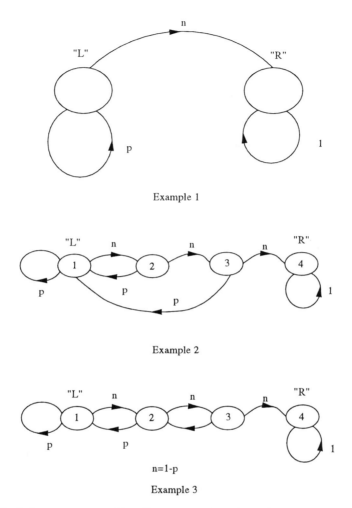

Figure 2.23 Lock detector state transition diagram. (© 1982 John Wiley & Sons, Inc. *From: Coherent Spread Spectrum Systems* by Jack Holmes. Reprinted by permission of John Wiley & Sons, Inc.)

The decision process is represented by a Markov absorbing chain with state "L"-lock, a number of additional transient states, and an absorbing state "R" representing resychronization. Whenever a negative decision is made (signal below the threshold probability n), the process is moved toward the right. In the first example, only one negative decision is needed to enter the resynchronization mode.

In the other two examples of the algorithm several additional check ups are made before entering "R" mode. Each positive decision (signal beyond the threshold probability p) will move the process on the left toward the "L" stage. In the second example, each positive decisions would move the process directly to the "L" state; whereas in the third example, the process follows an up-down counter type of operation. A performance analysis of these algorithms is based on the well-established theory of absorbing Markov chains [53]. As the first step, the state transition probability matrix should be arranged in a canonical form, which in the general case can be represented as

$$\mathbf{P} = \begin{Vmatrix} \mathbf{I} & \mathbf{O} \\ \mathbf{R} & \mathbf{Q} \end{Vmatrix} \qquad (2.82)$$

For a system with a total number of m-states, t-transient states, and $a = m - t$ absorbing states, these matrices represent the following:

I: a by a unity matrix;
O: a by t all-zero matrix;
Q: t by t matrix of transition probabilities between the transient states;
R: t by a matrix of transition probabilities between the transient and absorbing states.

For the three examples from Figure 2.23, the matrix **P** becomes

$$\mathbf{P}_1 = \begin{array}{c} \\ 2 \\ 1 \end{array} \begin{Vmatrix} 2 & 1 \\ 1 & 0 \\ \hline n & p \end{Vmatrix} \qquad (2.83)$$

$$\mathbf{P}_2 = \begin{array}{c} \\ 4 \\ 1 \\ 2 \\ 3 \end{array} \begin{Vmatrix} 4 & 1 & 2 & 3 \\ 1 & 0 & 0 & 0 \\ 0 & p & n & 0 \\ 0 & p & 0 & n \\ n & p & 0 & 0 \end{Vmatrix} \qquad (2.84)$$

$$\mathbf{P}_3 = \begin{array}{c} \\ 4 \\ 1 \\ 2 \\ 3 \end{array} \begin{Vmatrix} 4 & 1 & 2 & 3 \\ 1 & 0 & 0 & 0 \\ 0 & p & n & 0 \\ 0 & p & 0 & n \\ n & 0 & p & 0 \end{Vmatrix} \qquad (2.85)$$

The performance measure for these algorithms is the average lock in time, which is the average time for the process to reach state R.

If the dwell time in state i is T_i, then we introduce vector τ as

$$\tau = \{T_1, T_2, \cdots, T_t\}^T \tag{2.86}$$

We also define the fundamental matrix for an absorbing Markov chain as

$$\mathbf{N} = [\mathbf{I} - \mathbf{Q}]^{-1} \quad \text{dimension } (t \text{ by } t) \tag{2.87}$$

If we introduce

$$\xi = \{\xi_1, \xi_2, \cdots, \xi_t\}^T \tag{2.88}$$

as a vector with components ξ_i being the average time from the process to reach R given it starts from state "i", then the standard results from the absorbing Markov chain theory states [53]

$$\xi = \mathbf{N}\tau \tag{2.89}$$

If we additionally introduce notation

$$\mathbf{1} = \{1, 1, \cdots, 1\}^T \quad \text{dimension } (1 \text{ by } t) \tag{2.90}$$

$$\mathbf{T} = \begin{Vmatrix} T_1 & 0 & 0 \cdots 0 \\ 0 & T_2 & 0 \cdots 0 \\ 0 & 0 & \cdots T_2 \end{Vmatrix} \quad \text{dimension } (t \text{ by } t)$$

$$\| \; \|_{sq} = \text{square of each component of } \| \; \|$$

then the variance of each components ξ_i can be expressed as [53]

$$\text{Var } \xi_i = 2\mathbf{NTQN}\tau + \mathbf{N}(\tau_{sq}) - (\mathbf{N}\tau)_{sq} \tag{2.91}$$

For the three examples from Figure 2.23, (2.89) gives for the average lock in time $\overline{T} = \xi_1$

$$\overline{T}_1 = \frac{T}{n}$$

$$\overline{T}_2 = \frac{1 + n + n^2}{n^3} T$$

$$\overline{T}_3 = \frac{1 + 2n^2}{n^3} T \tag{2.92}$$

where T is the dwell time and $n = 1 - p$. For $p = 0.99$ we have

$$\overline{T}_1 = 100T$$
$$\overline{T}_2 \cong \overline{T}_3 \cong 10^6 T \qquad (2.93)$$

So, using a more sophisticated lock-in algorithm, the average lock-in time can be considerably improved.

References

[1] Van Trees, H. L., *Detection, Estimation and Modulation Theory, Part I*, New York: Wiley, 1968.
[2] Alem, W. K., G. K. Huth, J. K. Holmes, and S. Udalov, "Spread Spectrum Acquisition and Tracking Performance for Shuttle Communication Links," *IEEE Trans on Communications*, Vol. COM-26, Nov. 1978, pp. 1689–1703.
[3] Davisson, L. D., and P. G. Flikkema, "Fast Single-Element PN Acquisition for the TDRSS MA System," *IEEE Trans. on Communications*, Vol. 36, No. 11, Nov. 1988, pp. 1226–1235.
[4] Milstein, L. B., J. Gevargis, and P. K. Das, "Rapid Acquisition for Direct Sequence Spread Spectrum Communications Using Parallel SAW Convolvers," *IEEE Trans. on Communications*, Vol. 33, No. 7, July 1985, pp. 593–600.
[5] Sourour, E. A., and S. C. Gupta, "Direct-Sequence Spread-Spectrum Parallel Acquisition in a Fading Mobile Channel," *IEEE Trans. on Communications*, Vol. 38, No. 7, July 1990, pp. 992–998.
[6] Polydoros, A., and C. L. Weber, "A Unified Approach to Serial Search Spread-Spectrum Code Acquisition—Part I: General Theory," *IEEE Trans. on Communications*, Vol. 32, No. 5, May 1984, pp. 542–549.
[7] Sage, G. F., "Serial Synchronization of Pseudonoise Systems," *IEEE Trans. on Communications*, Vol. 12, Dec. 1964, pp. 123–127.
[8] Holmes, J. K., and K. T. Woo, "An Optimum PN Code Search Technique for a Given A Priori Signal Location Density," *NTC 78 Conf. Record*, Birmingham, Alabama, Dec. 3–6, 1978, Section 18.6, pp. 18.6.1–18.6.5.
[9] Gumacos, C., "Analysis of an Optimum Sync Search Procedure," *IRE Trans. on Communications Systems*, Vol. 11, March 1963, pp. 89–99.
[10] Posner, E. C., "Optimal Search Procedures," *IEEE Trans. on Information Theory*, Vol. IT-11, July 1963, pp. 157–160.
[11] Cheng, U., W. Hurd, and J. Statman, "Spread Spectrum Code Acquisition in the Presence of Doppler Shifts and Data Modulation," *IEEE Trans. on Communications*, Vol. 38, No. 2, Feb. 1990, pp. 241–250.
[12] Di Carlo, D. M., "Multiple Dwell Serial Synchronization of Pseudonoise Signals," Ph.D. Dissertation, Dept. of Electrical Engineering, University of Southern California, May 1979.
[13] Di Carlo, D. M., and C. L. Weber, "Statistical Performance of Single Dwell Serial Synchronization Systems," *IEEE Trans. on Communications*, Vol. 28, No. 8, Aug. 1980, pp. 1382–1388.
[14] Holmes, J. K., and C. C. Chen, "Acquisition Time Performance of PN Spread Spectrum Systems," *IEEE Trans. on Communications*, Special Issue on Spread Spectrum Communications, Vol. 25, Aug. 1977, pp. 778–784.
[15] Di Carlo, D. M., and C. L. Weber, "Multiple Dwell Serial Search: Performance and Application to Direct Sequence Code Acquisition," *IEEE Trans. on Communications*, Vol. 31, No. 5, May 1983, pp. 650–659.
[16] Wald, A., *Sequential Analysis*, New York: Wiley, 1947.
[17] Bussgang, J. J., and D. Middleton, "Optimum Sequential Detection of Signals in Noise," *Trans. IRE*, Vol. IT-1, Dec. 1955, pp. 5–18.

[18] Comparetto, G. M., "A General Analysis for a Dual Threshold Sequential Detection PN Acquisition Receiver," *IEEE Trans. on Communications*, Vol. 35, No. 9, Sept. 1987, pp. 956–960.
[19] Su, Y. T., and C. L. Weber, "A Class of Sequential Tests and Its Applications," *IEEE Trans. on Communications*, Vol. 38, No. 2, Feb. 1990, pp. 165–171.
[20] Hopkins, P. M., "A Unified Analysis of Pseudonoise Synchronization by Envelope Correlation," *IEEE Trans. on Communications*, Vol. 25, No. 8, Aug. 1977, pp. 770–778.
[21] Polydoros, A., and C. L. Weber, "A Unified Approach to Serial Search Spread-Spectrum Code Acquisition—Part II: A Matched-Filter Receiver," *IEEE Trans. on Communications*, Vol. 32, No. 5, May 1984, pp. 550–560.
[22] Polydoros, A., "On the Synchronization Aspects of Direct Sequence Spread Spectrum Systems," Ph.D. Dissertation, Dept. of Electrical Engineering, University of Southern California, Aug. 1982.
[23] Polydoros, A., and M. Simon, "Generalized Serial Search Code Acquisition: The Equivalent Circular State Diagram Approach," *IEEE Trans. on Communications*, Vol. 32, No. 12, Dec. 1984, pp. 1260–1268.
[24] Simon, M. K., J. K. Omura, R. A. Scholtz, and B. K. Levitt, *Spread Spectrum Communication III*, Rockville, MD: Computer Science, 1985.
[25] Braun, W. R., "Performance Analysis for the Expanding Search PN Acquisition Algorithm," *IEEE Trans. on Communications*, Vol. 30, No. 3, March 1982, pp. 424–435.
[26] Cheng, U., "Performance of a Class of Parallel Spread-Spectrum Code Acquisition Schemes in the Presence of Data Modulation," *IEEE Trans. on Communications*, Vol. 36, No. 5, May 1988, pp. 596–604.
[27] Davidovici, S., L. B. Milstein, and D. L. Schilling, "A New Rapid Acquisition Technique for Direct Sequence Spread-Spectrum Communications," *IEEE Trans. on Communications*, Vol. 32, No. 11, Nov. 1984, pp. 1161–1168.
[28] Elhakeem, A. K., G. S. Takbar, and S. C. Gupta, "New Code Acquisition Techniques in Spread Spectrum Communications," *IEEE Trans. on Communications*, Vol. 28, Feb. 1980, pp. 249–257.
[29] Jovanovic, V. M., "Analysis of Strategies for Serial Search Spread-Spectrum Code Acquisition—Direct Approach," *IEEE Trans. on Communications*, Vol. 36, No. 11, Nov. 1988, pp. 1208–1220.
[30] Jovanovic, V. M., "On the Distribution Function of the Spread-Spectrum Code Acquisition Time," *IEEE J. on Selected Areas Comm.*, Vol. 10, No. 4, May 1992, pp. 760–769.
[31] Lee, Y. H., and S. Tantaratana, "Sequential Acquisition of PN Sequences for DS/SS Communications: Design and Performance," *IEEE J. on Selected Areas in Communications*, Vol. 10, No. 4, May 1992, pp. 750–759.
[32] Mark, J. W., and I. F. Blake, "Rapid Acquisition Techniques in CDMA Spread-Spectrum Systems," *IEE Proc.*, Vol. 131, Part F, No. 2, April 1984, pp. 223–232.
[33] Meyr, H., and G. Polzer, "Performance Analysis for General PN-Spread Spectrum Acquisition Techniques," *IEEE Trans. on Communications*, Vol. 31, No. 12, Dec. 1983, pp. 1317–1319.
[34] Pan, S. M., D. E. Dodds, and S. Kumar, "Acquisition Time Distribution for Spread-Spectrum Receiver," *IEEE J. on Selected Areas in Communications*, Vol. 8, No. 5, June 1990, pp. 800–808.
[35] Pandit, M., "Mean Acquisition Time of Active and Passive-Correlation Acquisition Systems for Spread-Spectrum Communication Systems," *IEEE Proc.*, Vol. 128, Part F, No. 4, Aug. 1981, pp. 211–214.
[36] Putman, C. A., S. S. Rappaport, and D. L. Schilling, "A Comparison of Schemes for Coarse Acquisition of Frequency Hopped Spread Spectrum Signals," *Proc. ICC '81*, Denver, CO, June 1981, pp. 34.2.1–34.2.5.
[37] Rappaport, S. S., and D. Schilling, "A Two Level Coarse Code Acquisition Scheme for Spread Spectrum Radio," *IEEE Trans. on Communications*, Vol. 28, No. 9, Sept. 1980, pp. 1734–1742.
[38] Sourour, E. A., and S. C. Gupta, "Direct-Sequence Spread-Spectrum Parallel Acquisition in Nonselective and Frequency-Selective Rician Fading Channels," *IEEE J. on Selected Areas in Communications*, Vol. 10, No. 3, April 1992, pp. 535–544.

[39] Su, Y. T., "Rapid Code Acquisition Algorithm Employing PN Matched Filters," *IEEE Trans. on Communications*, Vol. 36., No. 6, June 1988, pp. 724–733.
[40] Weinberg, A., "Generalized Analysis for the Evaluation of Search Strategy Effects on PN Acquisition Performance," *IEEE Trans. on Communications*, Vol. 31, No. 1, Jan. 1983, pp. 37–49.
[41] Wilson, N. D., S. S. Rappaport, and M. M. Vasudevan, "Rapid Acquisition Scheme for Spread-Spectrum Radio in a Fading Environment," *IEEE Proc.*, Vol. 135, Part F, No. 1, Feb. 1988, pp. 95–104.
[42] Ziemer, R. E., and R. L. Peterson, *Digital Communications and Spread Spectrum Systems*, New York: MacMillan, 1985.
[43] Krebser, J., "Performance of FH-Synchronizers with Constant Search Rate in the Presence of Partial Band Noise," *Proc. 1980 Int. Zurich Seminar on Communications.*, 1980, pp. A9.1–A9.6.
[44] Miller, L. E., J. S. Lee, R. H. French, and D. J. Torrieri, "Analysis of an Antijam FH Acquisition Scheme," *IEEE Trans. on Communications*, Vol. 40, No. 1, Jan. 1992, pp. 160–170.
[45] Pawlowski, P., and A. Polydoros, "Optimization of a Matched Filter Receiver for FH Code Acquisition in Jamming," *Proc. IEEE 1993 Conf. Military Commun.*, Oct. 1985, pp. 1.1.1–1.1.7.
[46] Siess, E. W., and C. L. Weber, "Acquisition of Direct Sequence Signals with Modulation and Jamming," *IEEE J. on Selected Areas in Communications*, Vol. 4, No. 2, March 1986, pp. 254–272.
[47] Madhow, U., and M. B. Pursley, "Acquisition in Direct-Sequence Spread-Spectrum Communication Networks: An Asymptotic Analysis," *IEEE Trans. on Information Theory*, Vol. 39, No. 3, May 1993, pp. 903–912.
[48] Dlugos, D. M., and R. A. Scholtz, "Acquisition of Spread Spectrum Signals by an Adaptive Array," *IEEE Trans. on Acoustics, Speech and Signal Processing*, Vol. 37, No. 8, Aug. 1989, pp. 1253–1270.
[49] Kilgus, C. C., "Pseudonoise Code Acguisition Using Majority Logic Decoding," *IEEE Trans. on Communications*, Vol. 21, June 1973, pp. 772–774.
[50] Ward, R. B., "Acquisition of Pseudonoise Signals by Sequential Estimation," *IEEE Trans. on Communications*, Vol. 13, Dec. 1965, pp. 475–483.
[51] Ward, R. B., and K. P. Yiu, "Acquisition of Pseudonoise Signals by Recursion Aided Sequential Estimation," *IEEE Trans. on Communications*, Vol. 25, No. 8, Aug. 1977, pp. 784–794.
[52] Glisic, G. S., "Automatic Decision Threshold Level Control in Direct-Sequence Spread-Spectrum Systems," *IEEE Trans. on Communications*, Vol. 39, No. 2, Feb. 1991, pp. 187–192.
[53] Kameny, J., and J. Snell, *Finite Markov Chains*, New York: Van Hostrand, 1960.

Selected Bibliography

Aftelak, A., et al., "Design and Implementation of Spread Spectrum Demodulator for Data-Relay Systems," *IAF 92*, 0416, 1992.
Alem, W. K., "Advanced Techniques for Direct Sequence Spread Spectrum Acquisition," Ph.D. Dissertation, Dept. of Electrical Engineering, University of Southern California, Feb. 1977.
Anderson, B., and I. Moore, *Optimal Filtering*, Englewood Cliffs, NJ: Prentice-Hall, 1979.
Cahn, C. R., "Performance of Digital Matched Filter Correlator With Unknown Interference," *IEEE Trans. on Communications*, Vol. 19, No. 6, Dec. 1971, pp. 1163–1172.
Cheng, T., et al., "Single Dwell and Multi Dwell PN Code Acquisition in Multipath Rayleigh Fading Channel," *PIMRC'93*, Yokohama, Japan, Sept. 8–11, 1993, pp. 276–283.
Gilbert, E., "Capacity of a Burst-Noise Channel," *Bell System Tech. J*, Vol.39, 1960, pp. 1253–1265.
Glisic, G. S., "Automatic Decision Threshold Level Control (ADTLC) in Direct-Sequence Spread-Spectrum Systems Based on Matching Filtering," *IEEE Trans. on Communications*, Vol. 36, No. 4, April 1988, pp. 519–527.
Grieco, D. M., "The Application of Charge Coupled Devices to Spread Spectrum Systems," *IEEE Trans. on Communications*, Vol. COM-28, No. 9, Sept. 1980, pp. 1693–1705.

Grob, M., et al., "Microcellular Direct Sequence Spread Spectrum Radio System Using N-Path RAKE Receiver," *IEEE J. on Selected Areas in Comm.*, Vol. 8, No. 2, June 1990, pp. 772–780.

Hurd, W. J., et al., "High Dynamic GPS Receiver Using Maximum Likelihood Estimation and Frequency Tracking," *IEEE Trans. on Aerospace*, Vol. 23, Sept. 1987.

Kanal, L., et al., "Models for Channels with Memory and their Applications to Error Control," *Proc. IEEE*, Vol. 66, 1978, pp. 724–744.

Lehnert, I., and M. Pursley, "Multipath Diversity Reception of Spread Spectrum Multiple Access Communications," *IEEE Trans. on Communications*, Vol. COM-35, No. 11, Nov. 1987, pp. 1189–1198.

Miyagaki, Y., et al., "Double Symbol Error Rates of M-arg DPSK in a Satellite—Aircraft Multipath Channels," *IEEE Trans. on Communications*, Vol. COM-31, Dec. 1983, pp. 1285–1289.

Polydoros, A., and S. G. Glisic, "Code Synchronization: A Review of Principles and Techniques," *Code Division Multiple Access Communications*, S. Glisic and P. Leppanen (eds.), Amsterdam: Kluwer, 1995, pp. 225–266.

Rappaport, S. S., and D. M. Grieco, "Spread-Spectrum Signal Acquisition: Methods and Technology," *IEEE Comm. Magazine*, Vol. 22, No. 6, June 1984, pp. 6–21.

Soliman, S., and R. Scholtz, "Synchronization over Fading Dispersive Channels," *IEEE Trans. on Communications*, Vol. 36, No. 4, April 1988, pp. 499–505.

Stiffler, J. J., "Rapid Acquisition Sequences," *IEEE Trans. on Information Theory*, Vol. IT-14, No. 2, March 1968, pp. 221–225.

Stiffler, J. J., *Theory of Synchronous Communications*, Englewood Cliffs, NJ: Prentice-Hall, 1971.

Suzuki, R., et al., "Spread Spectrum Satellite Communication Terminal with Coherent Matched Filter," Conference record, *GLOBECOM'86*, Vol. 2, pp. 728–732.

Thompson, M., et al., "Non-Coherent PN Code Acquisition in Direct Sequence Spread Spectrum Systems Using a Neural Network," *Proc. IEEE 1993 Conf. Military Communications*, Vol. 1, Oct. 11–14, 1993, pp. 30–34.

Vajda, I., and G. Einarsson, "Code Acquisition for a Frequency-Hopping System," *IEEE Trans. on Communications*, Vol. COM-35, No. 5, May 1987, pp. 566–568.

Appendix A2 Code Acquisition Time Variance

For the analysis of the code acquisition time variance let us recall that the variance of a random variable x can be expressed as

$$\sigma_x^2 = E[(x - \bar{x})^2] = E[x^2] - E^2[x] \qquad (A2.1)$$

In addition to the mean of the synchronization time that has already been calculated, for the variance we need only the mean-square value. The synchronization time for a received phase that lies in the nth cell and is detected after j missed detections and k FAs is given by (2.38). The mean-square value is calculated using

$$\overline{T_s^2} = \sum_{n,j,k} T^2(n, j, k) \Pr[n, j, k] \qquad (A2.2)$$

The probability distribution $\Pr[n, j, k]$ is the same as that used to calculate the mean. Substituting (2.38) and (2.39) into (A2.2) yields

$$\overline{T_s^2} = \frac{1}{q} \sum_{n=1}^{q} \sum_{j=0}^{\infty} \sum_{k=0}^{K} [(n+jq)T_i + kT_{fa}]^2 \binom{K}{k} P_{fa}^k (1-P_{fa})^{K-k} P_d (1-P_d)^j \quad (A2.3)$$

where

$$A_2 = T_{fa}^2$$
$$A_1 = 2(n+jq)T_i T_{fa}$$
$$A_0 = (n+jq)^2 T_i^2 \quad (A2.4)$$

This expression can be simplified by recognizing that the sums over k are the moments of a binomially distributed random variable. The first sum is

$$\sum_{k=0}^{K} A_2 k^2 \binom{K}{k} P_{fa}^k (1-P_{fa})^{K-k} = A_2 [K^2 P_{fa}^2 + K P_{fa}(1-P_{fa})] \quad (A2.5)$$

and the other two terms were previously evaluated. Thus (A2.3) simplifies to

$$\overline{T_s^2} = \frac{P_d}{q} \sum_{n=1}^{q} \sum_{j=0}^{\infty} \{A_2 [K^2 P_{fa}^2 + K P_{fa}(1-P_{fa})] + A_1 [K P_{fa}] + A_0\}(1-P_d)^j \quad (A2.6)$$

This expression is evaluated by grouping like powers of j after substituting $K = n + jq - j - 1$, which results in

$$\overline{T_s^2} = \frac{P_d}{q} \sum_{n=1}^{q} \sum_{j=0}^{\infty} (B_2 j^2 + B_1 + B_0)(1-P_d)^j \quad (A2.7)$$

where

$$B_2 = q^2 T_i^2 + 2q(q-1)T_i T_{fa} P_{fa} + T_{fa}^2 P_{fa}^2 (q-1)^2$$
$$B_1 = 2nq T_i^2 + 2(2qn - n - q)T_i T_{fa} P_{fa}$$
$$\quad + 2T_{fa}^2 P_{fa}^2 (n-1)(q-1) + T_{fa}^2 (q-1)P_{fa}(1-P_{fa})$$
$$B_0 = n^2 T_i^2 + 2n(n-1)T_i T_{fa} P_{fa} + T_{fa}^2 (n-1)^2 P_{fa}^2 \quad (A2.8)$$
$$\quad + (n-1)P_{fa}(1-P_{fa})T_{fa}^2$$

The summations over j can be evaluated using (2.46) and the relation

$$\sum_{i=0}^{\infty} i^2 m^i = \frac{m(1+m)}{(1-m)^3} \quad (A2.9)$$

The result is

$$\overline{T_s^2} = \frac{1}{q}\sum_{n=1}^{q}\left[B_2\frac{(1-P_d)(2-P_d)}{P_d^2} + B_1\frac{1-P_d}{P_d} + B_0\right] \quad (A2.10)$$

The summation over n is evaluated by grouping like powers of n, which yields

$$\overline{T_s^2} = \frac{1}{q}\sum_{n=1}^{q}(D_2 n^2 + D_1 n + D_0) \quad (A2.11)$$

where

$$D_2 = (T_i + T_{fa}P_{fa})^2$$

$$D_1 = -2T_i T_{fa} P_{fa} - 2T_{fa}^2 P_{fa}^2 + T_{fa}^2 P_{fa}(1-P_{fa})$$
$$+ [2qT_i^2 + 2(2q-1)T_i T_{fa} P_{fa} + 2(q-1)T_{fa}^2 P_{fa}^2]\left(\frac{1-P_d}{P_d}\right)$$

$$D_0 = T_{fa}^2 P_{fa}^2 - T^2 P_{fa}(1-P_{fa})$$
$$+ [-2qT_i T_{fa} P_{fa} - 2(q-1)T_{fa}^2 P_{fa}^2 + (q-1)T_{fa}^2 P_{fa}^2(1-P_{fa})]\left(\frac{1-P_d}{P_d}\right)$$
$$+ [C^2 T_i^2 2q(q-1)T_i T_{fa} P_{fa} + (q-1)^2 T_{fa}^2 P_{fa}^2]\frac{(1-P_d)(2-P_d)}{P_d^2} \quad (A2.12)$$

The summations over n are evaluated using

$$\sum_{n=1}^{q} n = \frac{q(q+1)}{2}$$
$$\sum_{n=1}^{q} n^2 = \frac{q(q+1)(2q+1)}{6} \quad (A2.13)$$

The remainder of the calculation is straightforward. Using (2.47) to obtain the square of the mean value of T_s, the final result is (2.48).

CHAPTER 3

Code Tracking

3.1 PHASE-LOCKED LOOP FUNDAMENTALS

3.1.1 Analog PLL

The *phase-locked loop* (PLL) is a basic element of any configuration of code tracking systems. For this reason we shall first summarize the basic equations and performance of the PLL and then apply these general results to different code tracking system configurations.

Let us look at the problem where a phase of an input signal

$$s(t, \theta) = s(t) = \sqrt{2}A \sin\Phi(t) = \sqrt{2}A \sin(\omega_0 t + \theta(t)) \tag{3.1}$$

has to be estimated. Because we do not know the signal amplitude, including the frequency offset into the phase of the local signal, the MLT estimator from Figure 2.1 reduces to the scheme shown in Figure 3.1, known as the phase-locked loop, where

$$\hat{s}(t, \theta) = \frac{\partial s(t, \hat{\theta})}{\partial \hat{\theta}} = \hat{s}(t) = \sqrt{2}K_1\cos\hat{\Phi}(t) = \sqrt{2}K_1\cos(\omega_0 t + \hat{\theta}(t)) \tag{3.2}$$

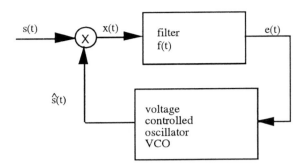

Figure 3.1 Phase-locked loop.

So, the PLL is an ML-motivated structure that is supposed to adjust $\Phi(t)$ and $\hat{\Phi}(t)$, so the phase error

$$\phi(t) = \Phi(t) - \hat{\Phi}(t) = \theta(t) - \hat{\theta}(t) \tag{3.3}$$

should be eliminated. The loop consists of a phase detector, implemented as a multiplier and filter with impulse response $f(t)$, and *voltage-controlled oscillator* (VCO). The integrator from Figure 2.1(b) is now replaced with a lowpass filter to allow slight time variations of the parameter to be synchronized. After multiplying $s(t)$ by $\hat{s}(t)$ and lowpass filtering this product, only the lowpass component

$$x(t) = AK_1 K_m \sin[\theta(t) - \hat{\theta}(t)] \tag{3.4}$$

will have an influence on the VCO output signal. The signal component with double frequency will be filtered out by the cascade of the filter and VCO. We will see later that the VCO has an integrating effect on the input control signal so that even if the loop filter is not used, the VCO will eliminate the double-frequency component of the product. The VCO control signal can be presented as

$$e(t) = f(t) * [K_m K_1 A \sin[\theta(t) - \hat{\theta}(t)]] \tag{3.5}$$

and its Laplace transform as

$$E(s) = F(s)X(s) \tag{3.6}$$

where $F(s)$ is the Laplace transform of $f(t)$. If the control law of the VCO is

$$\frac{d\hat{\Phi}(t)}{dt} = \omega_0 + K_0 e(t)$$

then by definition we have

$$\hat{\Phi}(t) = \omega_0 t + \hat{\theta}(t)$$

and

$$\frac{d\hat{\theta}}{dt} = K_0 e(t) = KA \int_0^t f(t-u) \sin[\theta(u) - \hat{\theta}(u)] \, du \qquad (3.8)$$

where $K = K = K_0 K_1 K_m$ is called loop gain. Using the phase error $\phi(t)$ defined by (3.3) the previous expression becomes

$$\frac{d\phi(t)}{dt} = \frac{d\theta(t)}{dt} - KA \int_0^t f(t-u) \sin \phi(u) \, du \qquad (3.9)$$

Based on these two equations the baseband model of the phase-locked loop is shown in Figure 3.2

3.1.2 The Linear Model

In the tracking mode, phase error $\phi(t)$ is assumed to be small so that $\sin \phi(t) \cong \phi(t)$. This results in a linear model of the PLL. Equation (3.9) now becomes

$$\frac{d\phi(t)}{dt} = \frac{d\theta(t)}{dt} - KA \int_0^t f(t-u) \phi(u) \, du \qquad (3.10)$$

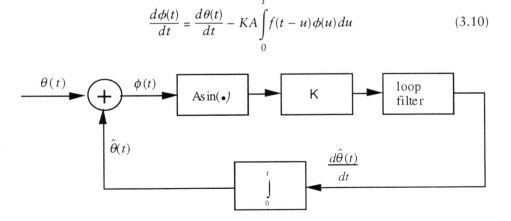

Figure 3.2 Baseband model of the phase-locked loop.

and its Laplace transform can be expressed as

$$s\phi(s) = s\theta(s) - KAF(s)\phi(s)$$

which results in

$$\phi(s) = \theta(s) - KAF(s)\phi(s)/s \tag{3.11}$$

Based on this equation, the linearized baseband model of the phase-locked loop is presented in Figure 3.3. In the next step we replace $\phi(s)$ by $\theta(s) - \hat{\theta}(s)$ and define the loop transfer function $H(s)$ as

$$H(s) = \frac{\hat{\theta}(s)}{\theta(s)} = \frac{KAF(s)}{s + KAF(s)} \tag{3.12}$$

One should note that the order of $H(s)$ is one higher than the order of the loop filter transfer function $F(s)$. Using this expression we also have

$$\phi(s) = \theta(s) - \hat{\theta}(s) = [1 - H(s)]\theta(s)$$
$$\frac{\phi(s)}{\theta(s)} = \frac{1}{1 + KAF(s)/s} = \frac{1}{1 + G_0(s)} \tag{3.13}$$

where $G_0(s) = KAF(s)/s$ is called the open-loop transfer function of the system.

3.1.3 Steady-State Phase Error

In order to get an initial insight into the loop behavior we will investigate its steady-state error $\phi(t \to \infty)$. Most of the time this value is more easily evaluated in the Laplace domain using the relation

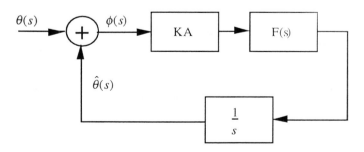

Figure 3.3 Linearized baseband model of the phase-locked loop.

$$\phi(t \to \infty) = \lim_{t \to \infty} \phi(t) = \lim_{s \to 0} s\phi(s) \qquad (3.14)$$

We will specify $\phi(t \to \infty)$ with respect to well-defined elementary signals such as

$$\theta(t) = \begin{cases} \Delta\theta \\ \Delta\omega t \\ \Delta\dot{\omega} t^2/2 \end{cases} \Rightarrow \theta(s) = \begin{cases} \Delta\theta/s & \text{phase step} \\ \Delta\omega/s^2 & \text{frequency step} \\ \Delta\dot{\omega}/s^3 & \text{frequency ramp} \end{cases} \qquad (3.15)$$

As a first example, let us consider the steady-state phase error resulting from a phase step of magnitude $\Delta\theta$. Using (3.13) to (3.15) we have

$$\phi(t \to \infty) = \lim_{s \to 0} \left\{ s \frac{\Delta\theta}{s} \frac{1}{1 + [KAF(s)/s]} \right\} = 0 \qquad F(0) \neq 0 \qquad (3.16)$$

In other words, even a zeroth-order filter, which results in a first-order loop, will compensate a phase step of the input signal.

Another example with $\theta(s) = \Delta\omega/s^2$ will result in

$$\phi(t \to \infty) = \lim_{s \to 0} \left\{ s \frac{\Delta\omega}{s^2} \frac{1}{1 + [KAF(s)/s]} \right\} = \lim_{s \to 0} \left[\frac{\Delta\omega}{s + KAF(s)} \right] \qquad (3.17)$$

If we now assume that the filter transfer function can be represented as

$$F(s) = \frac{1}{s^k} \cdot F_1(s)$$

and

$$0 < |F_1(0)| < \infty \qquad (3.18)$$

then we need at least $k = 1$ in order for (3.17) to reduce to zero. This means that the frequency step of the input signal can be eliminated by the second-order loop.

In the case of a frequency ramp we have

$$\phi(t \to \infty) = \lim_{s \to 0} \left\{ s \frac{\Delta\dot{\omega}}{s^3} \frac{1}{1 + [KAF(s)/s]} \right\} = \lim_{s \to 0} \left[\frac{\Delta\dot{\omega}}{s^2 + KAF(s)s} \right] \qquad (3.19)$$

If a filter with one pole at the origin was used to track this signal, we would have a phase error equal to

$$\phi(t \to \infty) = \frac{\Delta\dot{\omega}}{KAF_1(0)} \qquad (3.20)$$

As a general rule we can state that in order to track an input phase signal with a Laplace transform of the form s^{-k}, a filter with $(k-1)$ poles at the origin is needed if zero steady-state phase error is required.

3.1.4 Transient Response of the PLL Under Linear Conditions

For a given transfer function $H(s)$ and input phase $\theta(s)$, the transient behavior of a PLL under linear conditions can be analysed by evaluating $\phi(t)$ from (3.13) using an inverse Laplace transform. For the three forms of $\theta(s)$ specified by (3.15) and the two types of the second-order loop specified in Table 3.1, transient responses are shown in Figure 3.4. One should notice that for $2\zeta\omega_n \gg \omega_n^2/KA$ the two loops become identical. As was expected, the second-order loop will compensate for the phase step $\Delta\theta$ of the input signal. The transient response will be less oscillatory if the damping factor ζ is higher. The perfect second-order loop will also compensate for a step in frequency. Finally, for the frequency ramp at its input, even the perfect second-order loop will not be able to compensate for the error and the constant error $\phi = \Delta\dot{\omega}/\omega_n^2$ will remain even in the steady state. To track a frequency ramp with zero steady-state phase error, a loop filter with two poles at the origin is required. As a result, a third-order loop specified in Table 3.1 is obtained.

Transient responses of such a loop with $\alpha = \alpha_1$ for the three forms of the input signal specified by (3.15) are shown in Figure 3.5. One should notice that low α will compensate for phase step faster with less oscillations. At the same time, in order to compensate for frequency step and frequency ramp, a higher α is needed. For loop stability, all poles of $H(s)$ should be in the left-hand side of the complex plane. The stability conditions for the three types of the loop are also specified in Table 3.1.

3.1.5. Loop Operation in the Presence of Noise

In the next step we will assume that in addition to $s(t)$, narrowband noise of the form

$$n(t) = \sqrt{2}[n_1(t)\cos \omega_0 t + n_2(t)\sin \omega_0 t] \qquad (3.21)$$

Table 3.1
Specification of the Loop Parameters

loop order	F(s)	H(s)	parameters	stability conditions
2nd	$\dfrac{1+ST_2}{ST_1}$ active filter	$\dfrac{2\zeta\omega_n s + \omega_n^2}{s^2 + 2\zeta\omega_n s + \omega_n^2}$	$\omega_n = \sqrt{\dfrac{KA}{T_1}};\ \zeta = \dfrac{T_2\omega_n}{2}$	all KA
2nd	$\dfrac{1+ST_2}{1+ST_1}$	$\dfrac{\left[2\zeta\omega_n - \omega_n^2/KA\right]s + \omega_n^2}{s^2 + 2\zeta\omega_n s + \omega_n^2}$	$\omega_n = \sqrt{KA/T_1}$ $\zeta = \dfrac{T_2\omega_n}{2} + \dfrac{\omega_n}{2KA}$	all KA
3rd	$\dfrac{(1+ST_2)(1+ST_3)}{(ST_1)^2}$	$\dfrac{\omega_c(s+\alpha\omega_c)(s+\alpha_1\omega_c)}{s^3 + \omega_c(s+\alpha\omega_c)(s+\alpha_1\omega_c)}$	$\alpha = 1/\omega_c T_2$ $\alpha_1 = 1/\omega_c T_3$ $\omega_c = KA\left(\dfrac{T_2}{T_1}\right)\left(\dfrac{T_3}{T_1}\right)$	for $\alpha = \alpha_1$ $KA > \dfrac{1}{2T_2}\left(\dfrac{T_1}{T_2}\right)$

Code Tracking 111

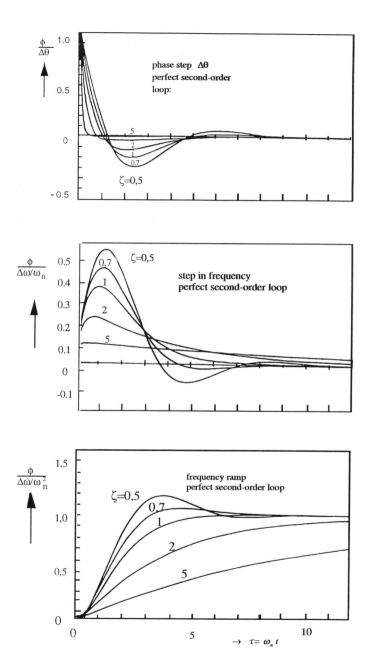

Figure 3.4 Transient second-order loop response under linear conditions. (© 1990 John Wiley & Sons, Inc. From: *Synchronization in Digital Communications, Volume 1* by H. Meyr. Reprinted by permission of John Wiley & Sons, Inc.)

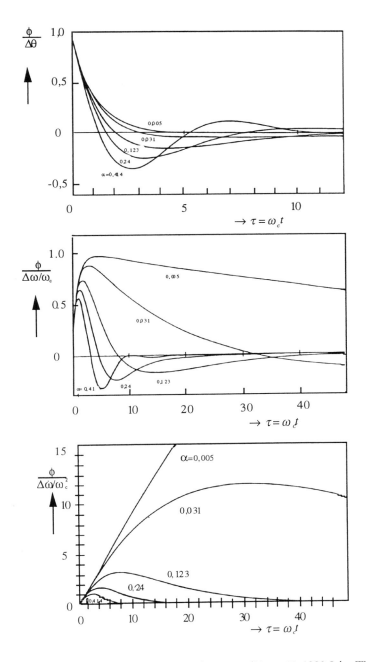

Figure 3.5 Transient third-order loop response under linear conditions. (© 1990 John Wiley & Sons, Inc. *From: Synchronization in Digital Communications, Volume 1* by H. Meyr. Reprinted by permission of John Wiley & Sons, Inc.)

is also present at the loop input. n_1 and n_2 are zero mean Gaussian variables. In this case, it can be shown that the baseband component of $x(t)$ has the form

$$x(t) = AK_1K_m\sin \phi(t) + n'(t) \qquad (3.22)$$

where

$$n'(t) = n_1(t)\sin \hat{\theta}(t) + n_2(t)\cos \hat{\theta}(t) \qquad (3.23)$$

is also a zero mean Gaussian variable. With these modifications, (3.9) now becomes

$$\frac{d\phi(t)}{dt} = \frac{d\theta(t)}{dt} - K\int_0^t [A \sin \phi(u) + n'(u)]f(t-u)\,du \qquad (3.24)$$

The model of this equation is given in Figure 3.6. In the linear regime block $A \sin(\ )$ is replaced by an amplifier with gain A. For the linear model the superposition principle is valid, so the response to $n'(t)$ can be analyzed separately. Putting $\theta_1(t) = 0$ the phase error due to $n'(t)$ is

$$\phi(t) = -\hat{\theta}(t) \qquad (3.25)$$

and the tracking error power density is

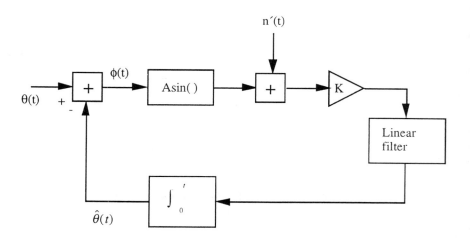

Figure 3.6 Model of phase-locked loop with additive noise.

$$S_\phi(\omega) = S_{\hat\theta}(\omega) = \left|\frac{KF(\omega)/i\omega}{1 + AKF(\omega)/i\omega}\right| S_{n'}(\omega) \qquad (3.26)$$

If the input noise one-sided spectral density is N_0, then $S_{n'}(\omega) = N_0/2$. Bearing in mind (3.12) we have

$$S_\phi(\omega) = \frac{N_0}{2A^2}|H(i\omega)|^2 \qquad (3.27)$$

The variance of the phase error due to noise is

$$\sigma_\phi^2 = \int_{-\infty}^{\infty} S_\phi(\omega)\,d\omega = \frac{N_0}{2A^2}\int_{-\infty}^{\infty}|H(i\omega)|^2 \frac{d\omega}{2\pi} = \frac{N_0}{A^2}\int_{0}^{\infty}|H(i\omega)|^2 \frac{d\omega}{2\pi} \qquad (3.28)$$

We define the loop-noise bandwidth as

$$B_L = \int_0^\infty |H(i\omega)|^2 \frac{d\omega}{2\pi} \qquad (3.29)$$

so that the phase-error variance in the steady state can be written as

$$\sigma_\phi^2 = \frac{N_0 B_L}{A^2} \qquad (3.30)$$

and is inversely proportional to the SNR in the loop.

3.1.6 Nonlinear Model of PLL

When the tracking error is large, a linear model of the loop cannot be used. So, for the analysis of the acquisition process where the initial error is large, a nonlinear model must be used. The second-order loop is the most interesting for practical applications. For the purpose of this analysis we will assume the normalized loop filter transfer function of the form

$$F(s) = \frac{s + a}{s} = \frac{E(s)}{X(s)} \qquad (3.31)$$

which results in

$$sE(s) = (s + a)X(s) \tag{3.32}$$

In the time domain this equation becomes

$$\frac{de(t)}{dt} = \frac{dx(t)}{dt} + ax(t) = \left(\frac{d}{dt} + a\right)x(t) \tag{3.33}$$

Differentiating (3.9) yields

$$\frac{d^2\phi(t)}{dt^2} + AKe'(t) = \frac{d^2\theta(t)}{dt} \tag{3.34}$$

or using (3.33)

$$\frac{d^2\phi(t)}{dt^2} + AK\left(\frac{d}{dt} + a\right)\sin\phi(t) = \frac{d^2\theta(t)}{dt^2} \tag{3.35}$$

From this point on we will use the notation

$$\theta(t) = \Delta\omega t + \theta_0 \Rightarrow d^2\theta/dt^2 = 0 \qquad t = \tau/AK$$
$$d\phi/dt = AK(d\phi/d\tau) = AK\dot\phi \qquad a' = a/AK$$
$$\dot\phi = d\phi/d\tau \qquad \ddot\phi/\dot\phi = (d\dot\phi/d\tau)/(d\phi/d\tau) = d\dot\phi/d\phi \tag{3.36}$$

Using this notation, (3.35) can be expressed as

$$\frac{d\dot\phi}{d\phi} = -\cos\phi - a'\frac{\sin\phi}{\dot\phi} \tag{3.37}$$

For an initial point $(\phi, \dot\phi)$ in the $\phi \to \dot\phi$ plane, a slope of the curve can be evaluated and a new infinitesimally close point $(\phi + \Delta\phi, \dot\phi + \Delta\dot\phi)$ can be found. By repeating the same procedure in each new point, the overall trajectory shown in Figures 3.7 and 3.8 for two different sets of parameters can be obtained. The overall trajectory is folded into the $(-\pi, \pi)$ region. Two different initial points are marked on the figure. When curve b reaches the point $(\phi - \pi)$, it continues at the point $(\phi + \pi)$. For curve a the direction is the opposite. From the figures one can see that for all four examples of the initial conditions the loop will end up in the synchrostate where $\phi = 0$ and $\dot\phi = 0$.

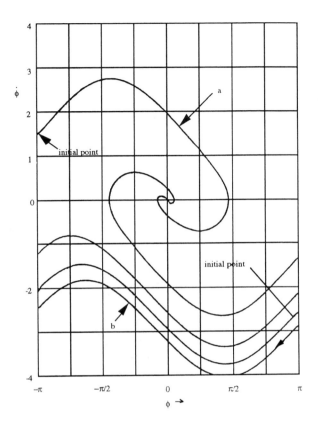

$$\xi = 1/2\sqrt{a/Ak} = 0,707, \ \omega_n = \sqrt{aAk} = 10^3 Hz:,$$
$$trajectory \ a; \ (\omega - \omega_0) = \dot{\phi}_0 Ak = 2 \bullet 10^3 Hz$$
$$t' = 1,363 \ 10^{-3}$$
$$trajectory \ b; \ (\omega - \omega_0) = \dot{\phi}_0 Ak = -4 \bullet 10^3 Hz.$$
$$t' = 2,348 \bullet 10^{-3}$$

Figure 3.7 Phase-locked loop trajectories.

3.1.7 Digital PLL

In this case the input signal $y(t)$ consisting of signal and additive noise will be sampled in regular time instances t_k, so we have

$$y_k = s_k + n_k$$

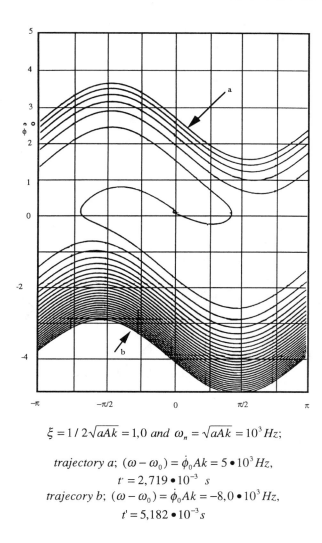

$\xi = 1/2\sqrt{aAk} = 1,0$ and $\omega_n = \sqrt{aAk} = 10^3 Hz$;

trajectory a; $(\omega - \omega_0) = \dot{\phi}_0 Ak = 5 \bullet 10^3 Hz$,
$t' = 2,719 \bullet 10^{-3}$ s
trajecory b; $(\omega - \omega_0) = \dot{\phi}_0 Ak = -8,0 \bullet 10^3 Hz$,
$t' = 5,182 \bullet 10^{-3} s$

Figure 3.8 Phase-locked loop trajectories.

and

$$s_k = \sqrt{2}\sin(\omega_0 t_k + \theta) \qquad (3.38)$$

where the time dependency of θ is dropped for simplicity. By analogy with Figure 3.1, the digital phase-locked loop (DPLL) is shown in Figure 3.9. If the control is performed in accordance with

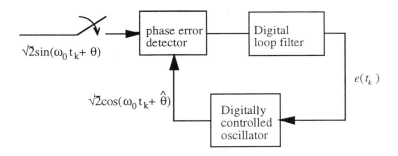

Figure 3.9 Digital PLL block diagram.

$$\hat{\theta}_{k+1} = \hat{\theta}_k + e(t_k) \tag{3.39}$$

then using a z-transform with z^{-1} representing unit delay (equal to the sampling interval) we have $z^{-1}\hat{\theta}_k = \hat{\theta}_{k-1}$ and the previous equation becomes

$$\hat{\theta}(t) = z^{-1}\hat{\theta}(z) + z^{-1}e(z)$$

and

$$\hat{\theta}(z) = \frac{z^{-1}}{1 - z^{-1}} e(z) \tag{3.40}$$

The equivalent baseband model for DPLL is shown in Figure 3.10 where g( ) is the phase detector nonlinearity function and $z^{-1}/(1 - z^{-1})$ is the digital-controlled oscillator transfer function. From this model we have directly

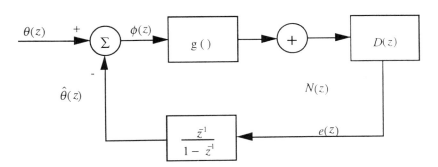

Figure 3.10 The baseband model for DPLL.

$$\hat{\theta}(z) = \frac{D(z)}{z-1}\{g[\phi(z)] + N(z)\}$$

and

$$\Phi(z) = \theta(z) - \hat{\theta}(z) \quad (3.41)$$

For the linear model with no noise we have

$$\Phi(z) = [1 - H(z)]\theta(z) \quad (3.42)$$

where

$$H(z) = \frac{D(z)}{(z-1) + D(z)} \quad (3.43)$$

If we choose $D(z) = G_1 + G_2/(1 - z^{-1})$ we will get the second-order loop with normalized parameters in the form

$$1 - H(z) = \frac{(z-1)^2}{(z-\alpha)^2 + \beta^2} \quad (3.44)$$

where α and β are functions of G_1 and G_2. The corresponding difference equation is

$$\phi_K - 2\alpha\phi_{K-1} + (\alpha^2 + \beta^2)\phi_{K-2} = \theta_K - 2\theta_{K-1} + \theta_{K-2} \quad (3.45)$$

For any form of the input phase sequence $\{\theta_k\}$ the phase error sequence $\{\theta_k\}$ is obtained from the recursion (3.45) for a given initial state $\phi_{-1} = \phi_{-2} = 0$.

3.2 CODE TRACKING LOOPS

3.2.1 Baseband Full-Time Early-Late Tracking Loop

In order to gain an initial understanding of the code tracking loop operation we will first consider a baseband loop. A conceptual block diagram of the loop is shown in Figure 3.11. This is an ELT motivated scheme defined by (2.8). In the code acquisition process, the normalized delay error between the input and local sequence $\delta = (\tau - \hat{\tau})/T_c$ is reduced to the range $\delta < 1$. The purpose of the code delay tracking loop is to further reduce this error down to zero and then to track any changes in

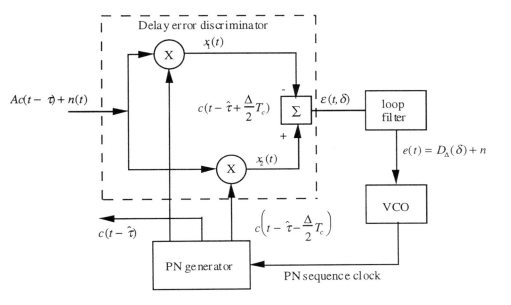

Figure 3.11 Baseband code tracking loop.

τ. In the literature this loop is known as a *delay-locked loop* (DLL). The input signal is correlated with two locally generated, mutually delayed replicas of the PN code. After filtering the useful component of the control signal $e(t)$ will be proportional to

$$D_\Delta(\delta) = R_c(\delta - \Delta/2) - R_c(\delta + \Delta/2) \qquad (3.46)$$

where $R_c(\delta)$ is the autocorrelation of the sequence. For different values of Δ the delay-locked discriminator characteristic $D_\Delta(\delta)$ is shown in Figure 3.12. Bearing in mind that $c = \pm 1$ and for $\Delta \geq 1$, one can show that noise characteristics will not be changed in the discriminator. So, for the analysis of the tracking error variance, results from Section 3.1 can be used directly.

3.2.2 Full-Time Early-Late Noncoherent Tracking Loop

Except in ranging, in most applications the input signal in DLL will be a complete DSSS signal. In order to eliminate information, a noncoherent structure, as shown in Figure 3.13, may be used with the simplest form of the input signal

$$r(t) = s(t) + n(t)$$

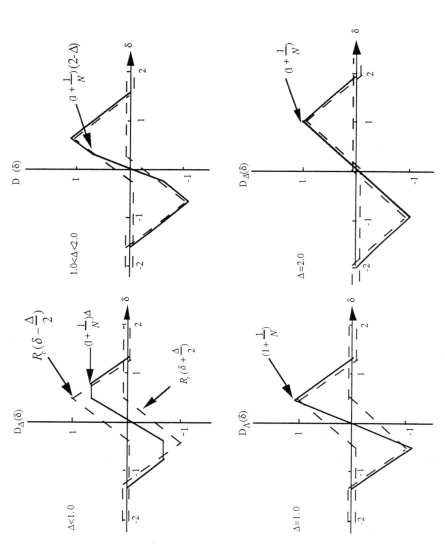

Figure 3.12 Delay-lock discriminator dc outputs for maximal-length sequence spreading codes and various values of Δ. (*From: Digital Communications and Spread Systems* by Ziemer/Peterson, © 1985. Reprinted by permission of Prentice-Hall, Inc., Upper Saddle River, NJ.)

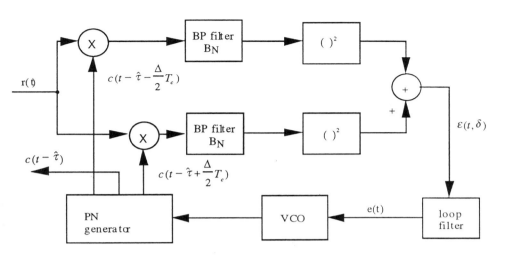

Figure 3.13 Full-time early-late noncoherent code tracking loop.

and

$$s(t) = Ab(t)c(t)\cos \omega_0 t \qquad (3.47)$$

It can be shown that the DC component of $\epsilon(t, \delta)$ is $A^2 D_\Delta(\delta)/2$, where

$$D_\Delta(\delta) \underline{\underline{\Delta}} R_c^2\left[\left(\delta - \frac{\Delta}{2}\right)T_c\right] - R_c^2\left[\left(\delta + \frac{\Delta}{2}\right)T_c\right] \qquad (3.48)$$

In order to keep the notation simple we are not using multiplication constants, so we lose track of signal dimensions. One should be aware of this in what follows. Evaluating the noise density at the loop filter output is a rather cumbersome although straightforward procedure. It can be shown that this density can be represented as

$$\eta = [(N_0)^2 B_N + N_0 A^2 D_\Sigma(\delta)]/2 \qquad (3.49)$$

where

$$D_\Sigma(\delta) = R_c^2\left[\left(\delta - \frac{\Delta}{2}\right)T_c\right] + R_c^2\left[\left(\delta + \frac{\Delta}{2}\right)T_c\right] \qquad (3.50)$$

For small δ we can use a linear model, which $\Delta = 1.0$ results in

$$D_\Delta(\delta) = 2\delta \qquad (3.51)$$

Using (3.39) we have

$$\sigma_\delta^2 = \frac{\eta B_L}{2A^4} = \frac{N_0 B_L}{2A^4}\left[N_0 B_N + \frac{A^2}{2}\right] = \frac{1}{2\rho_L}\left(1 + \frac{2}{\rho_{if}}\right) \quad (3.52)$$

where

$$\rho_L = \frac{2A^2}{N_0 B_L} \qquad \rho_{if} = \frac{A^2}{N_0 B_N} \quad (3.53)$$

and B_N is the bandpass filter bandwidth. Parameter ρ_L is the loop signal-to-noise power ratio and ρ_{if} is signal-to-noise power ratio at the output of IF (bandpass) filter.

3.2.3 Tau Dither Early-Late Noncoherent Tracking Loop

The loop operation is explained by using diagrams from Figure 3.14, and the block diagram of the loop is shown in Figure 3.15. The input signal plus noise is correlated with local PN sequence, and after bandpass filtering the signal amplitude will be proportional to the correlation function $R_c(\delta)$. Suppose that δ is in region 1^0 shown in Figure 3.14. In periodic time intervals T_d the phase of the local sequence is changed from $\hat{\tau} + \frac{\Delta}{2}T_c$ to $\hat{\tau} - \frac{\Delta}{2}T_c$ corresponding to the normalized delay error δ_{a1} and δ_{b1}, respectively. Whenever control signal $q(t) = \pm 1$ has a value of $+1$, δ will be lower and vice versa. Envelope variations of the bandpass filter output signal for the two regions are shown in Figure 3.15. If these envelope variations are detected (square-law device plus lowpass filter) and multiplied with the control signal $q(t)$, $\epsilon(t, \delta)$ is obtained. One can see that for the two regions, where δ is of the opposite sign, the DC components of $\epsilon(t, \delta)$ are also of the opposite sign, providing a correct control signal for VCO.

If $\delta = 0$, $q(t)$ will change δ from δ_{a3} to δ_{b3} and the envelope of $z(t)$ will be the same in both cases. This will result in a zero value control signal. In order to use (3.39) for the tracking error variance evaluation, the slope of the DC component at $\delta = 0$, and the power density function of the noise component of $e(t)$ must be evaluated. For the case when $\Delta = 1.0$, $N \gg 1$, and $1/T_d = B_N/4$, the result is

$$\sigma_\delta^2 = \frac{1}{2\rho_L}\left(1.811 + \frac{3.261}{\rho_{if}}\right) \quad (3.54)$$

where ρ_L and ρ_{if} are given again by (3.53). The advantage of the tau dither loop is low hardware complexity and the avoidance of arm gain imbalance problems, but

Code Tracking 125

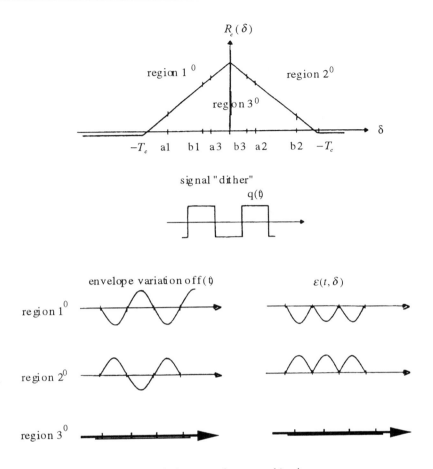

Figure 3.14 Operation of tau dither early-late noncoherent tracking loop.

the tracking error is almost twice as large as in the case of the DLL. One should be aware that unbalanced circuitry in the DLL would result into a constant bias in delay tracking. One way around this problem is a double dither *early-late* (E-L) noncoherent tracking loop, which is described in the next section.

3.2.4 Double Dither DLL

To explain the principle, we start with the baseband DLL whose delay discriminator characteristic is shown in Figure 3.16.

Early and late correlations might be different due to different gains in the two arms of the loop. If these two arms are periodically switched, $D(\delta)$ as shown in

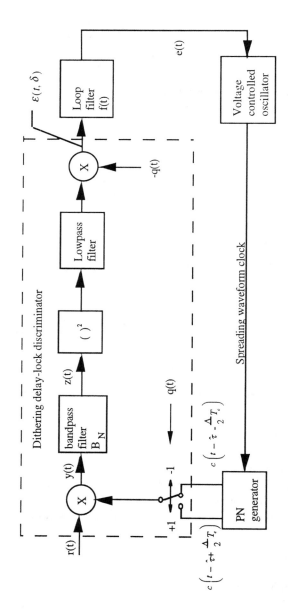

Figure 3.15 Tau dither early-late noncoherent tracking loop. (*From: Digital Communications and Spread Systems* by Ziemer/Peterson, © 1985. Reprinted by permission of Prentice-Hall, Inc., Upper Saddle River, NJ.)

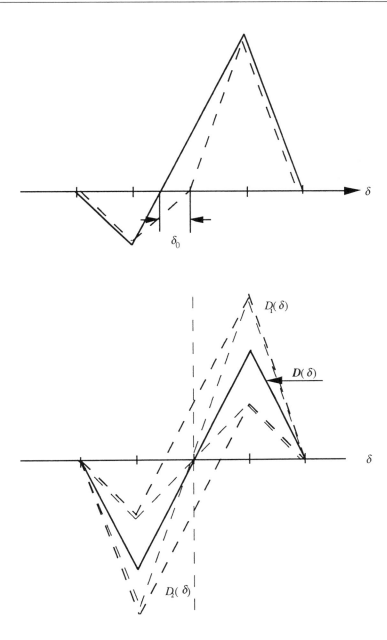

Figure 3.16 Average S-curve for double dither DLL.

Figure 3.16 will be obtained. The average value $D(\delta)$ will be an effective control signal and will not be biased. A noncoherent implementation of this idea is given in Figure 3.17.

3.2.5 Code Tracking Loop for Frequency Hopping Systems

An equivalent solution for DLL in FHSS systems is shown in Figure 3.18. An illustration of possible early channel mixer input and output signal spectra is shown in Figure 3.19. Signal waveforms leading to the creation of the VCO control signal are shown in Figure 3.20. For the opposite sign of the tracking error, the control signal is of the opposite sign too. The normalized S-curve of the tracking error discriminator is shown in Figure 3.21. One can show that (3.39) in this case results in

$$\sigma_\delta^2 = \frac{N_0^2 B_N/2 + N_0 A^2/2}{A^4} B_L = \frac{1}{\rho_L}\left(1 + \frac{1}{\rho_{if}}\right)$$

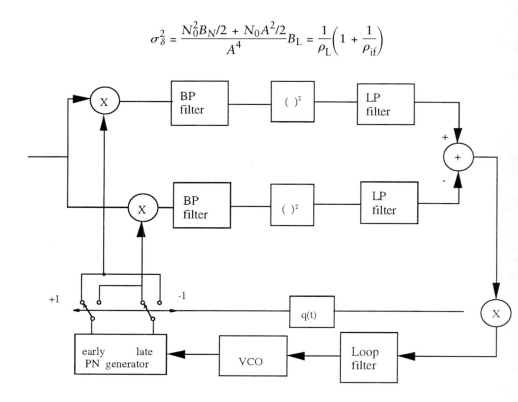

Figure 3.17 Block diagram of double dither DLL.

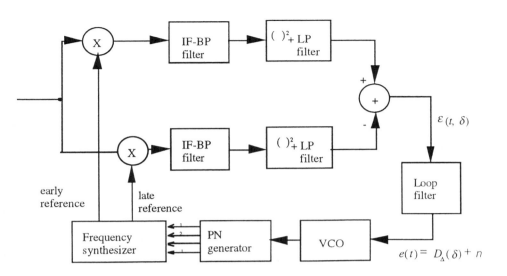

Figure 3.18 Full-time early-late noncoherent code tracking loop for FHSS.

3.2.6 Time-Shared Two-Channel Noncoherent Code

3.2.6.1 Tracking Loop for Slow FHSS Systems

A full-time E-L noncoherent code tracking loop for FHSS systems requires two mutually delayed reference signals. In most realizations, this means two synthesizers with some common elements. For this reason a code tracking loop with only one reference, which is shared by the early and late channels, might be more practical. A block diagram of such a loop is shown in Figure 3.22. Within one hopping interval T_h, each channel uses a reference signal during $T_h/2$. The frequency planes of the input and local reference signal for positive ($+\delta$) and negative ($-\delta$) tracking error are shown in Figure 3.23(a). Envelopes $q_e(t)$ and $q_l(t)$ are also shown in the figure. One can see from the figure that for positive δ, $l(\delta) > e(\delta)$ and vice versa, providing a correct sign of the control signal. The discriminator S-curve $V_F(\delta) = e(\delta) - l(\delta)$ is shown in Figure 3.23(b).

A further simplification of the scheme can be achieved by using the one-channel noncoherent loop shown in Figure 3.24. Frequency plane and signal waveforms are shown in Figure 3.25 for positive ($+\delta$) and negative ($-\delta$) tracking errors. The envelope detector signal is multiplied by the clock, alternating between +1 and −1 every $T_h/2$ sec. One can see that for the two different cases, positive and negative δ, the DC component of the opposite sign is obtained. The discriminator S-curve is the same shown in Figure 3.23(b).

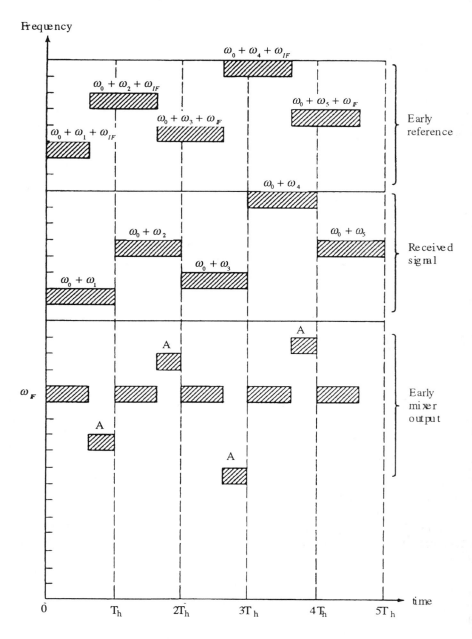

Figure 3.19 Illustration of typical early channel mixer input and output signal spectra. (*From: Digital Communications and Spread Systems* by Ziemer/Peterson, © 1985. Reprinted by permission of Prentice-Hall, Inc. Upper Saddle River, NJ.)

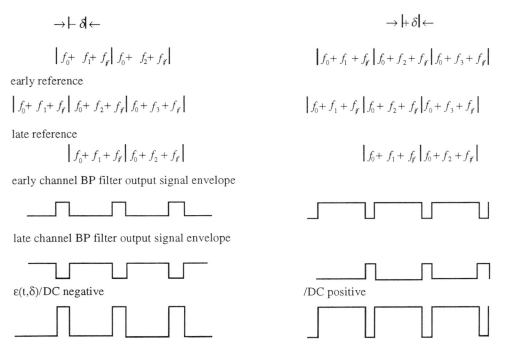

Figure 3.20 Frequency plane and signal waveforms in full-time early-late noncoherent code tracking loop for frequency hopping systems.

3.2.7 Discrete Tracking Process

Nowadays even the advanced *time division multiple access* (TDMA) schemes are using slow frequency hopping to improve the system performance in fading channels [1,2]. For these reason we additionally elaborate on the tracking problem in these systems.

Strictly speaking, the tracking process in spread spectrum systems cannot be analyzed independently of the acquisition process since these two processes interchange with each other continuously. The time that a system spends in each mode depends on the random noise. For these reasons, the signal tracking mode can be considered as a "renewal process" [3]. According to this theory, the timing error probability distribution function can be obtained by analyzing the general model with $\tau^{acq} \equiv 0$, where τ^{acq} is the average time spent in coarse acquisition. In accordance with the aforementioned explanation, it is assumed that in the tracking mode, the timing error takes on the values from a discrete set $\vec{\epsilon} = \{\epsilon_1, \epsilon_2, \cdots, \epsilon_h\}$, one of which is the sync state e_s (which implies $\epsilon(t) = 0$). The state e_0 is the reacquisition mode

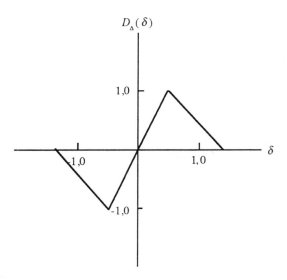

Figure 3.21 Normalized S-curve for full-time early-late noncoherent code tracking loop discriminator in FHSS system.

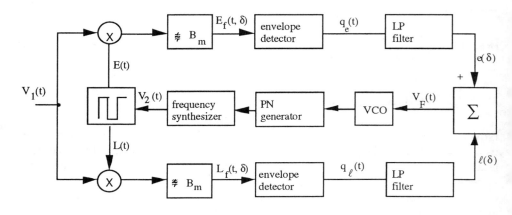

Figure 3.22 Time-shared two-channel noncoherent code tracking loop for slow FHSS systems.

(loss of synchronization). As noted previously, it is assumed that the process does not spend any time in state $\epsilon_0(\tau^{acq} \equiv 0)$. In any other state it spends $\tau_i = \tau$ sec, which is the time period between two decisions in the process (state dwell time). We define the parameter p_{ij} as the process state transition probability of going from state j to state i. Now assume that this process is observed for S successive state transitions.

Code Tracking 133

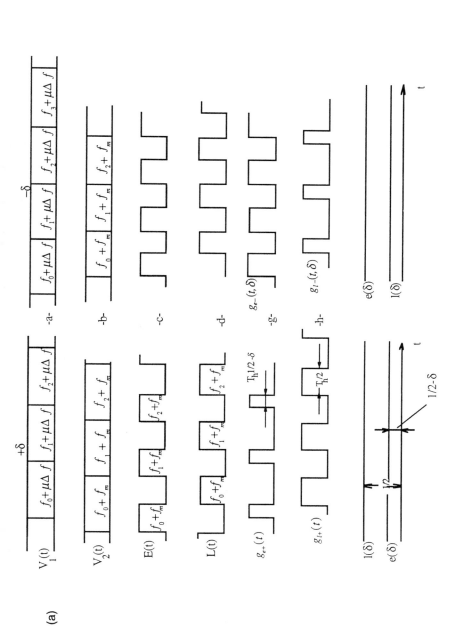

Figure 3.23 (a) Frequency plane and signal waveforms in time-shared two-channel noncoherent code tracking loop for slow FHSS system. (b) Discriminator S-curve.

(b)

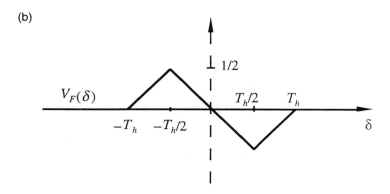

Figure 3.23 (continued).

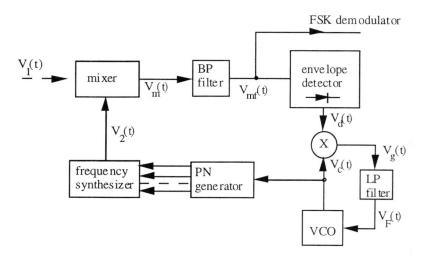

Figure 3.24 One-channel noncoherent code tracking loop for slow FHSS systems.

Let $S(i)$ be the number of times the system is observed to be in state i, and let $L(i)$ be defined by

$$L(i) = \lim_{S \to \infty} \frac{S(i)}{S} \qquad i = 0, 1, 2, \cdots, h \tag{3.55}$$

The process can be interpreted as a finite state Markov chain, where the $L(i)$ are the state probabilities of the Markov chain formed by observing the process only

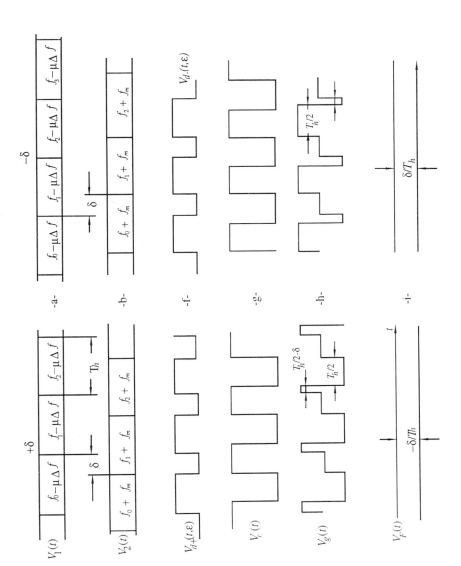

Figure 3.25 Frequency plane and signal waveforms in one-channel noncoherent code tracking loop for slow FHSS systems.

at state transition times. Since the time spent in each state is not the same (i.e., $\tau_0 \neq \tau_i$, $i \neq 0$), the Markov state probabilities $L(i)$ are not the absolute probabilities of the occurrence of the tracking system states at any instant in time but rather the probabilities of state selection, given that a state transition is occurring. The absolute state probabilities, designated $P(i)$, can be related to these state selection probabilities as

$$P(i) = \begin{cases} 0 & i = 0 \\ \dfrac{L(i)}{\sum_{i=1}^{h} L(i)} = \dfrac{L(i)}{1 - L(0)} & i = 1, 2, \ldots, h \end{cases} \quad (3.56)$$

Hence, the tracking error probabilities mass function can be expressed as

$$p(\epsilon/\vec{\epsilon}) = \sum_{i=1}^{h} P(i)\delta(\epsilon - \epsilon_i) \quad (3.57)$$

$P(0) = 0$ because, by definition, $\epsilon_0 \notin \vec{\epsilon}$. Also, the tracking error variance can be expressed as

$$\sigma_\epsilon^2(\vec{\epsilon}) = \sum_{i=1}^{h} \epsilon_i^2 p(\epsilon_i/\vec{\epsilon}) \quad (3.58)$$

In order to analyze the mean-time-to-lose lock, we consider state ϵ_0 of the process as an absorbing state. Using results from Section 2.7 the average time for the process to reach the absorbing state is given by

$$\mathbf{T} = (\mathbf{I} - \mathbf{Q})^{-1}\boldsymbol{\tau}$$

where $\mathbf{T}$ is the column vector with elements $T(i)$ representing the average time for the process to reach the absorbing state, given it starts from state ϵ_i, $\mathbf{I}$ is an identity matrix, $\mathbf{Q}$ is a matrix of the nonabsorbing (transient) state transition probabilities, and $\boldsymbol{\tau}$ is the column vector with element τ_i representing the process observation time at state ϵ_i. If we assume that the tracking process starts with the same probability $1/h$ from any state $\epsilon_i (i = 1, 2, \ldots, h)$, the mean-time-to-lose lock can be expressed as

$$T_{\text{lose}} = \frac{1}{h} \sum_{i=1}^{h} T(i) \quad (3.59)$$

Equations (3.55) to (3.59) completely describe the discrete renewal tracking process. In the next section, some examples of the algorithms for slow FH tracking are described. Besides the bits that carry information, in these systems, some additional bits are added into the binary stream for synchronization purposes. Depending on the way these bits are added to the information stream, we can talk about two different cases: (1) systems with a *distributed sync group* (DSG), where in each frame of h binary symbols, k bits are added for synchronization purposes; and (2) systems with a *concentrated sync group* (CSG), where every Mth frame is completely used for synchronization purposes.

These algorithms can be modeled as discrete renewal processes. After determining the transition probabilities P_{ij} for different variants of the algorithm, the general results from this section are used for the system performance analysis.

3.2.8 Discrete Tracking Systems for Slow FH

For the schemes described in Sections 3.2.5 and 3.2.6, the timing-error discriminator characteristic $S(\epsilon)$ (S-curve) is formed by the correlation of the input signal and a locally generated signal from a frequency synthesizer. Additional processing is used to provide a direct relationship between the correlator output signal and the VCO control signal, which is used to adjust the phase of the locally generated signal so that it tracks the phase of the input signal. As a result, this FH signal tracking system can be modeled as an analog PLL. The main drawback of this tracking system is loss of information due to the timing tracking error, since the absolute timing error ϵ increases as the hopping interval T_h increases. For a slow FH system where the information bit rate is much higher than the signal hopping rate, this means more and more bits lost (burst errors) due to the tracking error. For these applications, discrete tracking systems are more efficient. In what follows we describe some options available.

3.2.8.1 Systems With Distributed Sync Group

DSG(h, 1)-ML System

This is a system using minimum redundancy, namely, one where only a single sync pulse, situated in the middle of each h-bit frame, is used for tracking purposes. This sync pulse is referred to as an m-pulse, and the system is designated DSG(h, 1), or only (h, 1) for short. We assume that the m-pulse takes on the value "binary one" in each frame. Let us consider the possibility of using a ML estimate of the m-pulse position in an (h, 1) system with an uncertainty region $\epsilon \in \{\epsilon_1, \epsilon_2, \ldots, \epsilon_h\}$. After dehopping, the receiver extracts the binary signal in the demodulation process. In order not to have to limit results to a specific modulation/demodulation scheme, we characterize this process with a parameter P_e, which is the probability of bit error. In those time intervals where the frequencies of the input and the locally generated signal do not coincide due to the tracking error (α region in Figure 3.26), the receiver demodulator input signal is zero and the receiver outputs a binary one or a binary zero with probability 1/2. Having produced the binary signal, the receiver observes all h successive bit positions in a frame simultaneously. Let

$$\mathbf{B}(t) = \{b_1(t), b_2(t), \cdots, b_h(t)\} \qquad (3.60)$$

designate a vector whose elements are the values of the bits observed in one frame. In N successive frames, the receiver forms a matrix of the form

$$\mathbf{B} = [b_{ij}] = [b_j(t + (i-1)hT)] \qquad j = 1, 2, \cdots, h; i = 1, 2, \cdots, N \qquad (3.61)$$

where T is the bit interval. The columns of the matrix form new vectors

$$\vec{B}_j = \{b_j(t), b_j(t + hT), b_j(t + 2hT), \cdots, b_j(t + (N-1)hT)\} \qquad (3.62)$$

The receiver accepts the rth bit position as the sync position if vector $\vec{B}_r$ has the largest number of elements that have the value one. A possible position of the observed frame with respect to the real input signal frame is sketched in Figure 3.26. If the receiver accepts the rth bit position in the observed frame to be the sync position, then the synchronization will be lost if $r \in \alpha$ because the next time the phase position is updated, the input signal will not contain the real m-pulse (note that in each frame, $|\epsilon|/T$ bits will not be dehopped). In the analysis, we assume that the receiver synchronization control system will detect such an event with probability $P_{\text{ld}} = 1$. This assumption has always been used in the analysis of the renewal processes.

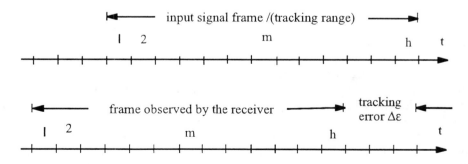

Figure 3.26 Position of the input signal frame and observed frame when the tracking error is $\Delta\epsilon$. (© 1991 IEEE. *From:* [4].)

(h, 1)-ML System State Transition Probabilities

An $(h, 1)$-ML system decision process can be modeled as a discrete renewal tracking process where the ith state of the process represents the case where the ith pulse position in the frame is the central bit position observed by the receiver. Due to the finite probability of finding two or more bit positions with the same maximum number of detected bit positions at the end of N-frames (observation interval), the general $(h, 1)$-ML algorithm described in the previous section must be specified in more detail. Depending on how we resolve the multiple ties ambiguity problem, various algorithms are possible. For example, when we have N_t ties (N_t positions with the same maximum number of detected positions), we could choose randomly (with probability $1/N_t$) one of these positions to be the sync state. This algorithm will be referred to as ML/RT (randomized ties) algorithm. Alternatively, if, at the end of the observation interval $N_t > 1$, the system does not change the state (i.e., we ignore these obeservables), the algorithm is referred to as ML/IT (observation interval with ties ignored). It can be shown that the IT system will perform better than the RT system. This can be explained by the fact that the decision process spends most of the time in the m-position. Hence, when ties occur, it is better to stay in the same state than to choose randomly any one of the states with ties. For this reason, in the rest of the section, we present the analysis of only the ML/IT system. If n_j is the number of positions detected in the jth bit position in N successive frames, then we define the following set of probabilities:

$$p_j(x) = P(n_j = x | j \neq m) = \binom{N}{x}(1/2)^x(1 - 1/2)^{N-x} = (1/2)^N \binom{N}{x} \quad (3.63)$$

and

$$P_j(x) = P(n_j < x | j = m) = \sum_{i=0}^{x-1} (1/2)^N \binom{N}{i} \quad (3.64)$$

and

$$p_m(x) = P(n_j = x | j = m) = \binom{N}{x}(1 - p_e)^x p_e^{N-x} \quad (3.65)$$

where p_e is the bit error probability. Strictly speaking, p_e depends on the FH signal tracking error, because due to this error the average bit synchronizer S-curve is modified as

$$S_{ba}(\epsilon) = \frac{1}{T_h} \int_0^{T_h} S_b(\epsilon, t)\, dt$$

$$= \frac{1}{hT} \int_{t \in T_{sp}} S_b(\epsilon)\, dt + \frac{1}{hT} \int_{t \in T_{sa}} 0 \cdot dt$$

$$= (1 - |\epsilon|/hT) S_b(\epsilon - kT) \tag{3.66}$$

where k is the integer portion of ϵ/T, T_{sp} is the portion of T_h where the FH signal is dehopped, and T_{sa} is the portion of the same time interval in which FH signal is not dehopped due to the tracking error ϵ. The subscripts "sp" and "sa" stand for "signal present" and "signal absent," respectively, because during T_{sa} the input signal to the system bit synchronizer is present and during T_{sa} it is not. Note that $\epsilon = $ const for $t = NhT$, where Nh is taken to be a large number. It should be also noted that even at the edge of the tracking range where $\epsilon_{max} = hT/2$, the modifying factor is not less than $1/2$. For most applications the bit synchronizer loop SNR is large enough so that even with the modified $S_b(\epsilon)$ curve the degradation in signal power is not excessive. If this is not the case, in each state j, p_e must be evaluated taking into account the modified SNR in the bit synchronization loop. We now evaluate the probability of having multiple ties ($N_t > 1$). It is convenient to express this probability as

$$p_t = P(t; x_0 = n_{max}) + P(t; x_0 < n_{max}) \tag{3.67}$$

where the first term refers to the case when the m position is one of the N_t positions having maximum number of ones n_{max}, and the second term refers to the case when only N_t nonsync positions have the maximum number of ones. In accordance with previous notation, this probability can be expressed as

$$p_t = \sum_{x_0=0}^{N} \sum_{N_t=2}^{h} \binom{h-1}{N_t-1} \cdot [p_j(x_0)]^{N_t-1} [P_j(x_0)]^{h-N_t} p_m(x_0)$$

$$+ \sum_{x_0=0}^{N-1} \sum_{n_{max}=x_0+1}^{N} \sum_{N_t=2}^{h-1} \binom{h-1}{N_t} \cdot \{p_j(n_{max})\}^{N_t} \{P_j(n_{max})\}^{h-N_t-1} p_m(x_0) \tag{3.68}$$

The probability of choosing the m-position as the sync state can be expressed as

$$p_0' = P(n_m > n_j \text{ for all } j \neq m)$$

$$= \sum_{x_0=0}^{N} P_j^{h-1}(x_0) p_m(x_0) \tag{3.69}$$

The probability of choosing any other bit position as the sync state is

$$p' = (1 - p'_0 - p_t)/(h - 1) \qquad (3.70)$$

Hence, the state transition probabilities can be now expressed as

$$p_{i0} = \pi(\epsilon_i) = 1/h \qquad i = 1, 2, \cdots, h \qquad (3.71)$$

$$p_{0j} = |m - j|p' \qquad j = 1, 2, \cdots, h \qquad (3.72)$$

and

$$p_{ij} = \begin{cases} p'_0 & i = m, j \neq m \\ p' & |i - j| \leq h \text{ and } j \neq i, i \neq m \\ 0 & |i - j| > h \\ p' + p_t & j = i \text{ and } i \neq m \\ p'_0 + p_t & j = i = m \end{cases} \qquad (3.73)$$

These p_{ij} are used to evaluate Markov state selection probabilities defined by (3.55). Then in accordance with (3.56) the steady-state probabilities are evaluated. The tracking error variance for the ML/IT system, defined by (3.58), is presented graphically in Figure 3.27. In order to make a rough comparison with a closed-loop tracking system, consider, for simplicity, a coherent E-L loop. A standard form for the variance of the tracking error, normalized to the hop time, can be approximated as

$$(\sigma_\epsilon/T_h)^2 \cong CN_0 B_L/S \qquad (3.74)$$

where T_h is the hop interval, N_0 is one-sided noise power density, S is signal power, and B_L denotes the equivalent loop bandwidth. For a coherent loop, we take $C = 1/2$; and in order to make a fair comparison, we set $B_L \cong (NT_h)^{-1}$, the inverse of total observation time in the discrete tracking system. Thus,

$$(\sigma_\epsilon/T_h)^2_{E/L} \cong \frac{1}{2} \cdot \frac{N_0}{SNT_h} = \frac{1}{2} \cdot \frac{N_0}{SNhT_b} = \frac{1}{2} \frac{(E_b/N_0)^{-1}}{hN} \qquad (3.75)$$

where E_b/N_0 gives the corresponding probability of error in the discrete tracking system. For coherent PSK and $p_e \simeq 10^{-2}$ we have $E_b/N_0 \simeq 2.5$.

Hence, for $h = 15$, we obtain $(\sigma_\epsilon^2/T_h)^2_{E/L} \cong 1/75N$. This relation is presented graphically in Figure 3.27 (dashed line). One can see that for $N > 12$ the discrete tracking system will be better than the standard E-L loop. Along similar lines, the

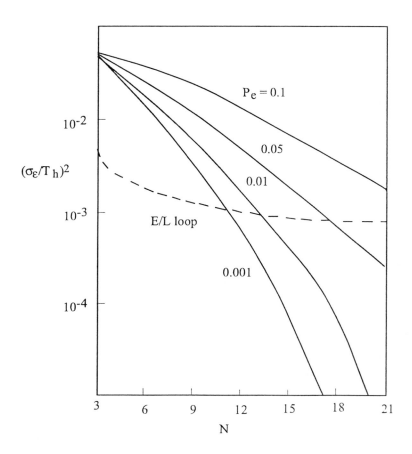

Figure 3.27 Tracking error variance $(\sigma_\epsilon/T_h)^2$ for ML/IT system versus the number of observables N. (© 1991 IEEE. *From*: [4].)

two methods can be compared for any p_e (corresponding to a given E_b/N_0). We should, however, keep in mind that, (3.58) is defined for the tracking range $(-T_h/2, T_h/2)$, as opposed to the range $(-3T_h/2, 3T_h/2)$ for the standard E-L loop. Finally, the mean-time-to-lose lock in an ML/IT system, defined by (3.59), is presented graphically in Figure 3.28 versus the number of observables N, with p_e a parameter. The main advantages of the $(h, 1)$-ML system are the good performance and low signal redundancy (one sync bit per frame). The main drawback of this system is, like in any other system using ML estimation of the sync parameters, its receiver hardware complexity. If the hardware complexity is the main concern, then some simplified solutions are possible. However, we must be prepared to accept either performance degradation or higher signal redundancy compared to the $(h, 1)$-ML system. An alternative discrete system for FH signal tracking is based on utilization of additional sync data (i.e., k sync pulses per frame) added to the binary

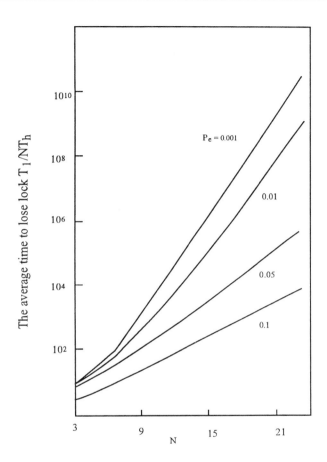

Figure 3.28 The mean-time-to-lose lock T_1/NT_h for ML/IT system versus the number of observables N. (© 1991 IEEE. *From*: [4].)

stream carrying the information. This system is referred to as a DSG(h, k) or simply (h, k) system. The receiver observes only one bit position in the frame instead of the h observables used simultaneously in the ML receiver. Thus, the receiver complexity is considerably reduced, whereas the signal redundancy is slightly increased. The system performance will depend on the amount of redundancy introduced for tracking purposes and could be even better than the performance of the (h, 1)-ML system. Details about such solutions can be found in [4].

3.2.8.2 Systems With Concentrated Synchro Group

This approach is based on the utilization of each Mth hopping interval completely for the transmission of additional data for tracking purposes and, hence, is referred to as a CSG. Each sync frame is of length h. The relative redundancy is now $r = h/Mh = 1/M$ and can be less than $1/h$. We thus have more flexibility to tradeoff system performance, complexity, and the amount of the redundancy introduced for synchronization (tracking) purposes. A PN sequence is used for the bits of the sync frame, and three slightly different tracking algorithms are described. For each of these algorithms the receiver front end is the same. The received FH signal is dehopped, and the receiver demodulator produces the binary data stream. Assuming that initial (coarse) acquisition of the hopping pattern is achieved, only a part of the sync frame is successfully demodulated, the remaining data being lost due to the tracking error. In those time intervals where the frequencies of the input and the locally generated signals do not coincide, the receiver demodulator outputs binary ONEs and binary ZEROs with probability 0.5. Each algorithm uses the demodulated binary stream in different ways to extract the information needed about the correct sync position of the input signal, and these are described separately in the following subsections.

ϵML Tracking System Algorithm

In this case, every Mth frame is composed of a PN sequence (of length h) used for synchronization purposes. In the receiver, the dehopped and demodulated binary stream is simultaneously correlated (bit by bit) with all possible cyclic shifts of the PN sequence during the time when the sync frame is expected. This might be realized with a bank of matched filters, where each filter is matched to a different cyclic shift of the sequence. The sequence shift with the largest value of the correlation at the sampling moment is declared to be the sync state. This system is called ϵML because all phase shifts of the sequence are correlated simultaneously and the phase shift ϵ with the maximum probability is chosen to be the sync-state. The maximum value of the correlation depends on both the synchronization error and the probability of bit error. A possible mutual phase relationship between the input signal and the locally generated sequence is sketched in Figure 3.29. The phase of the local sequence ϵ_j is defined as the position of its middle (m) bit with respect to the first bit of the input sequence. That is, ϵ_j means that the jth bit in the input sequence and the m-bit in the local sequence coincide in time. Hence, if the phase of the local sequence is ϵ_m, perfect synchronization is achieved ($\Delta\epsilon = \epsilon_j - \epsilon_m = \epsilon_m - \epsilon_m = 0$). As indicated above, due to the phase tracking error $\Delta\epsilon$, part of the input sequence will not be dehopped; specifically, in each hopping interval T_h, $|\Delta\epsilon|/T_b$ bits will not be dehopped. These bits will be replaced at the output of the receiver demodulator with a random binary stream having $p(0) = p(1) = 1/2$. The same a priori bit probabilities are assumed for the binary information stream. Hence, for a given phase of the local sequence, each matched filter will correlate a cyclic shift of the local sequence with the input sequence. The latter can be expressed as $R_{\overline{w}}(j) \oplus S_w(j) \oplus E(j)$ where $S_w(j)$

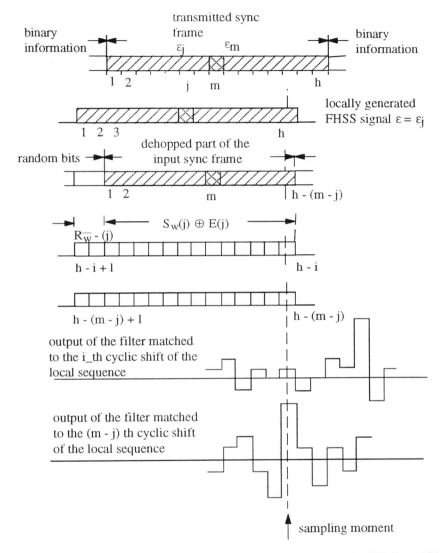

Figure 3.29 Signal waveforms in the system with ϵML tracking algorithm. (© 1991 IEEE. *From*: "Discrete Tracking System for Slow FH: Part II," by S. Glisic and L. Milstein in *IEEE Trans. on Communications*, Vol. 39, Feb. 1991.)

is the dehopped part of the input sequence, $E(j)$ is the bit error pattern in the time interval of the sequence $S_w(j)$, $R_{\overline{w}}(j)$ is a random sequence having $p(0) = p(1) = 1/2$, and $\oplus$ stands for modulo-2 summation. These parameters are sketched in Figure 3.29 and will be discussed in more detail later.

CML Tracking System Algorithm

The main drawback of the ϵML tracking algorithm is its complexity (a bank of h matched filters is used). Instead of this, we can use only one matched filter and declare the moment when its output reaches the maximum to be the sync state. In this case, for a given phase of the local sequence ϵ_j, at each instant lT_b, $l = 1, 2, \ldots, h$, the receiver correlates the local sequence S and the input sequence $R_{\overline{w}}(l, j) \oplus S_w(l, j) \oplus E(l, j)$ where $S_w(l, j)$ is the portion of the dehopped input sequence $S_w(j)$ that overlaps with the local sequence, $E(l, j)$ is the bit error pattern in the time interval of the sequence $S_w(l, j)$, and $R_{\overline{w}}(l, j)$ is a random sequence having $p(0) = p(1) = 1/2$. These parameters are sketched in Figure 3.30. This system is called CML because, for a given j, the complete correlation function $\rho(l, j)$ ($l = 1, 2, \ldots, h$) is formed and the value of l for which this correlation has the maximum value is chosen to be the sync state. If the maximum of $\rho(l, j)$ is in the region $l \in a$ (see Figure 3.30), the synchronization is lost because, when the next sync frame appears, the real correlation peak will lie outside the observed frame of correlation samples. Note that, for this system, a device for storage of the correlation value $\rho(l, j)$ must be provided since for each $l = 1, 2, \ldots, h$, the estimator output and associated timing phase are compared to the largest past output and its timing phase.

MF Tracking System Algorithm

A further simplification of the receiver can be achieved if the *matched filter* (MF) (correlator) output signal is compared with a threshold b (see Figure 3.30) and the first moment when it exceeds the threshold is accepted as the sync position. This algorithm, along with the previously described algorithms, are discussed in more detail in the following sections.

Tracking System Parameters

ML Systems On the basis of the decision rules described in the previous sections, both ϵML and CML systems can be modeled as a discrete renewal process. This model is based on the assumption that the system timing error tracking range is defined as a set of the local signal phases $\epsilon_t \in \{\epsilon_1, \epsilon_2, \ldots, \epsilon_m, \ldots, \epsilon_h\}$ where $-hT_b/2 \leq \epsilon_i - \epsilon_m \leq hT_b/2$, for any $i = 1, 2, \ldots, h$, and that the probability of detection of synchronization loss $P_{ld}(\epsilon \notin \epsilon_t) = 1$. This assumption is used in the analysis of the renewal processes. State j of the process represents the case when the phase of the local sequence is ϵ_j (the jth bit in the input sync frame and the mth bit in the local sequence coincide). State $\epsilon_j = \epsilon_m$ represents timing error $\Delta\epsilon = 0$ (correct sync position). We define p_{ij} as the probability that, for a given j, the largest value of the signal correlation function is in state i for the CML system. Equivalently, it

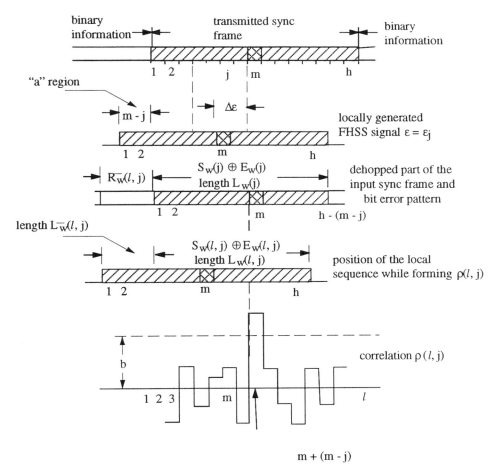

Figure 3.30 Mutual signal positions in the system with CML tracking algorithm. (© 1991 IEEE. *From*: "Discrete Tracking System for Slow FH: Part II," by S. Glisic and L. Milstein in *IEEE Trans. on Communications*, Vol. 39, Feb. 1991.)

is the probability that, for a given j, at the sampling moment, the output signal of the filter matched to the ith cyclic shift of the sequence is the largest for the ϵML system. For the purpose of analysis of the CML system, an additional parameter l, defined by $i = l - (m - j)$, is needed; so for each j, the correlation function $\rho(l, j)$ is formed in the range where $l = 1, 2, \ldots, h$. This corresponds to the range of the relative time shifts with respect to the local signal $\epsilon = \{-h - 1)T_b/2, (h - 1)T_b/2\}$. In order to explain this in more detail, assume that sequence S of length h is used as a sync frame. This sequence is defined by the bit pattern $S = \{S_1, S_2, \ldots, S_{h-1}, S_h\}$. In the receiver, some bits of the sequence are demodulated erroneously so that the input sequence to the correlator can be expressed as $S' = S \oplus E$, where E is the error pattern sequence of the form $E = \{E_1, E_2, \ldots, E_h\}$. The elements of this sequence E_i have value 1 with probability p_e and value 0 with probability $(1 - p_e)$, where p_e is the bit error probability.

Sequence Windowing

For the purpose of the ensuing presentation, we define a new operation called sequence windowing, denoted by $W(S)$. By definition, for the sequence S, we have

$$W^i(S) = \{l_i l_{i-1}, \cdots, l_1, S_1, S_2, \cdots, S_{h-i}\} = R_{\overline{w}}(i) \oplus S_w(i) \quad (3.76)$$

where $R_{\overline{w}}(i) = \{l_i, l_{i-1}, \cdots, l_1, 0, 0, \cdots, 0\}$, $S_{\overline{w}}(i) = \{0, 0, \cdots, 0, S_1, S_2, \cdots, S_{h-i}\}$, and l_r are binary symbols having a priori probability $p(l_r = 0) = p(l_r = 1) = 1/2$. The number of l_r bits in the sequence $R_{\overline{w}}(i)$ is designated as $L_{\overline{w}}(i)$, whereas the number of S_r bits in the sequence $S_w(i)$ is designated as $L_w(i)$. It should be noted that $L_{\overline{w}}(i) + L_w(i) = h$. This process can be thought of as a shifting of the sequence bits to the right for i positions (i unit shifts). For each unit shift, the last data bit leaves the frame (window) through the right frame boundary and one random bit enters the frame through the left frame boundary. Therefore, the sequence $S_w(i)$ is composed of the original sequence bits that did not reach the frame boundary in the process of the sequence shifting; whereas $R_{\overline{w}}(i)$ is composed of the random bits augmented to $S_w(i)$ in the same process. Also, by definition, we have

$$W^{-i}(S) = \{S_{i+1}, S_{i+2}, \cdots, S_h, r_1, r_2, \cdots, r_i\}$$
$$= S_w(-i) \oplus R_{\overline{w}}(-i) \quad (3.77)$$

where $S_w(-i) = \{S_{i+1} S_{i+2}, \cdots, S_h, 0, 0, \cdots, 0\}$, $R_{\overline{w}}(-i) = \{0, 0, \cdots, 0, r_1, r_2, \cdots, r_i\}$, and the r_r are binary symbols having equiprobable values 0 or 1 and entering the frame through the right frame boundary. It should be noted that, from the definitions given by (3.76) and (3.77), we have for $i > j$

$$W^{-j}\{W^i(S)\} = \{l_{i-j}, \cdots, l_1, S_1, S_2, \cdots, S_{h-i}, r_1, r_2, \ldots, r_j\}$$
$$= S_w(i, -j) \oplus R_{\overline{w}}(i, -j) \quad (3.78)$$

where $S_w(i, -j) = \{0, 0, \cdots, 0, S_1, S_2, \cdots, S_{h-i}, 0, 0, \cdots, 0\}$ and $R_{\overline{w}}(i, -j) = \{l_{i-j}, \cdots, l_1, 0, 0, \cdots, 0, r_1, r_2, \cdots, r_i\}$. Sequence $S_w(i, -j)$ is again composed of the original sequence bits that did not reach the frame boundaries in the process of the sequence shifting. If $i > j$, (3.78) becomes

$$W^{-j}\{W^i(S)\} = \{S_{j-i}, \cdots, S_{h-i}, r_1, r_2, \cdots, r_j\}$$
$$= S_w(i, -j) \oplus R_{\overline{w}}(i, -j) \quad (3.79)$$

where $S_w(i-j) = [S_{j-i}, \cdots, S_{h-i}, 0, 0, \cdots, 0]$ and $R_{\overline{w}}(i, -j) = \{0, \cdots, 0, r_1, r_2, \cdots, r_j\}$. Also, from the previous definitions, we have

$$W^j\{W^i(S)\} = W^{i+j}(S) = S_w(i, j) \oplus R_{\overline{w}}(i, j) \quad (3.80)$$

and

$$W^{-j}\{W^{-i}(S)\} = W^{-(i+j)}(S)$$
$$= S_w(-i, -j) \oplus R_{\overline{w}}(-i, -j) \quad (3.81)$$

In this case, $L_{\overline{w}}(i, j)$ will be used to designate the number of S_r bits in the sequence $S_w(i, j)$, whereas $L_{\overline{w}}(i, j) = h - L_w(i, j)$ will be used to designate the number of augmented bits (l or r) in the sequence $S_{\overline{w}}(i, j)$.

By analogy with the $W(S)$ operation, a sequence cyclic shifting operation can be defined in the same way, except that the windowing operation is now performed by shifting a window of length h along the sequence obtained by the periodic repetition of the original sequence, that is, $\ldots S_1, S_2, \ldots, S_h, S_1, S_2, \ldots, S_h, S_1, S_2, \ldots, S_h, \ldots$ rather than along the sequence obtained by augmenting the original sequence with the random bits, that is, $l_3, l_2, l_1, S_1, S_2, \ldots, S_h, r_1, r_2, r_3, \ldots$. This operation is designated as CYW(S), where CYW stands for cyclic windowing. For example, by analogy with (3.76),

$$\text{CYW}^i(S) = \{S_{h-i+1}, \cdots, S_h, S_1, S_2, \cdots, S_{h-i}\} \quad (3.82)$$

More details about the CYW(S) operation are given in Appendix A3.

State Transition Probabilities

For the given phase of the local sequence ϵ_j, for each $l = 1, 2, \ldots, h$, the correlation function $\rho(l, j)$ for the two systems is defined as (see Figures 3.29 and 3.30)

$$\rho(l, j) = \text{Cor}\{\text{CYW}^{-m+l}(S), W^{m-j}(S')\} \quad \text{for } \epsilon\text{ML} \quad (3.83)$$
$$\rho(l, j) = \text{Cor}\{S, W^{m+l}\{W^{m-l}(S')\}\} \quad \text{for CML} \quad (3.83a)$$

where Cor{ } stands for correlation defined as Cor$\{A, B\} = \Sigma_{i=1}^{h} a_i b_i$ and $a_i = \pm 1$, $b_i = \pm 1$. In (3.83), S is the locally generated sequence and $S' = S \oplus E$ is the demodulated sequence. It should be noted that the $\rho(l, j)$ are random variables because $W^{m-j}(S')$ and $W^{m-l}\{W^{m-j}(S')\}$ are composed of a number of the original sequence bits with possible random errors and a number of augmented bits (l or r) having random values 0 or 1 with equal probability. Due to the finite probability of finding two or more correlation points l with the same maximum value, the general ML algorithm, described in the previous sections, must be specified in more detail. As in Section 3.2.6.1, depending on how we resolve the multiple ties ambiguity problem, various algorithms are possible. For example, when we have N_t ties (i.e., the signal correlation function $\rho(l, j)$ has its maximum value for N_t different values of l), we might choose randomly (with probability $1/N_t$) one of these positions to be the sync state. Alternatively, one could have an algorithm whereby if, at the end of any observation interval, there are N_t ties, the system does not change state (these observables are ignored). This subalgorithm is referred to as the ML/IT (ignore ties) algorithm, and because it can be shown to outperform the former algorithm, it is the only one considered in what follows. As the first step in analyzing the ML/IT procedure, we evaluate the decision probability d_{lj}, which represents the probability that the lth correlation (recall $l = i + m - j$) corresponds to a single maximum for a given j (phase of the locally generated sequence ϵ_j). Now, it should be pointed out (see Figures 3.29 and 3.30) that in the process of forming h correlations $\rho(l, j)$ (for $l = 1, 2, \ldots, h$). L_A successive bits of the input signal are involved where

$$L_A = L_w(j) + L_{\overline{w}}(j) = h \qquad \text{for } \epsilon\text{ML system} \qquad (3.84)$$

$$L_A = L_{\overline{w}}(1, j) + L_w(j) + L_{\overline{w}}(h, j)$$
$$= 2h - 1 \qquad \text{for CML system} \qquad (3.85)$$

In the analysis to follow, we define a random sequence A, of length L_A as

$$A = R_{\overline{w}}(j) \oplus E_w(j) \qquad \text{for } \epsilon\text{ML system}$$
$$A = R_{\overline{w}}(1, j) \oplus E_w(j) \oplus R_{\overline{w}}(h, j) \quad \text{for CML system} \qquad (3.86)$$

For example, if $j < m$, these sequences can be represented as

$$A = \{l_{L_{\overline{w}}(j)}, \cdots, l_1, E_1, E_2, \cdots, E_{h-(m-j)}\} \qquad \text{for } \epsilon\text{ML}$$
$$A = \{l_{L_{\overline{w}}(1,j)}, \cdots, l_1, E_1, E_2, \cdots, E_{h-(m-j)}, r_1, r_2, \cdots, r_{L_{\overline{w}}(h,j)}\} \quad \text{for CML} \quad (3.87)$$

Once again, it should be pointed out that l and r bits take on equiprobable values 0 and 1, whereas E bits take on value 0 with probability $(1 - p_e)$ and the value 1 with probability p_e, where p_e is the bit error probability. For each particular combination of the bits in the sequence A, we have different values of the correlation functions $\rho(l, j)$. There are exactly $K = 2^{L_A}$ possible different realizations of the sequence A.

The probabilities of these realizations $p(A_k)$ are not the same because the a priori probabilities of the binary 1 and binary 0 in the sequence $E_w(j)$ are not the same. For the ϵML system, the probability of having a particular sequence A_k with $l_{\overline{w}}(j)$ "ones" in the first group of bits (so-called left or l-bits) and $l_w(j)$ in the second (E-bits) can be expressed as

$$p(A_k) = (1/2)^{l_{\overline{w}}(j)}(1/2)^{L_{\overline{w}}(j)-l_{\overline{w}}(j)} p_e^{l_w(j)}(1-p_e)^{L_w(j)-l_w(j)}$$
$$= (1/2)^{L_{\overline{w}}(j)} p_e^{l_w(j)}(1-p_e)^{L_w(j)-l_w(j)} \qquad (3.88)$$

For the CML system, the probability of having a particular sequence A_k with $l_{\overline{w}}(1, j)$ "ones" in the first group of bits (l-bits), $l_w(j)$ in the second (E-bits), and $l_{\overline{w}}(h, j)$ in the third (r-bits) can be expressed as

$$p(A_k) = (1/2)^{l_{\overline{w}}(1,j)}(1/2)^{L_{\overline{w}}(1,j)-l_{\overline{w}}(1,j)} p_e^{l_w(j)} \cdot (1-p_e)^{L_w(j)-l_w(j)}$$
$$\cdot (1/2)^{l_{\overline{w}}(h,j)}(1/2)^{L_{\overline{w}}(h,j)-l_{\overline{w}}(h,j)}$$
$$= (1/2)^{L_{\overline{w}}(1,j)+L_{\overline{w}}(h,j)} \cdot p_e^{l_w(j)}(1-p_e)^{L_w(j)-l_w(j)} \qquad (3.88a)$$

For each realization A_k of the sequence A, we define the parameter

$$n(l, j/A_k) = \begin{cases} 1 & \text{if } \rho(l, j/A_k) \text{ is the single maximum} \\ 0 & \text{elsewhere} \end{cases} \qquad (3.89)$$

Now, the probability that the correlation function has its single maximum at point l can be represented as

$$d_{lj} = \frac{1}{K} \sum_{k=1}^{K} n(l, j/A_k) p(A_k) \qquad (3.90)$$

The probability of having multiple ties can be expressed as

$$p_t = 1 - \sum_{l=1}^{h} d_{lj} \qquad (3.91)$$

Hence, the state transition probabilities can be expressed as

$$p_{i0} = \pi(\epsilon_i) = 1/h \qquad i = 1, 2, \cdots, h \qquad (3.92)$$

$$p_{00} = 0 \qquad (3.93)$$

$$p_{0j} = \begin{cases} \sum_{l=1}^{m-j} d_{lj} & j \le m \\ \sum_{l=h-(j-m)+1}^{h} d_{lj} & j > m \end{cases} \quad (3.94)$$

$$P_{ij} = \begin{cases} d_{lj} & j \le m \ \& \ 1 \le i \le h - (m-j) \ \& \ i \ne j \text{ or} \\ & j \ge m \ \& \ j - m \le i \le h \ \& \ i \ne j \\ d_{lj} + p_t & j \le m \ \& \ 1 \le i \le h - (m-j) \ \& \ i = j \text{ or} \\ & j \ge m \ \& \ j - m \le i \le h \ \& \ j = i \\ 0 & \text{elsewhere} \end{cases} \quad (3.95)$$

where $l = 1, 2, \ldots, h$ and $l = i + m - j$.

MF System. On the basis of the decision rule described earlier in this section, the MF tracking system can also be modeled as a discrete renewal process. The only exception is that the state transition probabilities are now different. In accordance with the algorithm description and by analogy with (3.89) we have

$$n(l, j|A_k) = \begin{cases} 1 & \text{if } \rho(l, j|A_k) \ge b \text{ and } \rho(c, j|A_k) < b; \ c < 1 \\ 0 & \text{elsewhere} \end{cases} \quad (3.96)$$

Hence, the probability d_{lj} can be expressed as

$$d_{lj} = \frac{1}{K} \sum_{k=1}^{K} n(l, j|A_k) p(A_k) \quad (3.97)$$

In other words, d_{lj} is the probability that the correlation function $\rho(l, j) \ge b$ and $\rho(c, j) < b$ for $1 \le c < l$. If none of the correlation functions exceed the threshold, the decision process stays in the same state. The probability of this event can be expressed as

$$P_n(j) = 1 - \sum_{l=1}^{h} d_{lj} \quad (3.98)$$

The state transition probabilities are again given by (3.95), except that the probabilities d_{lj} and p_t in (3.95) should be now replaced by d_{lj} (given by (3.97)) and $P_n(j)$ (given by (3.98)), respectively. Having derived the general expressions for the process transition probabilities P_{ij}, the system performance analysis is performed using the general results of discrete renewal process theory.

Approximations

Because the correlations $\rho(i, j)$ are not statistically independent, the calculation of the state transition probabilities, presented in the previous section, is quite difficult. To circumvent this, one can use approximations based on showing that the correlations between the $\rho(i, j)$ are small. Using this and a Gaussian approximation presented in Appendix A3, one can argue that we can indeed approximate the correlations as being independent of one another. It is shown in Appendix A3 that the correlation function $\rho(l, j)$ can be expressed as

$$\rho(l, j) = C(2m - j - l) + 2\Delta(l, j) + 2e(l, j) \qquad (3.99)$$

where $C(2m - j - l)$ is the periodic autocorrelation function of the sequence, whereas $\Delta(l, j)$ and $e(l, j)$ are random correction factors due to the presence of augmented and erroneously demodulated bits, respectively. Due to these factors, $\rho(i, j)$ is also a random variable. It is shown that if the sequence length h is large ($h \gg 1$) and $C(2m - j - l)$ is a two-level function that can be approximated as a Dirac impulse, then $\rho(l_p, j)$ and $\rho(l_q, j)$ are approximately independent variables for all $l_p \neq l_q$. This approximation is based on the fact that $\Delta(l, j)$ and $e(l, j)$ have binomial probability distribution functions (see Appendix A3). Using the Laplace approximation, these functions can be considered as Gaussian for $h \gg 1$. It is further shown that these variables are approximately uncorrelated; and so, being (approximately) Gaussian, they are taken to be independent. The expressions for the *probability mass function* (pmf) $P_{lj}(n)$ and cumulative probability $Q_{lj}(n)$ of the correlation function $\rho(l, j)$ are derived in Appendix A3. These probabilities are defined as

$$P_{lj}(n) = P(\rho(l, j) = n) \quad \begin{matrix} l, j = 1, 2, \cdots, h \\ n = -h, \cdots, 0, \cdots, h \end{matrix} \qquad (3.100)$$

and

$$Q_{lj}(n) = P(\rho(l, j) < n) \quad \begin{matrix} l, j = 1, 2, \cdots, h \\ n = -h, \cdots, 0, \cdots, h \end{matrix} \qquad (3.101)$$

Now (3.90) and (3.97) can be expressed (approximately) as

$$d_{lj} = \sum_{n=-h}^{h} P_{lj}(n) \prod_{\substack{k=1 \\ k \neq l}}^{h} Q_{kj}(n)$$

and

$$d_{lj} = (1 - Q_{lj}(b)) \prod_{k=1}^{l-1} Q_{kj}(b)$$

respectively. Numerical results indicate that, even for the case of a maximal length sequence with $h = 15$, this approximation is acceptable. For any larger h this approximation is better.

Performance

In order to gain an insight into the contribution of the dehopped part of the signal to the decision variable $\rho(l, j)$ defined by (3.99), the correlation function $C_{ws}(2m - l - j)$ for CML systems, defined by (A3.10d), is presented in Figure 3.31. In this example, a maximal length shift register sequence, defined by polynomial $D^4 \oplus D$, is used. The maximum of $C_{ws}(2m - l - j)$ occurs when $j = m = l$. This corresponds to the sync position. If the tracking error $\Delta\epsilon \neq 0$ ($j \neq m$), the maximum of $C_{ws}(2m - l - j)$ is at the point where $l = 2m - j$ and $C_{ws}(2m - l - j) < h$. The correlation function $C_{\overline{ws}}(2m - l - j)$, shown in the same figure, demonstrates how much of the useful information about the sync state has been lost due to the tracking error. The parameter $d_{l,j}$ (probability of choosing state l as the sync state when the tracking error is ϵ_j) for ϵML/IT, CML/IT, and MF systems is presented in Figures 3.32(a–e). From these figures the following conclusions can be drawn.

If $j \neq m$ (system is not in sync position m), algorithms ϵML/IT and CML/IT perform almost the same. Algorithm MF is worse because the peak value of $d_{l,j}$ (probability of choosing the sync-state) is considerably lower than in the case of the other two algorithms.

If $j = m$ (system is in sync position m), algorithm ϵML/IT will perform better because the sidelobes of $d_{l,j}$ are lower. This means that the probability of moving from the sync state to some other state is lower. Although the ϵML/IT algorithm is the best, it is not practical due to its hardware complexity. The phase error probability mass functions for CML/IT and MF algorithms are presented in Figure 3.33. From the figure, it can be seen by how much the CML/IT system outperforms the MF system.

3.3 CODE TRACKING IN FADING CHANNELS

The previously presented material on code tracking was based on the assumption that except for the additive white Gaussian noise the channel itself is not introducing any additional signal degradation. For some applications like land and satellite mobile communications, we have to take into account the presence of severe fading due to multipath propagation. In this section we will present one possible approach to code tracking in such an environment.

Code Tracking 155

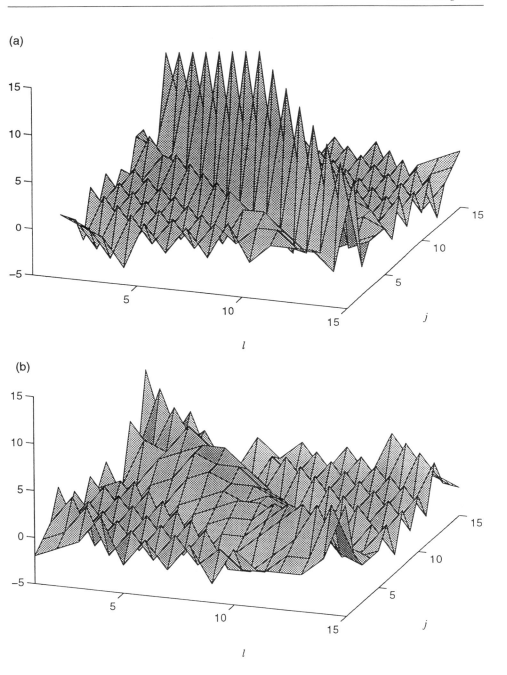

Figure 3.31 (a) Correlation function $C_{ws}(2m - l - j)$ for CML system and (b) correlation function $C_{\overline{ws}}(2m - l - j)$ for CML system.

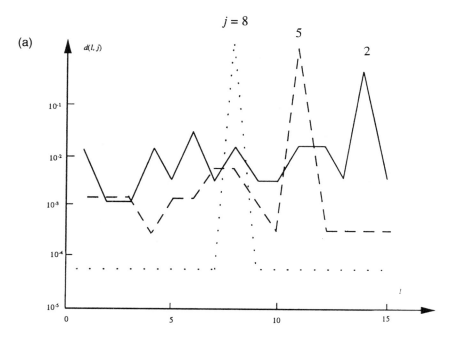

Figure 3.32 (a) Probability $d(l, j)$ for ϵML/IT system with $p_e = 0.1$ and $h = 15$, (b) probability $d(l, j)$ for CML/IT system with $h = 15$ and $p_e = 0.1$, (c) probability $d(l, j)$ for MF system with $h = 15$, $p_e = 0.1$, and $b = 7$, (d) probability $d(l, j)$ for MF system with $h = 15$, $p_e = 0.1$, and $b = 10$, and (e) probability $d(l, j)$ for MF system with $h = 15$, $b = 13$, and $p_e = 0.1$. (© 1991 IEEE. *From*: "Discrete Tracking System for Slow FH: Part II," by S. Glisic and L. Milstein in *IEEE Trans. on Communications*, Vol. 39, Feb. 1991.)

3.3.1 Channel Model

A channel with multipath propagation can be represented by a time-varying tapped-delay line, with impulse repose given by

$$h(\tau, t) = \sum_{l=0}^{N_\beta - 1} \beta_l(t) \delta(\tau - lT_s) \tag{3.102}$$

where T_s is the Nyquist sampling interval for the transmitted signal, N_β is the number of propagation paths, and $\beta_l(t)$ represent complex-valued time-varying channel coefficients. So, for the transmitted signal $s(t)$ where the received signal samples at $t = kT_s r(k)$ will consist of N_β mutually delayed replicas that can be represented as

$$r(k) = \sum_{l=0}^{N_\beta - 1} \beta_\ell(k) s[(k - l)T_s + \tau(k)] + n(k) \tag{3.103}$$

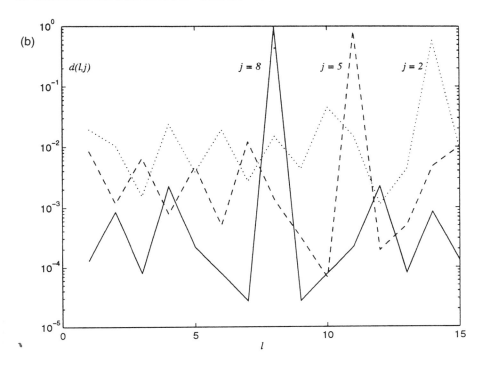

Figure 3.32 (continued).

In this equation $n(k)$ are samples of the noise with

$$E\{n(k-i)n*(k-j)\} = \sigma_n^2 \delta_{i,j} \quad (3.104)$$

In a Rake receiver each signal component is despread separately and then combined into a new decision variable for final decision. For the combining that provides maximum SNR, signal components are weighted with factors β_l^*. So the synchronization for the Rake receiver should provide a good estimate of delay τ and all channel intensity coefficients β_l, $l = 0, 1, \ldots, N_\beta - 1$. The operation of the Rake receiver will be elaborated upon later, and within this section we will concentrate on the joint channel coefficients (β_l) and code delay (τ) estimation using the extended Kalman filter.

For these purposes the channel coefficients and delays are assumed to obey the following dynamic model equations:

$$\begin{aligned} \beta_\ell(k+1) &= \alpha_\ell \beta_\ell(k) + w_l(k) \quad l = 0, 1, \ldots, N_\beta - 1 \\ \tau(k+1) &= \zeta \tau(k) + w_\tau(k) \end{aligned} \quad (3.105)$$

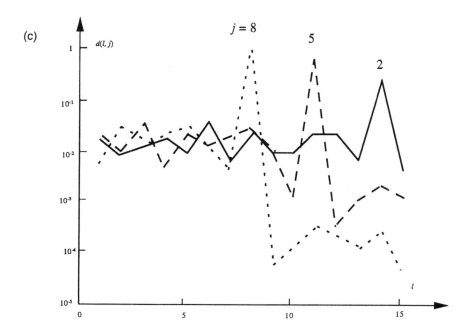

Figure 3.32 (continued).

where $w_l(k)$ and $w_\tau(k)$ are mutually independent white Gaussian processes with variances σ_{wl}^2 and σ_τ^2, respectively. In statistics these processes are called *autoregressive* (AR) processes of order K where K shows how many previous samples with indices $(k, k-1, k-2, \ldots, k-K+1)$ are included in the modeling of a sample with index $k+1$. In (3.105) the first-order ($K = 1$) AR model is used. The increased number of disturbances in signal are expected due to Doppler, the higher the variance of w_l, and the lower α_l should be used. Variance of w_τ will depend on Doppler but also on the oscillator stability. A comprehensive discussion of autoregressive modeling of wideband indoor radio propagation can be found in [5].

3.3.2 Joint Estimation of PN Code Delay and Multipath Using the Extended Kalman Filter

From the available signal samples $r(k)$ given by (3.103) we are supposed to find minimum variance estimates of β_l and τ. These will be denoted by

$$\hat{\beta}_l(k|k) = E\{\beta_l(k)|\mathbf{r}(k)\}$$
$$\hat{\tau}(k|k) = E\{\tau(k)|\mathbf{r}(k)\} \qquad (3.106)$$

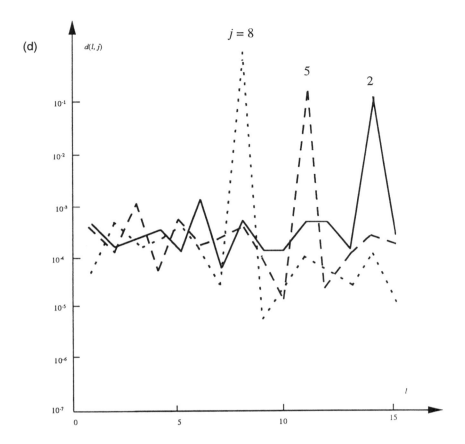

Figure 3.32 (continued).

where $\mathbf{r}(k)$ is a vector of signal samples

$$\mathbf{r}(k) = \{r(k), r(k-1), \cdots, r(0)\} \qquad (3.107)$$

From (3.103) one can see that $r(k)$ is linear in the channel coefficients $\beta_l(k)$, but it is nonlinear in the delay variable $\tau(k)$. A practical approximation to the minimum variance estimator in this case is the *extended Kalman filter* (EKF). This filter utilizes a first-order Taylor series expansion of the observation sequence about the predicted value of the state vector and will approach the true minimum variance estimate only if the linearization error is small. The basic theory of EKF is available in textbooks [6]. Keeping in mind that, in the delay tracking problem, the state model is linear while the measurement model is nonlinear, we have

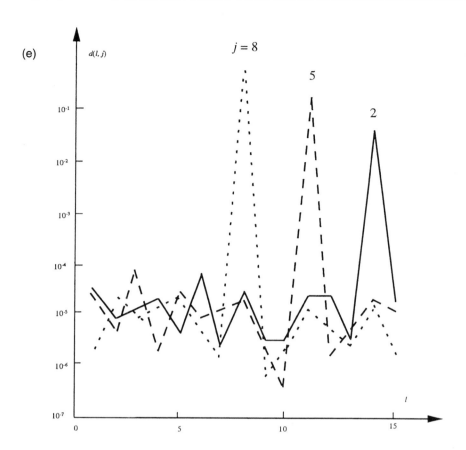

Figure 3.32 (continued).

$$\mathbf{x}(k+1) = \mathbf{F}\mathbf{x}(k) + \mathbf{G}\mathbf{w}(k)$$
$$z(k) = H(\mathbf{x}(k)) + n(k) \qquad (3.108)$$

In this equation $\mathbf{x}(k)$ represents the $(N_\beta + 1)$-dimensional state vector and $z(k)$ is the scalar measurement of $r(k)$. In terms of the previous notation we have

$$\mathbf{x}(k) = [\tau(k), \beta_0(k), \beta_1(k), \cdots, \beta_{N_\beta-1}(k)]^T$$

$$\mathbf{F} = \begin{bmatrix} \zeta & 0 & & \cdots & 0 \\ 0 & \alpha_0 & & \cdots & 0 \\ 0 & 0 & \alpha_1 & \cdots & 0 \\ \vdots & \vdots & \vdots & \vdots & \vdots \\ 0 & 0 & & \cdots & \alpha_{N_\beta-1} \end{bmatrix} \qquad (3.109)$$

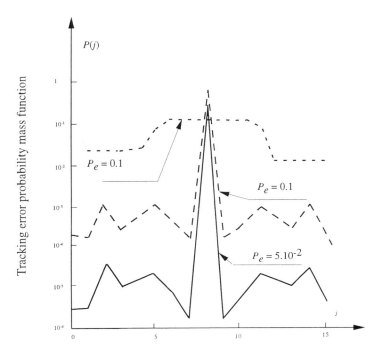

Figure 3.33 Tracking error probability mass function for CML/IT (——— and ---) and MF (···) systems. (© 1991 IEEE. *From*: "Discrete Tracking System for Slow FH: Part II," by S. Glisic and L. Milstein in *IEEE Trans. on Communications*, Vol. 39, Feb. 1991.)

$$\mathbf{w}(k) = [w_\tau(k), w_0(k), w_1(k), \cdots, w_{N-1}(k)]^T$$

$$H(\mathbf{x}(k)) = \sum_{l=0}^{N-1} \beta_l(k) s[(k-l)T_s + \tau(k)]$$

$$z(k) = r(k) = H[x(k)] + n(k)$$

$$\mathbf{G} = \mathbf{I}$$

Using general results of EKF theory [6] we have

$$\hat{\mathbf{x}}(k|k) = \hat{\mathbf{x}}(k|k-1) + \mathbf{K}(k)[z(k) - H(\hat{\mathbf{x}}(k|k-1))]$$

$$\mathbf{K}(k) = \mathbf{P}(k|k-1)\mathbf{H}'(k) \cdot [\mathbf{H}'(k)^H \mathbf{P}(k|k-1)\mathbf{H}'(k) + \sigma_n^2]^{-1}$$

$$\mathbf{P}(k|k) = [\mathbf{I} - \mathbf{K}(k)\mathbf{H}'(k)^H]\mathbf{P}(k|k-1) \qquad (3.110)$$

The matrix $H'(k)$ represents the time-varying gradient of the observation scalar with respect to the one-step prediction vector

$$\mathbf{H}'(k) = \left[\frac{\partial}{\partial x_1} H(\hat{\mathbf{x}}(k|k-1)), \frac{\partial}{\partial x_2} H(\hat{\mathbf{x}}(k|k-1)), \cdots, \frac{\partial}{\partial x_{N_\beta+1}} H(\hat{\mathbf{x}}(k|k-1)) \right]^H$$
(3.111)

The one-step predictions of the state vector and error covariance matrix are given as [6]

$$\hat{\mathbf{x}}(k+1|k) = \mathbf{F}\hat{\mathbf{x}}(k|k)$$
$$\mathbf{P}(k+1|k) = \mathbf{F}\mathbf{P}(k|k)\mathbf{F}^H + \mathbf{G}\mathbf{Q}\mathbf{G}^T$$

where

$$\mathbf{Q} = \text{diag}[\sigma_\tau^2, \sigma_{w0}^2, \cdots, \sigma_{w2}^2, \cdots, \sigma_{wN_\beta-1}^2]$$
(3.112)

After substituting the explicit forms of $\mathbf{x}(k)$ and $H(\mathbf{x}(k))$ into these equations, the following measurement update equation is obtained for the joint delay/multipath estimator:

$$\begin{bmatrix} \hat{\tau}(k|k) \\ \hat{\beta}_0(k|k) \\ \hat{\beta}_1(k|k) \\ \vdots \\ \hat{\beta}_{Nf-1}(k|k) \end{bmatrix} = \begin{bmatrix} \hat{\tau}(k|k-1) \\ \hat{\beta}_0(k|k-1) \\ \hat{\beta}_1(k|k-1) \\ \vdots \\ \hat{\beta}_{Nf-1}(k|k-1) \end{bmatrix}$$
(3.113)

$$+ \frac{1}{\sigma^2(k|k-1)} \mathbf{P}(k|k-1)$$

$$\cdot \begin{bmatrix} \frac{\partial}{\partial \hat{\tau}} s_{\hat{\beta}}^*(kT_s + \hat{\tau}(k|k-1)) \\ s * (kT_s + \hat{\tau}(k|k-1)) \\ s * ((k-1)T_s + \hat{\tau}(k|k-1)) \\ \vdots \\ s * ((k-N_\beta+1)T_s + \hat{\tau}(k|k-1)) \end{bmatrix}$$

$$\cdot [r(k) - s_{\hat{\beta}}(kT_s + \hat{\tau}(k|k-1))]$$

where $s_{\hat{\beta}}(t)$ denotes the estimate of the multipath distorted signal using the one-step predictions of $\beta_l(k)$.

$$s_{\hat{\beta}}(t) = \sum_{n=0}^{N_{\beta}-1} \hat{\beta}_n(k|k-1)s(t-nT_s) \quad (3.114)$$

The innovations variance $\sigma^2(k|k-1)$ is given by

$$\sigma^2(k|k-1) = [\mathbf{H}'(k)^H \mathbf{P}(k|k-1)\mathbf{H}'(k) + \sigma_n^2] \quad (3.115)$$

Note that $\sigma^2(k|k-1)$ is a scalar, thus a matrix inversion is not required, as would be the case for a vector measurement.

References

[1] European Telecommunications Standards Institute, GSM Specifications, ETSI TC-SMG, Sophia Antipolis, France.
[2] Rasky, P. D., G. M. Chiasson, and D. E. Borth, "Hybrid Slow Frequency-Hop/CDMA-TDMA as a Solution for High-Mobility, Widearea Personal Communications," *Proc. Fourth Winlab Workshop on Third Generation Wireless Information Networks*, East Brunswick, NJ, Oct. 19–20, 1993, pp. 199–215.
[3] Polydoros, A., and C. L. Weber, "Analysis and Optimization of Correlative Code-Tracking Loops in Spread Spectrum," *IEEE Trans. on Communications*, Vol. 33, Jan. 1985, pp. 30–43.
[4] Glisic, S., and L. Milstein, "Discrete Tracking System for Slow FH: Part I: Algorithms with Distributed Synchronization Group," *IEEE Trans. on Communications*, Vol. 39, No. 2, Feb. 1991, pp. 304–314.
[5] Holmes, J. K., *Coherent Spread Spectrum Systems*, New York: John Wiley, 1982.
[6] Andersson, B., and I. Moore, *Optimal Filtering*, Engelwood Clifts, NJ: Prentice-Hall, 1979.

Selected Bibliography

Bethel, R. E., and R. G. Rahikka, "Multisignal Time Delay Detection and Tracking," *IEEE Trans. on Aerospace and Electronic Systems*, Vol. 28, No. 3, July 1992, pp. 675–696.

Bethel, R. E., and R. G. Rahikka, "Optimum Time Delay Detection and Tracking," *IEEE Trans. on Aerospace and Electronic Systems*, Vol. 26, No. 5, Sept. 1990, pp. 700–712.

Bohmann, J., and H. Meyr, "An All-Digital Realization of a Baseband DLL Implemented as a Dynamical State Estimator," *IEEE Trans. on Acoustics, Speech, and Signal Processing*, Vol. ASSP-34, No. 3, June 1986, pp. 535–545.

Bourdreau, D., and P. Kabal, "Joint Time-Delay Estimation and Adaptive Recursive Least Squares Filtering," *IEEE Trans. on Signal Processing*, Vol. 41, No. 2, Feb. 1993, pp. 592–601.

Bowles, W. M., "GPS Code Tracking and Acquisition Using Extended-Range Detectors," Charles Stark Draper Lab., Inc., April 1979; see also *Proc. NTC'80*, Houston, Texas, Dec. 1980, pp. 24.1.1–24.1.5.

Champagne, B., M. Eizenman, and S. Pasupathy, "Exact Maximum Likelihood Time Delay Estimation for Short Observation Intervals," *IEEE Trans. on Signal Processing*, Vol. 39, No. 6, June 1991, pp. 1245–1257.

Chan, Y. T., J. M. F: Riley, and J. B. Plant, "Modelling of Time Delay and Its Applications to Estimation of Nonstationary Delays," *IEEE Trans. on Acoustics, Speech, and Signal Processing*, Vol. ASSP-29, No. 3, June 1981, pp. 577–581.

Chiang, H.-I., and C. L. Nikias, "A New Method for Adaptive Time Delay Estimation for Nongaussian Signals," *IEEE Trans. on Acoustics, Speech, and Signal Processing*, Vol. 38, No. 2, Feb. 1990, pp. 209–217.

Comparetto, G., "A Noncoherent Delay-Locked Loop Using the Exit-Time Criterion," *IEEE Trans. on Communications*, Vol. 35, No. 11, Nov. 1987, pp. 1240–1244.

Diaz, P., D. Henche, and R. Agusti, "A PN Code Delay Estimator Based on the Extended Kalman Filter for a DS/CDMA Cellular System," *The Fourth Int. Symp. on Personal, Indoor and Mobile Radio Communications*, Yokohama, Japan, Sept. 8–11, 1993.

Etter, D. M., and S. D. Stearns, "Adaptive Estimation of Time Delays in Sampled Data Systems," *IEEE Trans. on Acoustics, Speech, and Signal Processing*, Vol ASSP-29, No. 3, June 1981, pp. 582–587.

Feintuch, P. L., N. J. Bershad, and F. A. Reed, "Time Delay Estimation Using the LMS Adaptive Filter-Dynamic Behaviour," *IEEE Trans. on Acoustics, Speech, and Signal Processing*, Vol. ASSP-29, No. 3, June 1981, pp. 571–576.

Gill, W. J., "A Comparison of Binary Delay-Lock Loop Implementations," *IEEE Trans. on Aerospace and Electronic Systems*, Vol. 2, July 1966, pp. 415–424.

Glisic, S., and L. Milstein, "Discrete Tracking System for Slow FH: Part II: Algorithms with Concentrated Synchronization Group," *IEEE Trans. on Communications*, Vol. 39, No. 2, Feb. 1991, pp. 314–324.

Glisic, S., et al., "Efficiency of Digital Communication Systems," *IEEE Trans. on Communications*, Vol. COM-35, No. 6, June 1987, pp. 679–684.

Haas, W. H., and C. S. Lindquist, "A Synthesis of Frequency Domain Filters for Time Delay Estimation," *IEEE Trans. on Acoustics, Speech, and Signal Processing*, Vol. ASSP-29, No. 3, June 1981, pp. 540–548.

Hartmann, H. P., "Analysis of a Dithering Loop for PN Code Tracking," *IEEE Trans. on Aerospace and Electronic Systems*, Vol. 10, Jan. 1974, pp. 2–9.

Ho, K. C., Y. T. Chan, and P. C. Ching, "Adaptive Time-Delay Estimation in Nonstationary Signal and/or Noise Power Enviroments," *IEEE Trans. on Signal Processing*, Vol. 41, No. 7, July 1993, pp. 2289–2299.

Holmes, J. K., and L. Biederman, "Delay-Lock Loop, Mean-Time-to-Lose-Lock," *IEEE Trans. on Communications*, Vol. 26, Nov. 1978, pp. 1549–1556.

Hopkins, P. M., "Double Dither Loop for Pseudo Noise Code Tracking," *IEEE Trans. on Aerospace and Electronic Systems*, Vol. 13, Nov. 1977, pp. 644–650.

Howard, S., and K. Pahlavan, "Antoregressive Modelling of Wideband Indoor Radio Propagation," *IEEE Trans. on Communications*, Vol. 40, No. 9, Sept. 1992, pp. 1540–1552.

Iltis, R., "An EKF-Based Joint Estimator for Interference, Multipath and Code Delay in a DS Spread Spectrum Receiver," *IEEE Trans. on Communications*, 1994.

Iltis, R., "Joint Estimation of PN Code Delay and Multipath using the Extended Kalman Filter," *IEEE Trans. on Communications*, Vol. 38, No. 10, Oct. 1990, pp. 1677–1685.

Jacovitti, G., and G. Scarano, "Discrete Time Techniques for Time Delay Estimation," *IEEE Trans. on Signal Processing*, Vol. 41, No. 2, Feb. 1993, pp. 525–533.

Kosbar, K., "Open and Closed Loop Delay Estimation with Applications to Pseudo-Noise Code Tracking," Ph.D. Dissertation, Dept. of Electrical Engineering, University of Southern California, July 1988.

Kosbar, K., and A. Polydoros, "A Lower-Bounding Technique for the Delay Estimation of Discontinuous Signals, "*IEEE Trans. on Information Theory*, Vol. 38, No. 2, March 1992, pp. 451–457.

Kosbar, K. L., and J. L. Zaninovich, "Periodic PN Sequence Delay Estimation Using Phase Spectrum Data," *GLOBECOM'93*, pp. 1665–1669.

LaFlame, D. T., "A Delay-Lock Loop Implementation Which is Insensitive to Arm Gain Imbalance," *IEEE Trans. on Communications*, Vol. 27, Oct. 1979, pp. 1632–1633.

Layland, J. W., "On Optimal Signals for Phase-Locked Loops," *IEEE Trans. on Communications Technology*, Vol. 17, No. 5, Oct. 1969, pp. 526–531.

Lindsey, W. C., *Synchronization Systems in Communication and Control*, Englewood Cliffs, NJ: Prentice-Hall, 1972.

Lindsey, W. C., and H. Meyr, "Complete Statistical Description of the Phase-Error Process Generated by Correlative Tracking Systems," *IEEE Trans. on Information Theory*, Vol. 23, March 1977, pp. 194–202.

Lourtie, I. M. G., and J. M. F. Moura, "Multisource Delay Estimation: Nonstationary Signals," *IEEE Trans. on Signals Processing*, Vol. 39, No. 5, May 1991, pp. 1033–1048.

Meyr, H., "Delay-Lock Tracking of Stochastic Signals," *IEEE Trans. on Communications*, Vol. 24, March 1976, pp. 331–339.

Meyr, H., "Non-linear Analysis of Correlative Tracking Systems Using Renewal Process Theory," *IEEE Trans. on Communications*, Vol. 23, Feb. 1975, pp. 192–203.

Meyr, H., and G. Spies, "The Structure and Performance of Estimators for Real-Time Estimation of Randomly Varying Time Delay," *IEEE Trans. on Acoustics, Speech, and Signal Processing*, Vol. ASSP-32, No. 1, Feb 1994, pp. 81–94.

Miller, L. E., and J. S. Lee, "Error Analysis of Time Delay Estimation Using a Finite Integration Time Correlator," *IEEE Trans. on Acoustics, Speech, and Signal Processing*, Vol. ASSP-29, No. 3, June 1981, pp. 490–496.

Neilson, P. T., "On the Acquisition Behaviour of Binary Delay-Lock Loops," *IEEE Trans. on Aerospace and Electronic Systems*, Vol. 11, May 1975, pp. 415–418.

Polydoros, A., "On the Synchronization Aspects of Direct-Sequence Spread Spectrum Systems," Ph.D. Dissertation, Dept. of Electrical Engineering, University of Southern California, Aug. 1982.

Polydoros, A., and S. Glisic, "Code Synchronization: A Review of Principles and Techniques," *Code Division Multiple Access Communications*, S. Glisic and P. Leppänen (eds.), Boston, MA: Kluwer, 1995.

Pallas, M.-A., and G. Jourdain, "Active High Resolution Time Delay Estimation for Large BT Signals," *IEEE Trans. on Acoustics, Speech, and Signal Processing*, Vol. 39, No. 4, April 1981, pp. 781–788.

Raheli, R., A. Polydoros, and C-K. Tzou, "Per-Survivor Processing: A General Approach to MLSE in Uncertain Environments," *IEEE Trans. on Communications*, 1995.

Reed, F. A., P. L. Feintuch, and N. J. Bershad, "Time Delay Estimation Using the LMS Adaptive Filter—Static Behaviour," *IEEE Trans. on Acoustics, Speech, and Signal Processing*, Vol. ASSP-29, No. 3, June 1981, pp. 561–571.

Rodriques, M. A., R. H. Williams, and T. J. Carlow, "Signal Delay and Waveform Estimation Using Unwrapped Phase Averaging," *IEEE Trans. on Acoustics, Speech, and Signal Processing*, Vol. ASSP-29, No. 3, June 1981, pp. 508–513.

Scaraborough, K., N. Ahmed, and G. C. Carter, "On the Simulation of a Class of Time Delay Estimation Algorithms," *IEEE Trans. on Acoustics, Speech, and Signal Processing*, Vol. ASSP-29, No. 3, June 1981, pp. 534–540.

Segal, M., E. Weinstein, and B. R. Musicus, "Estimate-Maximize Algorithms for Multichannel Time Delay and Signal Estimation," *IEEE Trans. on Acoustics, Speech, and Signal Processing*, Vol. 39, No. 1, Jan. 1991, pp. 1–16.

Simon, M. K., "Noncoherent Pseudo Noise Code-Tracking Performance of Spread Spectrum Receivers," *IEEE Trans. on Communications*, Vol. 25, March 1977, pp. 327–345.

Simon, M. K., J. K. Omura, R. A. Scholtz, and B. K. Levitt, *Spread Spectrum Communications, Vol III*, Computer Science Press, 1985.

Spilker, J. J. Jr., and D. T. Magill, "The Delay-Lock Discriminator—An Optimum Tracking Device," *Proc. IRE*, Vol. 49, Sept. 1961, pp. 1–8.

Spilker, J. J. Jr., "Delay-Lock Tracking of Binary Signals," *IEEE Trans. on Space. Electronics Telemetry*, Vol. 9, March 1963, pp. 1–8.

Stein, S., "Algorithms for Ambiguity Function Processing," *IEEE Trans. on Acoustics, Speech, and Signal Processing*, Vol. ASSP-29, No. 3, June 1981, pp. 588–599.

Stiffler, J. J., *Theory of Synchronous Communications*, Englewood Cliffs, NJ: Prentice-Hall, 1971.

Tugnait, J. K., "On Time Delay Estimation with Unknown Spatially Correlated Gaussian Noise Using Fourth-Order Cumulates and Cross Cumulates," *IEEE Trans. on Signal Processing*, Vol. 39, No. 6, June 1991, pp. 1258–1267.

Tugnait, J. K., "Time Delay Estimation with Unknown Spatially Correlated Gaussian Noise," *IEEE Trans. on Signal Processing*, Vol. 41, No. 2, Feb. 1993, pp. 549–558.

Ward, R., "Optimization of Full-Time and Time-Shared Non Coherent Code Tracking Loops," Ph.D. Dissertation, Dept. of Electrical Engineering, University of Southern California, Aug. 1985.

Ward, R., and A. Polydoros, "Optimization of Full-Time and Time-Shared Noncoherent Code Tracking Loops," *Proc. MILCOM '85*, Boston, MA., Oct. 1985, pp. 1.6.1–5.

Wax, M., "The Estimate of Time Delay between Two Signals with Random Relative Phase Shift," *IEEE Trans. on Acoustics, Speech, and Signal Processing*, Vol. ASSP-29, No. 3, June 1981, pp. 497–501.

Welti, A. L., and B. Z. Bobrovsky, "Mean Time to Lose Lock for a Coherent Second-Order PN-Code Tracking Loop—The Singular Perturbation Approach," *IEEE J. on Selected Areas in Communications*, Vol. 8, No. 5, June 1985, pp. 809–818.

Yost, R. A., and R. W. Boyd, "A Modified PN Code Tracking Loop: Its Performance Analysis and Comparative Evaluation," *IEEE Trans. on Communications*, Vol. 30, May 1982, pp. 1027–1036.

Ziemer, R. E., and R. L. Peterson, *Digital Communications and Spread Spectrum Systems*, New York: Macmillan, Inc., 1985.

Appendix A3 Parameters of FH System Code Tracking

In this appendix, the general expressions for the probabilities $P_{lj}(n)$ and $Q_{lj}(n)$, defined by (3.100) and (3.101), respectively, are derived.

A3.1 Sequence Windowing

The well-known operation of sequence cyclic shifting is defined in terms of the windowing operation $W(S)$ and referred to as cyclic sequence windowing $CYW(S)$. By definition, for the sequence S we have

$$CYW^i(S) = \{S_{h-i+1}, \cdots, S_h, S_1, S_2, \cdots, S_{h-i}\} = S_{\overline{w}}(i) \oplus S_w(i) \quad (A3.1)$$

where $S_{\overline{w}}(i) = \{S_{h-i+1}, \cdots, S_h, 0, \cdots, 0\}$ and $S_w(i) = \{0, \cdots, 0, S_1, S_2, \cdots, S_{h-i}\}$. Also, by definition, we have

$$CYW^{-i}(S) = \{S_{i+1}, S_{i+2}, \cdots, S_h, S_1, S_2, \cdots, S_i\} = S_w(-i) \oplus S_{\overline{w}}(-i) \quad (A3.2)$$

where $S_w(-i) = \{S_{i+1}, S_{i+2}, \cdots, S_h, 0, 0, \cdots, 0\}$ and $S_{\overline{w}}(-i) = \{0, 0, \cdots, 0, S_1, S_2, \cdots, S_i\}$. By analogy with (3) and (4), for $i > j$,

$$CYW^{-j}\{CYW^i(S)\} = \{S_{i-j+1}, \cdots, S_h, S_1, S_2, \cdots, S_{h-i}, S_{h-i+1}, \cdots, S_{h-i+j}\}$$
$$= S_w(i, -j) \oplus S_{\overline{w}}(i, -j) \quad (A3.3)$$

where $S_w(i, -j) = \{0, 0, \cdots, 0, S_1, S_2, \cdots, S_{h-i}, 0, 0, \cdots, 0\}$ and $S_w(i, -j) = \{S_{i-j+1}, \cdots, S_h, 0, 0, \cdots, 0, S_{h-i+1}, \cdots, S_{h-i+j}\}$. If $i < j$, (A3.3) becomes

$$CYW^{-j}\{CYW^i(S)\} = \{S_{i-j+1}, \cdots, S_{h-i}, S_{h-i+1}, \cdots, S_h, S_1, \cdots, S_{j-i}\}$$
$$= S_w(i, -j) \oplus S_{\overline{w}}(i, -j) \quad (A3.4)$$

where $S_w(i, -j) = \{S_{j-i+1}, \cdots, S_{h-i}, 0, 0, \cdots, 0, 0\}$ and $S_{\overline{w}}(i, -j) = \{0, 0, \cdots, 0, S_{h-i+1}, \cdots, S_{j-i}\}$. From the previous definitions, we have

$$CYW^j\{CYW^i(S)\} = CYW^{i+j}(S) = S_w(i, j) \oplus S_{\overline{w}}(i, j)$$

$$CYW^{-j}\{CYW^{-i}(S)\} = CYW^{-(i+j)}(S) = S_w(-i, -j) \oplus S_{\overline{w}}(-i, -j) \quad (A3.5)$$

A3.2 General Expressions for $L_w(j)$, $L_w(l, j)$, and $L_{\overline{w}}(l, j)$

These parameters are illustrated graphically in Figure 3.30. According to the definition for both ϵML and CML systems, we have

$$L_w(j) = h - |m - j| \quad \text{and} \quad L_{\overline{w}}(j) = h - L_w(j) \quad (A3.6)$$

Also, for the CML system

$$L_w(l, j) = \begin{cases} L_w(j) - (m - l) & j \le m \ \& \ l \le m \\ L_w(j) & j \le m \ \& \ m \le l \le 2m - j = h - j + 1 \\ L_w(j) - (l - 2m + j) & j \le m \ \& \ h - j + 1 \le l \le h \\ L_w(j) - (2m - l - j) & j \ge m \ \& \ 1 \le l < 2m - j \\ L_w(j) & j \ge m \ \& \ 2m - j \le l < m \\ L_w(j) - (l - m) & j \ge m \ \& \ l \ge m \end{cases}$$
(A3.7)

and $L_{\overline{w}}(l, j) = h - L_w(l, j)$.

A3.3 The Correlation Function $\rho(l, j)$

For the CML system, we have

$$\begin{aligned}\rho(l, j) &= \text{Cor}\{S, R_{\overline{w}}(m - l, m - j) \oplus S'_w(m - l, m - j)\} \\ &= \text{Cor}\{S, R_{\overline{w}}(m - l, m - j) \oplus S_w(m - l, m - j) \\ &\quad \oplus E_w(m - l, m - j)\} \\ &= \text{Cor}\{S, D_{\overline{w}}(m - l, m - j) \oplus S_{\overline{w}}(m - l, m - j) \\ &\quad \oplus S_w(m - l, m - j) \oplus E_w(m - l, m - j)\}\end{aligned} \quad (A3.8)$$

where $R_{\overline{w}}$ is represented as $D_{\overline{w}} \oplus S_{\overline{w}}$. In other words, $D_{\overline{w}}$ is the "disagreement sequence" between $R_{\overline{w}}$ and $S_{\overline{w}}$. Using the fact that $S(2m - l - j) = S_w(m - l, m - j) \oplus S_{\overline{w}}(m - l, m - j)$, we have $\rho(l, j) = \text{Cor}\{S, S(2m - l - j) \oplus H(2m - l - j)\}$ where $H = D_{\overline{w}} \oplus E_w$ is the so-called "correction sequence." That is, it accounts for both errors due to noise and disagreements from the sync sequence due to the random data. Using the facts that the bipolar sequence $\tilde{H} = 2H - 1$ and $S \oplus H = -\tilde{S} \cdot \tilde{H}$, the correlation function can be expressed as

$$\rho(l, j) = \sum_{x=1}^{h} \tilde{S}_x \cdot (S \oplus H)_{x+(2m-l-j)}$$
$$= C(2m - j - l) + 2\Delta(l, j) + 2e(l, j) \qquad (A3.9)$$

where $\Delta(l, j) = \Sigma_{x=1}^{h} \tilde{S}_x \cdot \tilde{S}_{x+(2m-l-j)} \cdot D_{\overline{w}(x+2m-l-j)}$, $e(l, j) = -\Sigma_{x=1}^{h} \tilde{S}_x \cdot \tilde{S}_{x+(2m-l-j)} \cdot E_{w(x+2m-l-j)}$, and $C(x)$ is the sequence periodic correlation function defined as $C(x) = \Sigma_{y=1}^{h} \tilde{S}_y \cdot \tilde{S}_{y+x}$ for the ϵML system,

$$\rho(l, j) = \text{Cor}\{S(-m + l), R_{\overline{w}}(m - j) \oplus S'_w(m - j)\}$$
$$= C(2m - l - j) + 2\Delta(l, j) + 2e(l, j) \qquad (A3.9a)$$

where $\Delta(l, j) = -\Sigma_{x=1}^{h} D_{\overline{w}(x+m-j)} \tilde{S}_{x-m+l} \cdot \tilde{S}_{x+m-j}$ and $e(l, j) = -\Sigma_{x-1}^{h} E_{w(x+2m-l-j)} \tilde{S}_{x-m+l} \cdot \tilde{S}_{x+m-j}$.

A3.4 Additional Parameters

Finally, two additional parameters are defined:

$$C_{ws}(x) = \sum_{S_y \in ws} \tilde{S}_y \tilde{S}_{y+x} \qquad (A3.10a)$$

$$C_{\overline{ws}}(x) = \sum_{S_y \in \overline{ws}} \tilde{S}_y \tilde{S}_{y+x} \qquad (A3.10b)$$

where ws is set of the sequence S bits that are included in S_w and $\overline{ws}$ is the set of sequence S bits that are not included in S_w. Bearing in mind the definition of ws and $\overline{ws}$, we have $C(x) = C_{ws}(x) + C_{\overline{ws}}(x)$. For the ϵML system, $C_{ws}(2m - l - j)$ can be expressed as

$$C_{ws}(2m - l - j) = \begin{cases} \sum_{x=1}^{L_w(j)} \tilde{S}_x \tilde{S}_{x+(2m-l-j)} & j \leq m \\ \sum_{x=h-L_w(j)+1}^{h} \tilde{S}_x \tilde{S}_{x+(2m-l-j)} & j > m \end{cases} \qquad (A3.10c)$$

For the CML system, the same function can be expressed as

$$C_{ws}(2m - l - j) = \begin{cases} \sum_{x=1}^{L_w(l,j)} \tilde{S}_x \tilde{S}_{x+(2m-l-j)} & j < m \ \& \ l \le m \\ \sum_{x=1}^{L_w(j)} \tilde{S}_x \tilde{S}_{x+(2m-l-j)} & j < m \ \& \ m \le l \le 2m - j = h - j + 1 \\ \sum_{x=L_w(j)-L_w(l,j)+1}^{L_w(j)} \tilde{S}_x \tilde{S}_{x+(2m-l-j)} & j < m \ \& \ h - j + 1 \le l \le h \end{cases}$$

$$= \begin{cases} \sum_{x=L_{\overline{w}}(j)+1}^{L_{\overline{w}}(j)+L_w(l,j)} \tilde{S}_x \tilde{S}_{x+2m-l-j} & j \ge m \ \& \ 1 \le l \le 2m - j \\ \sum_{x=L_{\overline{w}}(j)+1}^{h} \tilde{S}_x \tilde{S}_{x+2m-l-j} & j \ge m \ \& \ 2m - j \le l \le m \\ \sum_{x=L_{\overline{w}}(l,j)+1}^{h} \tilde{S}_x \tilde{S}_{x+2m-l-j} & j \ge m \ \& \ l \ge m \end{cases}$$

(A3.10d)

It should be noted that (A3.10c) can be obtained from (A3.10d) by replacing $L_{\overline{w}}(l, j)$ with $L_{\overline{w}}(j)$ and $L_w(l, j)$ with $L_w(j)$ (bearing in mind the relation $L_w(j) + L_{\overline{w}}(j) = h$). The parameter $\Delta(l, j)$ in (A3.9) is the correction factor due to the disagreements of the $\rho_{\overline{ws}}$ and $C_{\overline{ws}}$ functions. These disagreements are a consequence of the sequence windowing. The parameter $e(l, j)$ is the correction factor due to the bit errors in the $S_w(l, j)$ sequence. If there are no errors, $e(l, j) \equiv 0$, whereas $\Delta(l, j)$ equals zero if the random bits of the $R_{\overline{w}}(l, j)$ sequence (introduced by the windowing process) happens to coincide with the corresponding sequence bits of $S_{\overline{w}}(l, j)$ (in which case $D_{\overline{w}} \equiv \{0\}$). Both parameters $\Delta(l, j)$ and $e(l, j)$ are random variables and can be either negative or positive. As the next step, we define parameters $N_{ws}^+(x)$ = the number of the positive products ($\tilde{S}_y \cdot \tilde{S}_{y+x} = 1$) in the sum defined by (A3.10a) and $N_{ws}^-(x)$ = the number of the negative products ($\tilde{S}_y \cdot \tilde{S}_{y+x} = -1$) in the same sum. $N_{\overline{ws}}^+(x)$ and $N_{\overline{ws}}^-(x)$ are the corresponding products in the sum defined by (A3.10b). On the basis of the previous definitions, we have

$$C_{ws}(x) = N^+_{ws}(x) - N^-_{ws}(x)$$

$$C_{\overline{ws}}(x) = N^+_{\overline{ws}}(x) - N^-_{\overline{ws}}(x)$$

$$L_w(x) = N^+_{ws}(x) + N^-_{ws}(x)$$

$$L_{\overline{w}}(x) = h - L_w(x) = N^+_{\overline{ws}}(x) + N^-_{\overline{ws}}(x)$$

Hence $N^+(x)$ and $N^-(x)$ can be expressed as

$$N^+_{ws}(x) = \{C_{ws}(x) + L_w(x)\}/2,$$

$$N^-_{ws}(x) = \{L_w(x) - C_{ws}(x)\}/2$$

$$N^+_{\overline{ws}}(x) = \{C_{\overline{ws}}(x) + L_{\overline{w}}(x)\}/2$$

$$N^-_{\overline{ws}}(x) = \{L_{\overline{w}}(x) - C_{\overline{ws}}(x)\}/2$$

A3.5 Probability Mass Functions

Consider the probability mass function of the parameter $e(l, j)$, which will be denoted $p_e(e)$. If a is the number of bit errors in ws where $\tilde{S}_y\tilde{S}_{y+x}$ should be $+1$ and b the number of bit errors in ws where $\tilde{S}_y\tilde{S}_{y+x}$ should be -1, then we have $e(l, j) = e = b - a$ and $n_e = b + a$ where n_e is the overall number of bit errors in ws. Hence, the probability $p_e(e)$ can be expressed as

$$p_e(e) = \sum_{a,b} \binom{N^+_{ws}}{a} p_e^a (1 - p_e)^{N^+_{ws}-a} \cdot \binom{N^-_{ws}}{b} p_e^b (1 - p_e)^{N^-_{ws}-b}$$

or

$$p_e(e) = \sum_{n_e=0}^{L_w(x)} \binom{[C_{ws}(x) + L_w(x)]/2}{(n_e - e)/2} \cdot \binom{[L_w(x) - C_{ws}(x)]/2}{(n_e + e)/2} p_e^{n_e} (1 - p_e)^{L_w(x)-n_e}$$

(A3.11)

Similarly, if a is now the number of bit disagreements between $S_{\overline{w}}(i, j)$ and $R_{\overline{w}}(i, j)$ in the $\overline{ws}$ region where $\tilde{S}_y\tilde{S}_{y+x}$ should be $+1$ and b is now the same parameter where $\tilde{S}_y\tilde{S}_{y+x}$ should be -1, then $\Delta(l, j) = \Delta = b - a$ and $n_\Delta = b + a$ where n_Δ is the overall number of disagreements between $S_{\overline{w}}(i, j)$ and $R_{\overline{w}}(i, j)$ in $\overline{ws}$ region. Hence, $p_\Delta(\Delta)$ can be expressed as

$$p_\Delta(\Delta) = \sum_{a,b} \binom{N_{\overline{ws}}^+}{a}(1/2)^a(1-1/2)^{N_{\overline{ws}}^+ - a} \cdot \binom{N_{\overline{ws}}^-}{b}(1/2)^b(1-1/2)^{N_{\overline{ws}}^- - b}$$

or

$$p_\Delta(\Delta) = (1/2)^{L_{\overline{w}}(x)} \sum_{n_\Delta = 0}^{L_{\overline{w}}(x)} \binom{[C_{\overline{ws}}(x) + L_{\overline{w}}(x)]/2}{(n_\Delta - \Delta)/2} \cdot \binom{[L_{\overline{w}}(x) - C_{\overline{ws}}(x)]/2}{(n_\Delta + \Delta)/2}$$

(A3.12)

Finally, the probability mass function of $\rho = \rho(l, j)$ can be expressed as

$$p_\rho(\rho) = \sum_{e=e_{\min}}^{e_{\max}} p_\Delta\left(\frac{\rho - C(x) - 2e}{2}\right) p_e(e) \quad (A3.13)$$

On the basis of the definition of parameter e, the parameters $e_{\min}$ and $e_{\max}$ can be expressed as $e_{\min} = -a_{\max} = -N_{ws}^+$ and $e_{\max} = b_{\max} = N_{ws}^-$. Hence the probabilities $P_{lj}(n)$ and $Q_{lj}(n)$ defined by (3.100) and (3.101), respectively, can be expressed as

$$P_{lj}(n) = P(\rho(l, j) = n) = p_\rho(n) \quad (A3.14a)$$

$$Q_{lj}(n) = P(\rho(l, j) < n) = \sum_{\rho = \rho_{\min}}^{n-1} p_\rho(\rho) \quad (A3.14b)$$

where

$$\rho_{\min} = C(x) + 2\Delta_{\min} + 2e_{\min}$$
$$e_{\min} = -N_{ws}^+ = -\{C_{ws}(x) + L_w(x)\}/2$$
$$\Delta_{\min} = -N_{\overline{ws}}^+ = -\{C_{\overline{ws}}(x) + L_{\overline{ws}}(x)\}/2$$

Upon combining these relations, we have $\rho_{\min} = C(x) - [C(x) + h] = -h$.

A3.6 Approximations

In what follows, we would like to show that $\rho(l_p, j)$ and $\rho(l_q, j)$ are approximately independent. To do this, we will show that they are approximately Gaussian and also approximately uncorrelated. We are interested in the pmf of ρ (A3.13); and from (A3.9) we see that ρ is determined by the sum of two random variables, Δ and e. These variables are each, in turn, the difference of two other random variables, a and b, in the regions $\overline{ws}$ and ws, respectively.

Since a and b have binomial distributions, using the so-called Laplace approximation, these distributions can be approximated as Gaussian; thus, we approximate both e and Δ (and consequently ρ) as Gaussian. Some results of numerical evaluation of the function $P_{lj}(n) = P(\rho(l, j) = n) = p_\rho(n)$, defined by (A3.13) with $h = 15$ and $p_e = 10^{-1}$, are presented in Figure A3.1, with l and j as parameters. As a reference, in the same figure, a Gaussian pdf with variance $\sigma^2 = 8$ and average value $m = 1$ is also presented. As the next step, we must show that the variables $\rho(l_p, j)$ and $\rho(l_q, j)$ are approximately uncorrelated for any $l_p \ne l_q$ or equivalently

$$r = \frac{C_{pq}}{\sqrt{C_p C_q}} \ll 1 \qquad (A3.15)$$

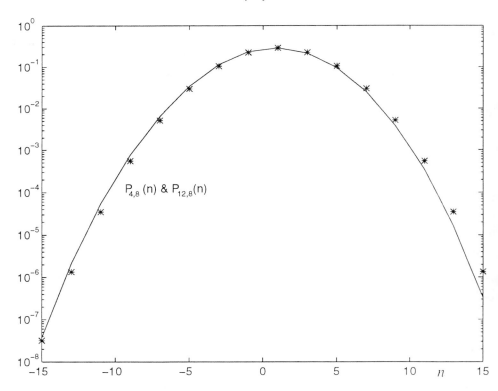

Figure A3.1 Probabilities $P_{lj}(n)$ for $p_e = 10^{-1}$ and Gaussian curve with $2\sigma^2 = 16$, and $m = 1$.

where

$$C_{pq} = \text{Cor}\{\rho(l_p, j), \rho(l_q, j)\} = E\{[\rho(l_p, j) - \overline{\rho}(l_p, j)][\rho(l_q, j) - \overline{\rho}(l_q, j)]\}$$
$$C_p = C_q = C_{pp} = C_{qq}$$

Using (A3.9), $\rho(l, j)$ can be expressed as $\rho(l, j) = C(2m - j - l) + 2\delta(l, j)$ where $\delta(l, j) = \{\Delta(l, j) + e(l, j)\}$ and, for any $l = l_p$, can be expressed as

$$\delta(l_p, j) = \delta_p = -\sum_{i=1}^{b} H(i)\tilde{S}(i)\tilde{S}(i \neq k_p) \quad (A3.16)$$

where $k_p = 2m - l_p - j$. In (A3.16), $H(i) = D_{\overline{w}}(i) \oplus E_w(i)$ is a unipolar sequence having $p(0) = p(1) = 1/2$ for $D_{\overline{w}}(i)$ ($i \in \overline{ws}$) and $p(0) = 1 - p(1) = 1 - p_e$ for $E_w(i)$ ($i \in ws$) where p_e is the channel bit error rate. In other words, if $H(i) = 1$, the contribution to the correction factor will depend on the sign of the product $\tilde{S}(i)\tilde{S}(i + k_p)$; whereas if $H(i) = 0$, there will be no contribution to the correction factor. For a random (and approximately for a pseudorandom) sequence the product $\tilde{S}(i)\tilde{S}(i + k_p)$ will be equiprobably +1 or −1, so the average value of $\delta(l_p, j)$ can be approximated as zero. C_{pq} can be expressed as

$$C_{pq} = 4E\{\delta_p \delta_q\} = 4\sum_i \sum_j E\{H(i)H(j)\}$$
$$\cdot \tilde{S}(i)\tilde{S}(i + k_p)\tilde{S}(j)\tilde{S}(j + k_q)$$
$$= 4\sum_i \sum_j \rho_H(i - j)\tilde{S}(i)\tilde{S}(i + k_p)\tilde{S}(j)\tilde{S}(j + k_q) \quad (A3.17)$$

which, in general, cannot be evaluated analytically in closed form. Even so, it is possible to estimate the value of the correlation $E(\delta_p \delta_q)$ for the case of interest where $k_p \neq k_q$, which implies $l_p \neq l_q$. If the probability $P(H(i) = 1) = p_e$ was the same over the entire range of the summation, the correlation function $\rho_H(\tau)$ would be

$$\rho_H(\tau) = \begin{cases} p_e & \tau = 0 \\ p_e^2 & \tau \neq 0 \end{cases} \quad (A3.18)$$

so that for each $p \neq q$, (A3.15) can be expressed as

$$r = \frac{C_{pq}}{C_{pp}} = \frac{(p_e - p_e^2)C(k_p - k_q) + p_e^2 C(k_p)C(k_q)}{(p_e - p_e^2)C(0) + p_e^2 C^2(k_p)} \quad (A3.19)$$

where $C(x)$ is the sequence correlation function. Bearing in mind that $C(0) \gg C(x)$ for any $x \neq 0$, for $p_e \ll 1$, we see that

$$r \sim \frac{C(k_p - k_q)}{C(0)} \ll 1 \qquad (A3.20)$$

Equations (A3.18) to (A3.19) are valid if $P(H(i) = 1) = p_e$ is the same over the entire summation range. Unfortunately, this is not true in our case. Rather, we have

$$P(H(i) = 1) = \begin{cases} p_e & i \in \text{ws} \\ 1/2 & i \in \overline{\text{ws}} \end{cases} \qquad (A3.21)$$

As a result, the correlation function $\rho_H(i - j)$ in (A3.18) has different values for $i - j \neq 0$. Consider then rewriting (A3.17) as

$$C_{pq} = 4\{\rho_H(0)\sum_i \tilde{S}^2(i)\tilde{S}(i + k_p)\tilde{S}(i + k_q) \\ + \sum_i \sum_{\substack{j \\ i \neq j}} \rho_H(i - j)\tilde{S}(i)\tilde{S}(i + k_p)\tilde{S}(j)\tilde{S}(j + k_q)\} \qquad (A3.22)$$

Bearing in mind that $\tilde{S}^2(i) = 1$, we have

$$C_{pq} = 4\left\{\rho_H(0)C(k_p - k_q) + \sum\sum\right\} \qquad (A3.23)$$

where $C(x)$ is the sequence correlation function and $\sum\sum$ is the double sum from (A3.22). If an m-sequence is used, then, using the shift-and-add property, we have $\tilde{S}(j) \cdot \tilde{S}(j + k_q) = \tilde{S}(j + k_q^1)$ and $\tilde{S}(i) \cdot \tilde{S}(i + k_p) = \tilde{S}(i + k_p^1)$ where k_q^1 and k_p^1 are integers that depend on k_q and k_p and the generating polynomial of the sequence, so the parameter $\sum\sum$ can be expressed as

$$\sum\sum = \sum_i \tilde{S}(i + k_p^1)\sum_{j \neq i}\rho_H(i - j)\tilde{S}(j + k_q^1) \qquad (A3.24)$$

The summation $I(i) = \sum_j \rho_H(i - j)\tilde{S}(j + k_q^1)$ can be thought of as the averaging of the product of a slowly varying positive function (ρ_H) and a rapidly varying carrier. Since $\rho_H(i - j) < 1$ and the carrier $S(j + k_q^1) = \pm 1$, if $h \gg 1$, we have $|I(i)| \ll h$. Now the remaining summation in (A3.24) can be expressed as

$$\sum\sum = \sum_i I(i)S(i + k_p^1) \ll h \qquad (A3.25)$$

Using (A3.23) to (A3.25) in (A3.19), one can see that inequality (A3.20) still holds. Therefore, once again, we see that there is relatively little correlation between δ_p and δ_q; hence, the assumption that they are approximately uncorrelated appears to be justified.

CHAPTER 4
▼▼▼

SPREAD SPECTRUM SYSTEM TECHNOLOGY

For packetized transmission in CDMA radio networks, most often the packet period is too short for standard code acquisition techniques, and some form of PN-matched filter techniques is preferable. In this chapter we will present a number of different receiver structures based on using PN-matched filters. These filters may be implemented either in standard *digital signal processing* (DSP) technology or using special technologies like *charge coupled devices* (CCD) or *surface acoustic wave* (SAW) components. DSP is widely used and documented in many other fields of communications, so in this book we emphasize CCD and SAW technologies, which are very specific for spread spectrum systems. We do not discuss the technology itself from the fabrication point of view but rather those aspects that influence the receiver structure. In Sections 4.1 and 4.2 basic parameters of a matched filter implemented in CCD technology will be discussed. This should explain the main limitations imposed on receiver design. In Section 4.4 a number of receiver schemes based on CCD PN-matched filters are described. Principles of SAW technology and its main applications are discussed in Sections 4.5 and 4.6, respectively.

4.1 CCD TECHNOLOGY

A block diagram of a PN-matched filter, matched to a sequence of length M, is shown in Figure 4.1(b). It consists of M cells tapped delay line where with each

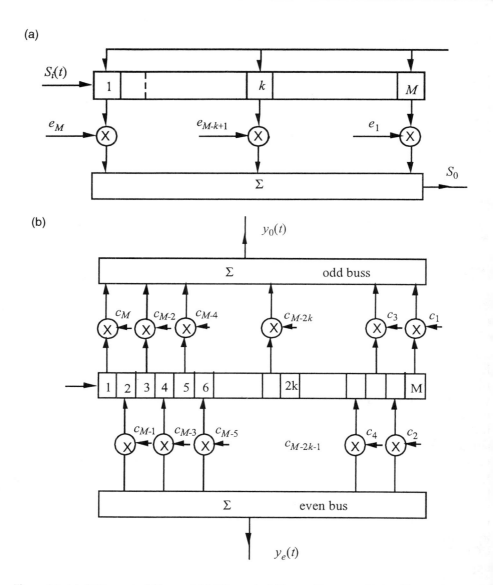

Figure 4.1 (a) CCD-matched filter and (b) PN-matched filter architecture with 2 cell/tap and even odd buses.

clock pulse ($T = T_c$) analog signal sample is transferred from one cell to the next. At a given moment, signals from each tap are multiplied by the corresponding sequence chip, as shown in the figure, and all these products are summed, resulting in signal S_o, that can be represented as

$$S_i = b(n)c(m)$$
$$S_o(l) = b(n-1)C_{c,c}(l-M) + b(n)C_{c,c}(l) \tag{4.1}$$

where $b(n)$ is the bit in the nth bit interval, $c(m)$ is the sequence mth chip, and $C_{c,c}(l)$ is the aperiodic correlation function defined as

$$C_{c,c}(l) = \begin{cases} \sum_{m=0}^{M-1-l} c(m)c(m+l) & 0 \leq l \leq M-1 \\ \sum_{m=0}^{M-1+l} c(m-l)c(m) & 1-M \leq l < 0 \\ 0 & |l| \geq M \end{cases} \tag{4.2}$$

At the moment when the input sequence in the line overlaps completely with the coefficients of the filter ($l = 0$) we have from (4.1) and (4.2)

$$S_o(0) = b(n)M \cdot \theta_{cc}(0) \tag{4.3}$$

where $\theta_{cc}(\ )$ is the periodic autocorrelation function of code c. If we introduce the notation

$$\mathbf{b} = |b(n-1), b(n)| \quad , \quad \mathbf{C}_{c,c}(l) = \begin{vmatrix} C_{c,c}(l-M) \\ C_{c,c}(l) \end{vmatrix} \tag{4.2a}$$

then (4.1) becomes

$$S_o(l) = \mathbf{b}\mathbf{C}_{c,c}(l) \tag{4.1a}$$

which for $l = 0$ reduces to (4.3). In practice, the CCD-matched filter from Figure 4.1(a) can be extended to include q cells per tap to allow higher sampling rates for better time resolution. Figure 4.1(b) shows a PNMF with $q = 2$ cells per tap. The odd PN code chips are used to control taps of an odd code bus, while the even PN code chips are associated with an even bus. This structure not only simplifies fabrication but also allows the PNMF to handle a number of different modulation formats, which will be discussed later in this chapter. In these schemes, instead of using odd and even chips of the same code, two different uncorrelated codes will be used.

For $e = 0$ two outputs from Figure 4.1(b) can be represented as

$$Y_o(0) = b(n)(M/2)\theta_o(0)$$
$$Y_e(0) = b(n)(M/2)\theta_e(0) \qquad (4.3a)$$

where $\theta_o(0)$ and $\theta_e(0)$ are correlation functions of odd and even PN sequence chips, respectively. Summing these two outputs results in (4.3).

The operation of a tapped delay line can be considered as an "analog shift register" where at each clock pulse arrival, signal samples are moved for one position to the right. The physical implementation of this function in CCD technology is explained in Figure 4.2. Different layers of the device are shown in Figure 4.2(a). By applying a signal at ϕ, a potential dwell is created and a free minority charge is generated proportional to the signal strength. Then by applying a signal to ϕ_2 the potential dwell is extended to the next cell, enabling the free charge to spill over as shown in Figure 4.2(c). Removing the signal from ϕ_2 will shrink the potential dwell and the equivalent effect is the transfer of charge from one cell to another. So, in essence, a CCD is a nearly ideal semiconductor analog shift register. It is fabricated in N-MOS semiconductor technology, allowing large-scale integration of many support circuits on to the integrated circuit. The CCD has the following desirable attributes:

1. Silicon fabrication simplicity;
2. High packing density;
3. High reliability;
4. Low power consumption;
5. Intrinsically low-noise analog signal processing.

In order to get better insight into the last statement, some sources of the noise generated in CCD components will be discussed in the next section.

4.2 CCD PN-MATCHED FILTER IMPERFECTIONS

In order to be able to understand the limitations of CCD technology, a number of sources contributing to the imperfections of these components will be discussed.

4.2.1 DC Zero Offset

To enable bipolar signaling, all CCD cells are maintained at the same quiescent charge, representing a DC zero offset sometimes called the fat zero. For maximum

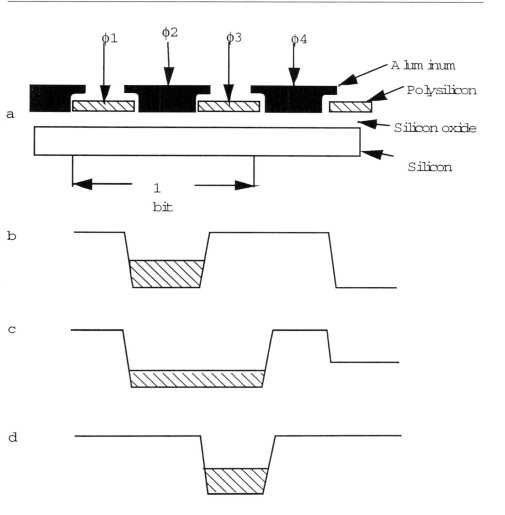

Figure 4.2 Schematic representation of charge transfer in a CCD device.

dynamic range, this reference is set near $Q_{SAT}/2$ where Q_{SAT} is the maximum charge packet that can be processed without saturation. In the general case, for these purposes, each sample of the signal, x_k, can be expressed as

$$x_k = n_k + c_k m_k \tag{4.4}$$

where n_k is the quiescent number of electrons in the first cell at the kth sample time and m_k is the additional number added or subtracted, depending on the sign of phase code c_k. Both n_k and m_k are assumed to be independent Poisson random variables, each with equal mean and variances $\bar{n}$ and $\bar{m}$, respectively. At the correlation peak, the output is given by

$$y_0 = \sum_{k=1}^{M} c_k(n_k + c_k m_k) \qquad (4.5)$$

The expected value of y_0 for a given code is given by

$$\bar{y}_0 = M\bar{m} + \bar{n}\Sigma c_k \qquad (4.6)$$

The second term is a bias that is dependent upon the balance of the code. For balanced codes, this term vanishes so that $\bar{y}_0 = M\bar{m}$. The variance of y_0 is given by

$$\text{var}(\bar{y}_0) = \text{var}(\Sigma c_k n_k + \Sigma m_k) = M(\bar{n} + \bar{m}) \qquad (4.7)$$

since

$$\text{var}(c_k n_k) = c_k^2 \bar{n} = \bar{n} \qquad (4.8)$$

The SNR is the squared mean over the variance

$$\text{SNR} = M\bar{m}^2/(\bar{m} + \bar{n}) \qquad (4.9)$$

Usually $\bar{m} \cong \bar{n}$, so the variability of the PNMF output due to code unbalance is down by $M\bar{n}/2$. With $\bar{n} = n_e \cong 10^{-6}$ and $M = 512$, the SNR would be 84 dB.

4.2.2 Charge Transfer Inefficiency

The charge transfer mechanism presented in the previous section is not perfect. In each transfer, a certain small amount of charge will not be transferred to the next cell. In order to better understand this phenomenon suppose that a unit pulse of unit charge is propagating in the filter. If ϵ is the portion of the charge that is not transferred, then the propagation of the unit pulse can be represented as shown in Figure 4.3.

With the first clock pulse, $(1 - \epsilon)$ of the charge will be transferred to the next cell and ϵ of it will remain in cell one. The second clock pulse will transfer $(1 - \epsilon)$ of ϵ from the first cell to the second one and ϵ out of $(1 - \epsilon)$ of the charge of the second cell will remain in cell two due to imperfect transfer, resulting in an overall $2\epsilon(1 - \epsilon)$ charge. The amplitude of the unit pulse after two clock pulses (two transfers) will be $(1 - \epsilon)^2$. From Figure 4.3 one can see that the charge transfer inefficiency

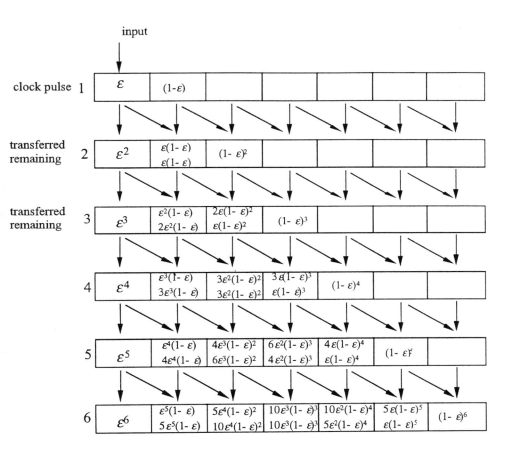

Figure 4.3 Pulse propagation in CCD-matched filter.

can be accounted for if the filter coefficients c_k are replaced by an effective set of weights given by

$$h_k = \sum_{j=0}^{k-1} c_{k-j} \binom{k-1}{j} (\epsilon)^j (1-\epsilon)^{k-j} \qquad k = 1, 2, \cdots, M \qquad (4.10)$$

If double sampling per tap (two transfers per tap) is used this will be modified with $\epsilon \rightarrow 2\epsilon$ resulting in

$$h_k = \sum_{j=0}^{k-1} c_{k-j} \binom{k-1}{j} (2\epsilon)^j (1-2\epsilon)^{k-j} \qquad (4.11)$$

The output of the effective PNMF is

$$y_0 = \sum_{k=1}^{M} h_k x_{M+1-k} \qquad (4.12)$$

For a BPSK signal passed through a chip-matched filter, the CCD samples are given by $x_{M+1-k} = c_k A T_c$ for an amplitude A and chip width T_c. Hence,

$$y_0 = A T_c \sum_{k=1}^{M} c_k h_k = A T_c \sum_{k=1}^{M} \sum_{j=0}^{k-1} c_k c_{k-j} \binom{k-1}{j} (2\epsilon)^j (1-2\epsilon)^{k-j} \qquad (4.13)$$

Taking the expected value and assuming that $E(c_k c_{k-j}) = \delta_{j0}$ yields

$$\bar{y}_0 = A T_c \sum_{k=1}^{M} (1-2\epsilon)^k = A T_c (1-2\epsilon) \frac{1-(1-2\epsilon)^M}{1-(1-2\epsilon)}$$
$$\cong M A T_c (1 - M\epsilon) \qquad (4.14)$$

for $M\epsilon \ll 1$.

Since the maximum output is equal to MAT_c, the decibel loss in peak signal due to charge transfer inefficiency can be defined as

$$L_s \triangleq 10 \log(\bar{y}_0/MAT_c)^2 \text{ dB}$$
$$\cong -8.68 M\epsilon + 4.34 (M\epsilon)^2 \text{ dB} \qquad (4.15)$$

L_s is plotted versus $M\epsilon$ in Figure 4.4. A 3-dB loss occurs at $M\epsilon \cong 0.3$. This plot can be considered the filter insertion loss versus $M\epsilon$.

Next consider a sequence of zero-mean noise samples $\langle z_k \rangle$ uncorrelated from tap to tap. The output noise variance is then given by

$$\sigma_0^2 = E\left(\sum_{k=1}^{M} h_k z_{M+1-k}\right)^2 = \sigma_n^2 \sum_{k=1}^{M} E(h_k^2) \qquad (4.16)$$

where the noise variance $\sigma_n^2 = N_0 T_c/2$ for white noise with double-sided power density $N_0/2$ is passed through an ideal matched filter (a unit pulse of width T_c). Based on code orthogonality,

$$E(h_k^2) = \sum_{j=0}^{k-1} \binom{k-1}{j} \binom{k-1}{j} (2\epsilon)^{2j} (1-2\epsilon)^{2(k-j)} \qquad (4.17)$$

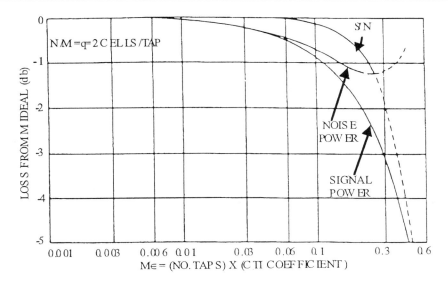

Figure 4.4 Effects of CCD CTI on PNMF performance. (© 1980 IEEE. *From*: "Inherent Signal to Noise Ratio Limitations of Charge-Coupled Devices," by D. M. Grieco in *IEEE Trans. on Communications*, Vol. COM-28, May 1980.)

Keeping only terms up to ϵ^2 and summing gives

$$\sigma_0^2 \cong M[1 - 2M\epsilon + (2M\epsilon)^2]\sigma_n^2 \qquad (4.18)$$

for large M but small $M\epsilon$.

The loss in output noise variance due to charge transfer inefficiency is defined as

$$L_N \underline{\underline{\Delta}} \ 10 \ \log(\sigma_0^2/M\sigma_n^2) \ \text{dB}$$

$$\cong -8.68 M\epsilon(1 - 2M\epsilon) \ \text{dB} \qquad (4.19)$$

This loss is plotted in Figure 4.4. At about $M\epsilon = 0.3$, the SNR begins to increase. After this point, higher order $M\epsilon$ terms cannot be ignored. The dashed line indicates this region of $M\epsilon$. The output SNR is given by dividing (4.18) into the square of (4.14)

$$\text{SNR} \cong (2MA^2 T_c/N_0)[1 - 3(M\epsilon)^2 - 6(M\epsilon)^3 + 24(M\epsilon)^5 \pm \cdots] \qquad (4.20)$$

When $\epsilon \to 0$, the SNR is equal to $2E_s/N_0$, the usual formula for a matched filter with a signal energy $E_s = MA^2 T_c$. The loss in SNR due to *charge transfer inefficiency* (CTI) is defined as

$$L_{\text{SNR}} \triangleq 10 \log(\text{SNR}/(2E_s/N_0)) \text{ dB} \qquad (4.21)$$

This loss is also plotted in Figure 4.4. Note that (4.21) falls off more gradually than (4.14) or (4.18) because (4.21) varies as $(M\epsilon)^2$ while (4.14) and (4.18) vary linearly with $M\epsilon$. In fact, the SNR loss can be approximated by

$$L_{\text{SNR}} \cong -13(M\epsilon)^2(1 + 2\epsilon) \quad \text{for } M\epsilon \lesssim 0.3 \qquad (4.22)$$

The SNR loss is less than 1 dB as long as $M\epsilon < 0.23$. Since for buried-channel CCDs $10^{-4} < \epsilon < 10^{-5}$, PNMFs of 2300 taps to 23,000 taps (and twice as many cells) can be used. In reality, IC size usually limits the number of cells to about 500. However, multiple-CCD PNMFs may be connected in tandem or recursive configurations for longer integration lengths.

4.2.3 Additional Impairments

Some additional sources of noise in CCD components will be summarized in this subsection.

4.2.3.1 Input Circuit Noise

During sampling of the analog input signal, a number of thermally generated electrons will be stored in the CCD cell. As shown in [1], the mean-squared number of thermal electrons collected on a capacitance C is kTC/q_0^2, where k is Boltzmann's constant, T is temperature, and q_0 is the charge per electron. At room temperature, the root mean square number of electrons is equal to $400\sqrt{C'}$ where C' is the capacitance in picofarads. For summing bus with M taps, the mean output signal level is given by Mn_e. Hence, the output SNR is given by

$$\text{SNR}_0 = (Mn_e)^2/[M(kTC/q_0^2)] = M(n_e/400\sqrt{C'})^2 \qquad (4.23)$$

at room temperature. A typical value for the input capacitance is 0.1 pF. Thus, with $M = 255$ and $n_e = 10^6$, we have $\text{SNR}_0 = 102$ dB. Therefore, the fat-zero analysis can serve as a worst case analysis of input noise.

4.2.3.2 Dark Current

During each dwell time, a number of electrons are generated thermally. This is called dark current. The number of dark charges increases from cell to cell as the charge packets traverse the CCD. It can be shown that the signal-to-dark current noise ratio can be expressed as

$$\text{SNR} = \frac{(M\overline{m})^2}{M^3 \overline{q}^{-2}/3} = \frac{3}{M}\left(\frac{\overline{m}}{\overline{q}}\right)^2 \qquad (4.24)$$

The variables can be expressed as $\overline{m} = \rho_0 A_c$ where ρ_0 is the fat-zero electron density ($\sim 10^{11}$ electrons/cm^2) and A_c is the cell area, while $\overline{q} = J_D A_c/q_0 f_c$ where J_D is the dark current density (~ 50 nA/cm^2), f_c is the clock frequency, and q_0 is the charge per electron (1.6×10^{-19} C/electron). Substitution yields

$$\text{SNR} = \frac{3}{M}\left(\frac{\rho_0 q_0 f_c}{J_D}\right)^2 \qquad (4.25)$$

J_D increases exponentially with temperature, approximately doubling every 9°C increase in device temperature [2]. Using the above values and $f_c = f_0 \times 10^6$ yields

$$\text{SNR} = 149 + 20 \log f_0 - 10 \log M - 2T/3 \text{ dB} \qquad (4.26)$$

High temperatures, low frequencies, and long devices will tend to decrease the SNR. A typical case might be $f_0 = 5$ MHz and $M = 500$. Then at $T = 21°C$, we have $SNR = 122$ dB; while at $T = 99°C$, we have $SNR = 63$ dB. A worst case might be $J_D = 100$ nA/cm^2 at 21°C, $f_0 = 1$ MHz, $M = 1000$, and $T = 99°C$, resulting in $SNR = 47$ dB. While generally that value would be satisfactory, cooling might be required in certain circumstances. CCDs can be efficiently air cooled and do not require elaborate cooling mechanisms.

4.2.3.3 Reset Noise

During each sampling interval, the bus output lines are reset to a known dc level. An output capacitance is charged through a resistor to this dc level, with additional thermal noise introduced by the resistor. For a typical set of the circuit parameters it can be shown that this so-called reset noise can be neglected at the output of a sample and hold circuit.

4.3 REPORTED SIGNAL PROCESSORS BASED ON CCD COMPONENTS

4.3.1 Analog-Analog Correlators

For full analog-analog correlator *programmable transversal filter* (PTF) capability, two distinct CCD approaches have been developed. In the first approach direct mixing of the stored analog signal samples with the analog filter weight can be accomplished in a single *metal-oxide-semiconductor* (MOS) transistor multiplier at each tapping point. This processor is reported in [3].

Alternative designs use a four-*field-effect transistor* (FET) bridge circuit for increased accuracy. When combined with the CCD tapped delay line, these compact realizations of the transversal filter cell allow 256-point PTFs to be designed in a single integrated device [4]. The dynamic range at each filter point of 90 dB to 108 dB in a 1Hz noise bandwidth has been reported with a total power dissipation of 0.25W [5]. Further cascading to achieve 2048-point filter capability has subsequently been demonstrated. This approach thus offers compact, low-power PTFs, but the difficulties in performing analog-analog operation with sufficient accuracy limit the device to a clock rate of 1 MHz to 5 MHz and, hence, to usable signal bandwidths of several hundred kilohertz. An alternative method to obtain the full analog-analog operation combines the CCD delay function with an array of multiplying digital-to-analog converters [6]. The analog signal samples are fed into each MDAC, which comprises an array of charge sources and transfer gates whose areas are related by powers of two. The filter weight values are stored as digital words where each bit turns the source gate either fully on or off so that the adjacent holding well is either filled to a level proportional to the analog signal or held empty of charge. The summed charge in the MDAC output is thus proportional to the analog signal sample and the multibit digital reference values. A schematic representation of this operation is given in Figure 4.5.

4.3.2 Analog-Binary Correlators

In spread spectrum communications, the information is encoded as a digital (binary) phase-modulated signal, and hence in the matched-filter receiver only an analog-binary correlator is required. This enables the CCD tapped delay line to be combined with a set of binary polarity switches and an output sum bus. Such devices have been considered for deployment in spread spectrum receivers for the SEEK COMM, and they have also been considered for developments in the Joint Tactical Information Distribution System. More about these components can be found in [7–9].

4.3.3 Double Sampling

CCD analog-binary correlators have been engineered for military temperature range operation with CMOS-compatible (complementary MOS) interfaces and provide dual-channel 256-sample 128-tap filter capability with an input sample rate that is comparable with the CCD MDAC approach. The double-sampling single-tapped structure (i.e., sampling twice per code chip but sensing only one sample per chip internally at a given time) provides considerable flexibility [10].

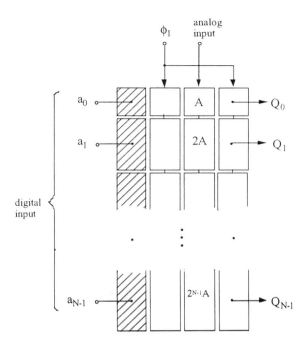

Figure 4.5 Schematic for N-bit MDAC used in CCDs to multiply analog signal samples by digital tap weight value.

Only alternate samples participate in the correlation each clock cycle. This mode of operation allows the device to operate by double sampling a single input waveform or single sampling of two independent waveforms (or the I and Q waveforms from a single channel). Single sampling of one waveform can also be achieved with interspersed zeros to provide an auto-zero facility at the output, for improved accuracy. Magnitude summation of the two filter channels is included on-chip if desired.

4.3.4 Simultaneous/Parallel Search

An alternative design of the matched-filter receiver is the time-integrating correlator [11] which multiplies the incoming signal by a binary reference and integrates the resulting product within the device structure. If a set of time-delayed signal replicas are provided by the CCD pipe-organ structure, the correlation can be extended to cover multiple offset time differences in the received and reference codes. This permits the correlator to simultaneously search over several chips of code in a manner similar to the FIR filter design, but the code length is no longer limited by the device delay but rather by the (presumably much longer) integration time. Such CCD designs have been operating at 20 MHz clock rates performing 32 simultaneous correlations with integration over 1,023 chip codes. These designs are applicable to spread spectrum navigation systems such as the NAVSTAR *global positioning system* (GPS) [12].

4.3.5 High-Speed Correlators

Improved CCD operating speed can be achieved by using high-electron mobility *gallium arsenide* (GaAs) semiconductor material. Schottky-barrier GaAs CCDs have recently demonstrated efficient charge transfer at 4.2 GHz and transversal-filter operation at 1 GHz [13].

4.3.6 Wide Bandwidth Devices

Another GaAs charge device utilizes surface-acoustic waves on a piezoelectric semiconductor to provide an electrodeless, sinusoidal, traveling-wave clock. These *acoustic charge transport* (ACT) devices, first investigated in a lithium niobate/silicon gap-coupled form [14] and further developed in a monolithic GaAs realization [15], offer bandwidths of several hundred megahertz. However, they sacrifice the clock agility of capacitive gate CCDs, and accurate tapping for high dynamic range operation has not yet been demonstrated.

4.4 RECEIVER CONFIGURATIONS BASED ON MATCHED FILTERS

In this section a number of receiver structures based on the utilization of a PN-matched filter will be discussed. One should be aware that the filter may be implemented either in CCD technology or using standard DSP components. The only difference would be that in the case of DSP, signal samples will propagate through the delay line in digital form rather than as analog samples. As the first step we will assume a general modulation format so that in a simplified form the transmitted signal can be represented as

$$s(t) = b_o c_o p(t)\cos(\omega t + \phi) + b_e c_e p(t - T_c/2)\sin(\omega t + \phi) \qquad (4.27)$$

where chip pulse shape $p(t)$ is limited to a one chip interval. We use indices "o" for odd and "e" for even channels in accordance with the previous discussion about the odd/even structure of CCD filters. This will be extended to DSP as well in order to unify the presentation. So, in a I-channel binary stream, b_o and sequence c_o are used to create a DS signal. In a Q-channel binary stream, b_e is transmitted using sequence c_e, which is uncorrelated with c_o and time shifted for a half-chip interval ($T_c/2$). The general form of the receiver front end is shown in Figure 4.6.

After chip-matched filtering and frequency down conversion, S_i and S_q are produced

$$S_i = b_o c_o \rho(t) \cos\phi + b_e c_e \rho(t - T_c/2) \sin\phi$$
$$S_q = -b_o c_o \rho(t) \sin\phi + b_e c_e \rho(t - T_c/2) \cos\phi \quad (4.28)$$

where $\rho(t)$ is the autocorrelation of chip pulse $p(t)$. If we assume that phase ϕ does not change much in period $\Delta t = MT_c$, the outputs of the matched filters can be represented as

$$R_{io}(t) = \text{cor}(S_i, c_o)$$
$$= \mathbf{b}_o \mathbf{C}_{o,o}(t) \cos\phi + \mathbf{b}_e \mathbf{C}_{e,o}(t - T_c/2) \sin\phi \quad (4.29)$$

where autocorrelation $\mathbf{C}_{o,o}$ and crosscorrelation $\mathbf{C}_{e,o}$ are given with appropriate modification of (4.2) where $c(m)c(m + t)$ should be replaced by $c(m)c(m + t)\rho(t - [m - 1]T_c)$ to account for the chip pulse correlation. For low crosscorrelations we can represent all four matched filter outputs as

$$R_{io}(t) = \mathbf{b}_o \mathbf{C}_{o,o}(t) \cos\theta + n_{io}$$
$$R_{ie}(t) = \mathbf{b}_e \mathbf{C}_{e,e}(t - T_c/2) \sin\theta + n_{ie}$$
$$R_{qo}(t) = -\mathbf{b}_o \mathbf{C}_{o,o}(t) \sin\theta + n_{qo}$$
$$R_{qe}(t) = \mathbf{b}_e \mathbf{C}_{e,e}(t - T_c/2) \cos\theta + n_{qe} \quad (4.30)$$

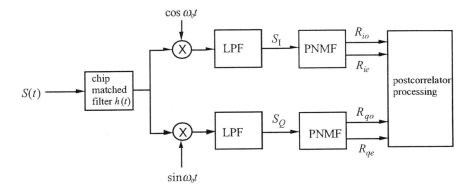

Figure 4.6 Receiver front end. (© 1980 IEEE. *From:* [8].)

The noise components represent terms proportional to the crosscorrelation functions. These four signals are used in the postcorrelator processor for data demodulation.

4.4.1 Coherent Digital Demodulation of BPSK Data and BPSK Spread Spectrum Modulation

The receiver block diagram for this case is shown in Figure 4.7(a).

Carrier and code (epoch) synchronization are provided by a PPL and ETL, respectively. Block diagrams of these loops are given in Figures 4.7(b,c), respectively. In this case (4.30) has only the first two terms and a data estimate is obtained from the first term as

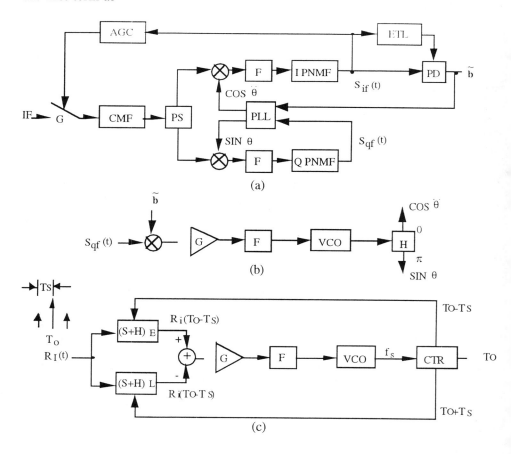

Figure 4.7 Coherent digital demodulator block diagram. (© 1980 IEEE. From: [8].)

$$\hat{b} = \text{sgn } R_i(MT_s + T_\Delta)$$
$$= \text{sgn } R_i(\Delta T) = \text{sgn}(Mb\rho(T_\Delta) + n_i) \quad (4.31)$$

where ΔT is the sampling error and the signum function is implemented as a polarity detector PD.

4.4.2 BPSK Data and OQPSK Spread Spectrum Modulation

In this case all terms of (4.30) are used as shown in Figure 4.8. Using (4.30) we get

$$R_{io} + R_{qe} = MD\cos\theta + n_{io} + n_{qe}$$
$$R_{ie} - R_{qo} = MD\sin\theta + n_{ie} + n_{qo} \quad (4.32)$$

where

$$D = b_o\rho_o(\tau) + b_e\rho_e(\tau - T_c/2)$$

The input to the polarity detector is $MD \cos \Delta\theta$ where $\Delta\theta = \theta - \tilde{\theta}$. The form of $D(t)$ is shown in Figure 4.9. For double sampling, samples of b_e and b_o will be obtained simultaneously.

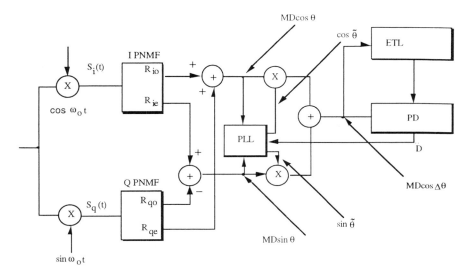

Figure 4.8 Receiver structure for BPSK data and OQPSK spread spectrum modulation. (© 1980 IEEE. From: [8].)

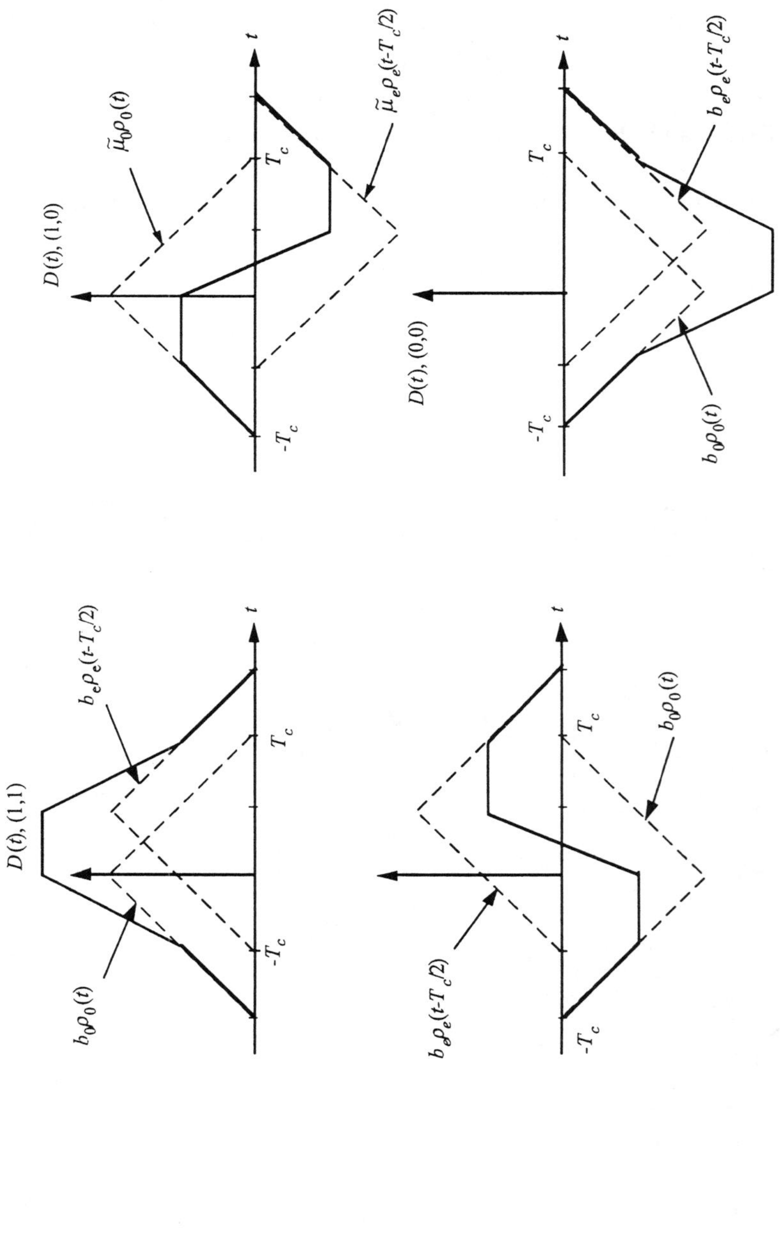

Figure 4.9 Signal waveforms.

4.4.3 Differentially Coherent Demodulation of QPSK Data Modulation

By ignoring noise components, (4.30) in two successive bit intervals become

$$R_{io}(t) = b_{o1}C_{o,o}(t)\cos\theta$$
$$R_{ie}(t) = b_{e1}C_{e,e}(t)\sin\theta$$
$$R_{qo}(t) = -b_{o1}C_{o,o}(t)\sin\theta$$
$$R_{qe}(t) = b_{e1}C_{e,e}(t)\cos\theta \quad (4.33)$$

and T_b seconds later

$$R_{io}(t + T_b) = b_{o2}C_{o,o}(t)\cos\theta'$$
$$R_{ie}(t + T_b) = b_{e2}C_{e,e}(t)\sin\theta'$$
$$R_{qo}(t + T_b) = -b_{o2}C_{o,o}(t)\sin\theta'$$
$$R_{qe}(t + T_b) = b_{e2}C_{e,e}(t)\cos\theta' \quad (4.34)$$

Bits with index 1 and 2 are of the same sign if a binary one is transmitted and of the opposite sign if a binary zero is transmitted.

Phase θ' is given as $\theta' = (\omega + \Delta\omega)T_b + \theta$ and for constant frequency $\Delta\omega = 0$ it becomes $\theta' = \omega T_b + \theta = 2k\overline{u} + \theta = \theta$. If $\Delta\omega \neq 0$, $\theta' = \Delta\omega T_b + \theta$. Postcorrelator signal processing consists of creating two decision variables of the form

$$S_{di}(t) = R_{io}(t)R_{io}(t + T_b) + R_{qo}(t)R_{qo}(t + T_b)$$
$$S_{dq}(t) = R_{ie}(t)R_{ie}(t + T_b) + R_{qe}(t)R_{qe}(t + T_b) \quad (4.35)$$

At the sampling moments these signals become

$$S_{di}(t) = M^2 b_{o1} b_{o2} \rho_o^2(0)\cos\Delta\theta \cong M^2 b_{o1} b_{o2}$$
$$S_{dq}(t) = M^2 b_{e1} b_{e2} \rho_e^2(0)\cos\Delta\theta \cong M^2 b_{e1} b_{e2} \quad (4.36)$$

which provides proper demodulation of differentially modulated data in I- and Q-channels. Based on these equations, the receiver block diagram is given in Figure 4.10. The main drawback of this structure is a need for a large number of analog multipliers that does not lend the overall scheme to all-digital VLSI implementation. In order to get rid of these multipliers we can use slight modifications of the overall system, which is described in the next section.

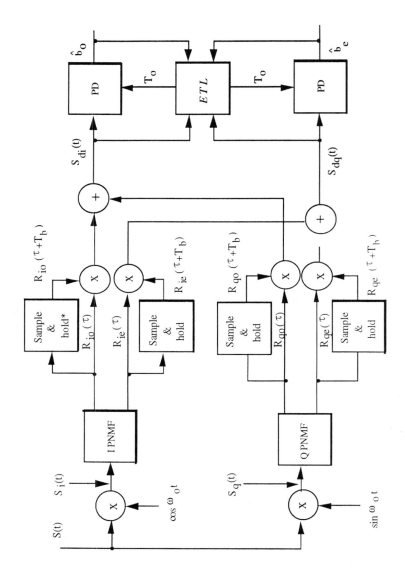

Figure 4.10 Differentially coherent demodulation of QPSK data modulation.

4.4.4 Sum and Difference Demodulator

For simplicity assume only a BPSK signal so that, for example, only the first and the third terms from (4.33) and (4.34) are generated. These terms are now combined as

$$|\Sigma| = |R_i(t) + R_i(t + T_b)| + |R_q(t) + R_q(t + T_b)|$$
$$|\Delta| = |R_i(t) - R_i(t + T_b)| + |R_q(t) - R_q(t + T_b)| \quad (4.37)$$

The data decision variable is now formed as

$$D = |\Sigma| - |\Delta| \quad (4.38)$$

For $\theta = \theta'$ this becomes

$$D = k|b_1 + b_2| - k|b_1 - b_2| \quad (4.39)$$

with

$$k = M\rho(\tau)\{|\cos\theta| + |\sin\theta|\} \geq M\rho(\tau) \quad (4.40)$$

For binary one $b_1 = b_2$ and $D = 2k$, and for binary zero $b_1 = -b_2$ so that $D = -2k$. The final detector makes a decision as

$$\hat{b} = \text{sgn } D \quad (4.41)$$

The circuit block diagram is given in Figure 4.11.

4.4.5 Noncoherent Pulse Amplitude Modulation

Pulse amplitude modulation (PAM) can be achieved by inverting the phase of a subset of consecutive PN chips. If k such chips are inverted, we have at the correlation peak instead of $M\rho(0)$ a correlation peak $M\rho^k(0)$ given as

$$A(k) = M\rho^{(k)}(0) = \sum_{i=1}^{M-k} c_i^2 - \sum_{i=M-k+1}^{M} c_i^2 = M - 2k \quad (4.42)$$

which can have value in the range $(-M, M)$. Suppose now that the transmitted signal given by (4.27) is modified as

$$s(t) = b_o c_o p(t)\cos(\omega t + \phi_o) + c_e^{(k)} p(t)\sin(\omega t + \phi_o) \quad (4.43)$$

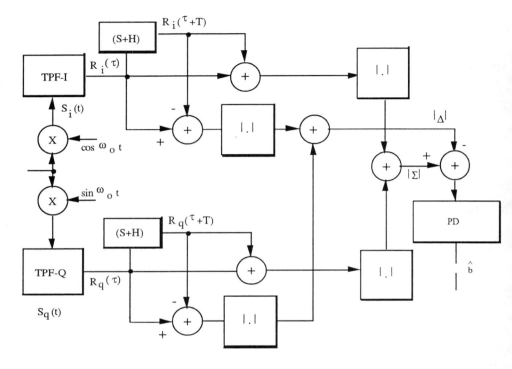

Figure 4.11 Difference demodulator block diagram.

where $c_e^{(k)}$ stands for the sequence c_e with k-chips inverted. Equation (4.30) in the sampling moment with noise ignored becomes

$$R_{io} = Mb_o\cos\theta$$
$$R_{ie} = M\rho_e^{(k)}(0)\sin\theta$$
$$R_{qo} = -Mb_o\sin\theta$$
$$R_{qe} = M\rho_e^{(k)}(0)\cos\theta \qquad (4.44)$$

If the sign and amplitude of the signal sample are encoded as

$$b_o = \text{sgn } A(k) \qquad M\rho_e^{(k)}(0) = |A(k)| \qquad (4.45)$$

then (4.44) becomes

$$R_{io} = M \text{ sgn } A(k) \cdot \cos\theta$$
$$R_{ie} = |A(k)|\sin\theta$$
$$R_{qo} = -M \text{ sgn } A(k)\sin\theta$$
$$R_{qe} = |A(k)|\cos\theta \tag{4.46}$$

The final receiver signal output is formed as

$$\hat{S} = R_{io}R_{qe} - R_{ie}R_{qo}$$
$$= M \text{ sgn } A(k)|A(k)|(\cos^2\theta + \sin^2\theta)$$
$$= M \text{ sgn } A(k)|A(k)| \tag{4.47}$$

which is an amplified (factor M) version of the transmitted signal sample. The receiver block diagram is given in Figure 4.12.

4.4.6 Coherent PAM

In this case (4.43) is further modified so that k_1 chips of sequence c_o and/or k_2 chips of sequence c_e are inverted. The signal sample amplitude is encoded as

$$A(k) = A_o(k_1) - A_e(k_2) = M[\rho_o^{(k_1)}(0) - \rho_e^{(k_2)}(0)] \tag{4.48}$$

which now vary in the range $(-2M, 2M)$. Equation (4.44) now becomes

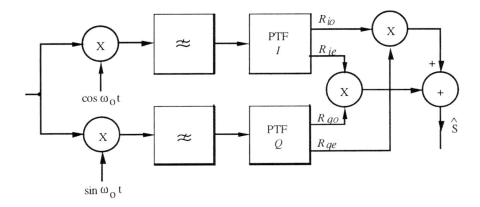

Figure 4.12 Noncoherent amplitude demodulator.

$$R_{io} = M\rho_o^{(k_1)}(0)\cos\theta = A_o(k_1)\cos\theta$$
$$R_{ie} = M\rho_e^{(k_2)}(0)\sin\theta = A_e(k_2)\sin\theta$$
$$R_{qo} = -M\rho_o^{(k_1)}(0)\sin\theta = -A_o(k_1)\sin\theta$$
$$R_{qe} = M\rho_e^{(k_2)}(0)\cos\theta = A_e(k_2)\cos\theta \qquad (4.49)$$

The signal sample estimate is formed as

$$\hat{S} = R_{io} - R_{qe} = [A_o(k_1) - A_e(k_2)]\cos\theta \qquad (4.50)$$

which for a coherent receiver with $\theta = 0$ results in $A(k)$. For estimated phase $\hat{\theta}$ the real receiver will form

$$\hat{S} = (R_{io} - R_{qe})\cos\hat{\theta} - (R_{qo} + R_{ie})\sin\hat{\theta}$$
$$= A(k)\cos(\theta - \hat{\theta}) \hat{=} A(k) \qquad (4.51)$$

4.5 SURFACE ACOUSTIC WAVE TECHNOLOGY

4.5.1 Convolution Function

CCD technology described in the previous section provides components for the baseband signal processing. In this section we discuss a different approach that provides signal processing components at IF frequencies. This approach is based on using a nonlinear interaction between two acoustic waves propagating in an acoustic material (e.g., $LiNbO_3$ plus silicon). Without going into the physical processes of this phenomenon, we model this interaction as shown in Figure 4.13. Two signals, $f(t)\exp(j\omega_o t)$ and $g(t)\exp(j\omega_o t)$, are transformed in an *electrical/acoustic* (E/A) wave transducer into acoustic substrate waves that propagate in the material in opposite directions. This is taken into account using different signs x and $-x$ for the spatial axis. At each point of the acoustic the interaction between the two waves is approximated with square law nonlinearity producing three components

1. $(k/2)f^2(t - x/v)\exp 2j(\omega_0 t - kx)$
2. $(k/2)g^2(t + x/v)\exp 2j(\omega_0 t + kx)$
3. $kf(t - x/v)q(t + x/v)\exp(2j\omega_0 t)$ \qquad (4.52)

where v is the signal propagation velocity in the acoustic material. Each of the three components is collected on the bottom electrode so that the overall output signal is a spatial integral with respect to x. The first two components of (4.52) are periodic

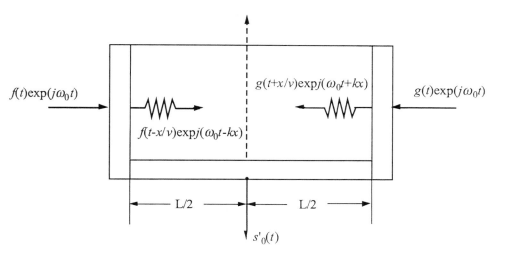

Figure 4.13 Signal propagation in a SAW component.

functions of x, and its integral in the range $(-L/2, L/2)$ will have a value close to zero. After integration, the third component will result in

$$k\exp(2j\omega_0 t) \int_{-L/2}^{L/2} f(t - x/v)g(t + x/v)\,dx = S'_0(t) \quad (4.53)$$

The envelope of this signal can be rearranged as

$$S_0(t) = k \int_{-L/2}^{L/2} f(t - x/v)g(t + x/v)\,dx$$
$$= kv \int_{t-L/2v}^{t+L/2v} f(\tau)g(2t - \tau)\,d\tau \quad (4.54)$$

which represents the convolution of the two signals $f(t)$ and $g(t)$ with time compression (factor 2). Based on these few relations, the block diagram of a SAW convolver is shown in Figure 4.14. If instead of $g(t)$ its time-inverted function $g(-t)$ is used, the same component will perform a correlation function.

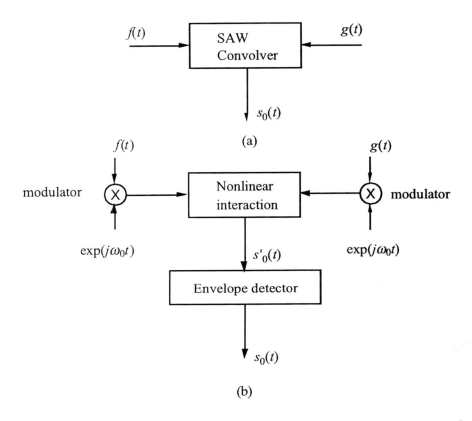

Figure 4.14 Block diagram of SAW convolver: (a) formal representation and (b) block diagram of the signal processing.

4.5.2 Fourier Transform

The component from Figure 4.14 can be used to produce real-time Fourier transform. We start with the definition of a Fourier transform of function $f(t)$

$$F(\omega) = \int_{-\infty}^{\infty} f(\tau)\exp(-j\omega\tau)\,d\tau \qquad (4.55)$$

At the output of a real-time Fourier transformer the frequency is proportional to time, which will be represented as

$$F(\omega) = F(\mu t) = \int_{-\infty}^{\infty} f(\tau)\exp(-j\mu t\tau)\,d\tau \qquad (4.56)$$

Using a formal transformation $-t\tau = [(t - \tau)^2 - t^2 - \tau^2]/2$, the previous equation becomes

$$F(\mu t) = \exp(-j\mu t^2/2)\int_{-\infty}^{\infty} d\tau\{f(\tau)\exp(-j\mu\tau^2/2)\}\exp[j\mu(t - \tau)^2/2] \qquad (4.57)$$

The block diagram of this equation is shown in Figure 4.15.

4.5.3 Inverse Fourier Transform

The inverse Fourier transform is obtained in a similar way. We start with the definition

$$f(t) = \frac{1}{2\pi}\int_{-\infty}^{\infty} F(\omega)\exp(j\omega t)\,d\omega \qquad (4.58)$$

Using a similar formal transformation $t\tau = [(t + \tau)^2 - t^2 - \tau^2]/2$ we have

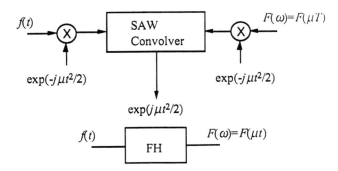

Figure 4.15 Real-time Fourier transform.

$$f(t) = \int_{-\infty}^{\infty} d\tau F(\tau) \exp j[(t + \tau)^2 - t^2 - \tau^2]/2$$

$$= \exp(-jt^2/2) \int_{-\infty}^{\infty} d\tau \{F(\tau)\exp(-j\tau^2/2)\} \exp(t + \tau)^2 \quad (4.59)$$

The block diagram of this processing is shown in Figure 4.16. One should notice that in this scheme a SAW correlator instead of convolver is used.

4.5.4 Transform Domain Signal Filtering

An example of signal processing based on using components from the previous sections is shown in Figure 4.17.

Input signal plus interference, represented as $f(t) + n(t)$, is first passed through a *Fourier Transform* (F.T.) component to produce $F(\omega) = F(\mu t)$. If $n(t)$ is strong narrowband interference, $F(\mu t)$ will have a high value in the time period corresponding to the frequency of the interfering signal. If the time notch $w(t)$ is used to shut down the receiver in this time period, that would be equivalent to producing a notch in the frequency domain in the frequency band occupied by the interference. The despreading process is realized in the frequency domain where the correlation corresponds to multiplication with conjugated signal spectrum. The process is completed by an additional inverse Fourier transform.

If the frequency of the interfering signal is not known it can be detected using a threshold detector after a F.T. processor. Any time the signal becomes higher than the threshold is an indication that strong interfering signal spectral components are present and the receiver is shut down. This is shown in Figure 4.17(b).

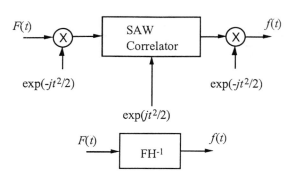

Figure 4.16 Inverse Fourier transform.

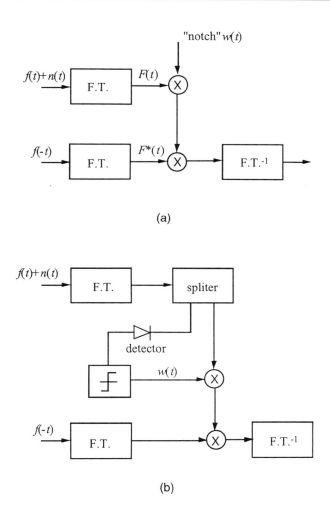

Figure 4.17 Transform domain signal filtering: (a) nonadaptive receiver and (b) adaptive receiver.

4.6 REPORTED SURFACE ACOUSTIC WAVE DEVICES

As we already explained, the SAW convolver [16,17] relies on two colinear surface acoustic waves propagating in opposite directions through the device. A nonlinear interaction produces the local sum-frequency product signal, which is subsequently integrated over the interaction region of the device. The SAW propagation medium is usually lithium niobate. In the simplest acoustic convolver, inherent nonlinearities in the lithium niobate are utilized and a metal plate electrode collects the output signal [16]. Considerable recent development has gone into improving the efficiency of this device with acoustic-beamwidth compressors using multistrip couplers [18], waveguide horns [19], and focused chirp transducers [20] and into reducing the self-convolution (foldover) spurious signals with dual channel operation and multistrip couplers for artifact cancellation [17,21,22].

Other convolver realizations produce the convolution in a semiconductor layer that is electrostatically coupled to a SAW propagating either in a lithium niobate crystal on the opposite side of an air gap [16,23] or in a piezoelectric film on the semiconductor surface [24]. All convolvers are attractive as they use a simple construction, without any structurally imposed coding, and the programmability is achieved by varying the modulation on the time-reversed reference waveform. Thus they can function as either an analog-binary correlator or fully *programmable transversal filter* (PTF) that can be programmed in the time it takes a new reference waveform to propagate into the device (1 μs to 50 μs). Further, the lack of structurally imposed reference-waveform coding makes the device adjustable to minimize Doppler effects. However, because of the counterpropagation of the input signals, the output signal has twice the center frequency and bandwidth of the inputs, and it also suffers a time compression relative to the input signals. This plus the practical difficulty in generating an accurate high-speed analog reference waveform and the limited dynamic range of the device severely limits its application to programmable frequency filtering (i.e., PTF operation). For programmable matched filter operation, convolvers can be relatively easily applied to spread spectrum waveform demodulation in communication and radar [21,25,26]. The processing gain from correlation also helps overcome the dynamic range limitation. The SAW device that most closely resembles the CCD FIR filter is the SAW programmable transversal filter (which often only possesses a binary reference capability). Here a delay line has uniformly spaced taps that are connected through an array of RF switches to a summing bus. The switches are programmed from a shift register with associated latches so that the present code vector can be held while the next one is being loaded. Various realizations of an SAW tapped delay line and associated switching circuits have been studied for almost 15 years [27,28]. The first realizations used hybrid wire-bond interconnects between the SAW and silicon *large-scale integration* (LSI) switching circuits. A more sophisticated approach is to use MOSFET switches [27] and gap-couple the SAW [29,30] or propagate it in a piezoelectric overlay such as zinc oxide [29,31]. An effort has also been directed toward a full monolithic implementation [32]. A typical component [28] with hybrid wire-bonded devices is a 256-point matched filter for a 64-MHz chip rate, *minimum-shift-keyed* (MSK) waveform [33]. It consists of eight separate 32-tap switching circuits, which use UHF bipolar transistor switches providing 20-dB on-off isolation at the 240-MHz device center frequency. The switching circuits are mounted on a thick-film hybrid and wire-bonded through a fan-in interconnect to the SAW tapped delay line. These devices offer <0.5 dB loss in processing gain from ideal, with spurious signals at −47 dB relative to peak output. Direct RF and clock feedthrough levels are presently slightly higher than this value and limit the current performance. This hybrid approach involves a complex assembly process with many wire-bonded interconnections. These can be reduced at the expense of an increase in SAW device fabrication complexity if the SAW is propagated directly on the semiconductor material [27] using, for example, the piezoelectric zinc oxide overlay. Hybrid approaches are generally limited to switching of the signals tapped from the SAW. However, if several parallel switching circuits are coupled to a single SAW tapped delay line, then binary weighting of the switched outputs can provide a limited analog-analog transversal filter capability [28]. Thus, using an approach that is architecturally similar to the CCD MDAC-based PTF [6], SAW PTFs can also be designed. Another method to obtain the full analog-analog correlator (PTF) capability without incurring the interconnection problem is to capacitively couple the electric fields associated with a propagating SAW into a silicon integrated circuit, where they interact with a stored reference. One of the devices, the SAW/FET [29], has its origins in earlier SAW memory correlators and convolvers. In these devices the memory capability was achieved by sampling the electric field pattern associated with a propagating SAW and storing the modulation information in a Schottky [34] or PN diode array [35].

CCD devices have also been used to read out the modulation information from the diode array [36] or to program the diode array [37]. CCD-based reference waveform programming does not offer wide tap weight control range, and the alternative integrated or monolithic correlators [24] have to be programmed with a modulated IF waveform. These problems motivated the development of alternative PTF solutions such as the SAW/FET [29]. In the resent realization of the SAW/FET [30], the propagating electric field is sampled capacitively with a metal finger that is coupled to the output bus and weighted by a MOS variable capacitor, providing analog tap-weight control. Each tap has an associated MOSFET which, during programming, allows a bias voltage to be sequentially placed on each of the MOS variable capacitors (metal fingers). Tap-weight control and output signal summation circuitry are located on a silicon integrated circuit that is coupled through an air-gap to the lithium niobate SAW delay line. Typical device parameters [30] are 350 taps, a 100-MHz programmable bandwidth, 1.5-μs interaction region, and 30-dB tap-weight control range. A single SAW/FET can function as either an analog-binary correlator (with binary tap-weight control), a PTF for arbitrary bandshape filters, or a programmable waveform generator.

A hybrid SAW/FET device has also been reported [38] that combines a low-insertion-loss SAW lithium niobate delay line with a separate silicon FET array. GaAs FET arrays are also available [39]. The necessity for wire-bond interconnects restricts the capability to 16 taps, but the use of FET tap weighting [27] gives improved tap-weight control range over other approaches, including monolithic GaAs devices. The low number of taps precludes the achievement of sophisticated bandpass filters with high out-of-band rejection; but as all the tap weights are available separately, the filter response of these SAW/FET devices can be made time varying under adaptive feedback control [40]. In addition to performing frequency or matched filtering, this allows the insertion of stopbands to reduce nonstationary interference; and as it is accomplished in a closed-loop feedback processor, this will automatically compensate for nonlinearities in the tap weight control circuits. The $(\sin x)/x$ response can be modified to reject signals close to the main filter lobe. Similar SAW-based adaptive processor results have been obtained with a SAW convolver [41] that achieves 15-dB suppression of a spurious trailing pulse [42]. The best performance in terms of cancellation depth has been obtained from adaptive filters utilizing hybrid SAW PTFs. These combine a conventional tapped delay line with hybrid tap weighting and control networks [23] to achieve 200-tap capability with 2.5-MHz bandwidth and 50 dB of CW suppression. As it was explained in Section 4.5.2, SAW techniques permit the realization of wideband Fourier transforms in compact, low-power consumption units [43,44]. When configured for programmable transversal filtering, these processors give direct access to the frequency spectrum [45,46] permitting, in addition, spectral modification for interference suppression. The resulting processor has performance capabilities that are controlled by the SAW Fourier transform processors described in Section 4.5.2. They are restricted by the phase and amplitude errors introduced by analog Fourier transformation and the problems of block rather than continuous processing to typically 20 dB to 30 dB of cancellation. In addition, they have a lower bandwidth than former SAW PTF approaches [30], but the constituent SAW chirp filters are much simpler to fabricate. Other SAW-based PTFs have been reported, in particular, ones using piezoelectric GaAs substrates. This material offers a monolithic implementation, but with the penalties imposed by a low electromechanical coupling material. Two distinct designs have been reported: one based on acoustoelectric interaction with metal-semiconductor FET (MESFET) taps in the beam path [32,47], the other a memory correlator structure with a capacitively coupled Schottky-diode array [48]. These devices do not offer the 350-tap capability of the SAW/FET, but they promise high bandwidth (50 MHz to 200 MHz) processing. A promising new class of device harnesses the photorefractive effect in iron-doped lithium niobate to optically write acoustically active gratings [49]. Because bulk acoustic waves are used, this technique could produce transversal filters and dispersive delay lines with bandwidths exceeding 1 GHz and delays of 10 μs as well as simpler high-Q filter devices.

References

[1] Carnes, J. E., and W. F. Kosonocky, "Noise Sources in Charge-Coupled Devices," *RCA Rev.*, Vol. 33, June 1972, pp. 327–343.

[2] Grove, A. S., *Physics and Technology of Semiconductor Devices*, New York: Wiley, 1967.

[3] Denyer, P. B., J. Mavor, and J. W. Arthur, "Miniature Programmable Transversal Filter Using CCD/MOS Technology," *Proc. IEEE*, Vol. 67, No. 1, Jan. 1979, pp. 42–50.

[4] Sage, J. P., and A. M. Cappon, "CCD Analog-Analog Correlator with Four-FET Bridge Multipliers," *Japanese J. Appl. Phys.*, Vol. 19, Supplement 19-1, 1980, pp. 256–268.
[5] Dix, J. F., and J. W. Widdowson, "CCD Processors for Sonar," *Advanced Signal Processing*, D. J. Creasy (ed.), London: Peter Peregrinus, 1985, Chap. 10.
[6] Chiang, A. M., and B. E. Burke, "A High Speed CCD Digitally Programmable Transversal Filter," *IEEE J. Solid-State Circuits*, Vol. SC-18, No. 6, Dec. 1983, pp. 745–753.
[7] Gandolfo, D. A., J. R. Tower, J. I. Pridgen, and S. C. Munroe, "Analog-Binary CCD Correlator," *IEEE J. Solid-State Circuits*, Vol. SC-14, April 1979, pp. 518–525.
[8] Greico, D. M., "The Application of CCDs to Spread-spectrum Systems," *IEEE Trans. on Communications*, Vol. COM-28, No. 9, Sept. 1980, pp. 1693–1705.
[9] Munroe, S. C., "The Use of CCD Correlations in the SEEK COMM Spread-Spectrum Receivers," *Proc. IEEE MILCOM Conf.*, 1982, paper 6.3.
[10] Munroe, S. C., "CCD-Based Signal Processing for Spread-Spectrum Systems," *Proc. AFCEA Tactical Communication Conf.*, April 22–24, 1986.
[11] Burke, B. E., and D. L. Smythe, Jr., "A CCD time integrating correlator," *IEEE J. Solid-State Circuits*, Vol. SC-18, No. 6, Dec. 1983, pp. 736–744.
[12] Blair, P. K., "Receivers for the Navstar System," *IEEE Proc.*, Vol. 127, Pt. F, No. 2, April 1980, pp. 163–167.
[13] Sovero, E. A., R. Sahai, W. A. Hill, and J. A. Higgins, "Microwave Frequency GaAs Charge-Coupled Devices," *IEEE GaAs Integrated Circuits Symp.*, 1984, pp. 101–104.
[14] Gaalema, S. D., R. J. Schwartz, and R. L. Gunshor, "Acoustic Surface Wave Interaction with a Charge-Coupled Device" *Appl. Phys. Lett.*, Vol. 29, No. 2, July 1976, p. 82.
[15] Hoskins, M. J., E. G. Bogus, and B. J. Hunsinger, "Experimental Performance of the Buried Channel Acoustic Charge Transport Device," *IEEE Electron Device Lett.*, Vol. 4, No. 11, Nov. 1983, p. 396.
[16] Kino, G. S., C. Ludvik, H. J. Shaw, W. R. Shreve, J. M. White, and D. K. Winslow, "Signal Processing by Parametric Interactions in Delay Line Devices," *IEEE Trans. on Microwave Theory Technology*, Vol. MTT-21, April 1973, pp. 244–255.
[17] Wang, W. C. (Ed.) Special Issue on SAW Convolvers and Correlators. *IEEE Trans. on Sonics and Ultrasonics*, Vol. SU-32, Sept. 1985, p. 5.
[18] Maerfeld, C., P. Defranould, and G. W. Farnell, "Some Applications of a Non-Symmetrical Multistrip Coupler." *Proc. IEEE Ultrasonic Symp.*, 1973, pp. 155–158.
[19] Yao, I., "High Performance Electric Convolver with Parabolic horns," *Proc. IEEE Ultrasonic Symp.*, 1980, pp. 37–42.
[20] Grassl, H. P., and H. Engan, "Small-Aperture Focussing Chirp Transducers vs. Diffraction-Compensated Beam Compressors in Elastic SAW Convolvers," *IEEE Trans. on Sonics and Ultrasonic*, Vol. SU-32, 1985, p. 5.
[21] Yao, I., and J. F. Cafarella, "Applications of SAW Convolvers to Spread-Spectrum Communication and Wideband Radar," *IEEE Trans. on Sonics and Ultrasonics*, Vol. SU-32, No. 5, Sept. 1985, pp. 76–770.
[22] Lewis, M. F., C. L. West, J. M. Deacon, and R. F. Humphries, "Recent Developments in SAW Devices," *IEEE Proc.*, Vol. 131, Pt. A, No. 4, June 1984, pp. 186–215.
[23] Reible, S. A., "Acoustoelectric Convolver Technology for Spread-Spectrum Communications," *IEEE Trans. on Microwave Theory Technology*, Vol. MTT-19, May 1981, pp. 463–473.
[24] Kino, G. S., "Zinc Oxide on Silicon Acousto-Electronic Devices." *Proc. IEEE Ultrasonic Symp.*, 1979, pp. 900–910.
[25] Grant, P. M., "The Potential Application of Analogue Matched and Adaptive Filters in Spread-Spectrum Communications," *Radio and Electronics Engrg.*, Vol. 52, No. 5, May 1982, pp. 24–258.
[26] Kowatsch, M., "Application of SAW Technology to Burst-Format Spread-Spectrum Communications," *IEEE Proc.*, Vol. 131, Pt. F, Dec. 1984, pp. 731–741.

[27] Staples, E. J., and L. T. Claiborne, "A Review of Device Technology for Programmable Surface Wave Filters," *IEEE Trans. on Microwave Theory Technology*, Vol. MTT-21, No. 4, April 1973, pp. 279–287.
[28] Lattanza, J., F. G. Herring, P. M. Krencik, and A. F. Clerihew, "Programmable RF Signal Processors Demonstrated for Spread-Spectrum Communication Systems," *Microwave Systems News and Comm. Technology*, Vol. 15, No. 4, April 1985, pp. 76–91.
[29] Green, J. B., and G. S. Kino, "The SAW-FET Signal Processor," *IEEE Trans. on Sonics Ultrasonics*, Vol. SU-32, No. 5, Sept. 1985, pp. 734–744.
[30] Oates, D. E., D. L. Smythe, and J. B. Green, "SAW/FET Programmable Transversal Filter with 100 MHz Bandwidth and Enhanced Programmability," *Proc. IEEE Ultrasonic Symp.*, 1985, pp. 124–129.
[31] Duquesncly, J. Y., P. C. Doncot, H. Gautier, J. P. Do Huu, J. M. Uro, and M. Peltier, "A Monolithic 7 Tap-Programmable Transversal Filter on Gallium Arsenide," *Proc. IEEE Ultrasonic Symp.*, 1984, pp. 303–307.
[32] Grudkowski, T. W., G. K. Montress, M. Gilden, and J. F. Black, "GaAs Monolithic Devices for Signal Processing and Frequency Control," *Proc. IEEE Ultrasonic Symp.*, 1980, pp. 88–97.
[33] Amoroso, F., and J. A. Kivett, "Simplified MSK signalling technique," *IEEE Trans on Communications*, Vol. COM-25, 1977, pp. 433–441.
[34] Ingebrigtsen, K. A., "The Schottky Diode Memory and Correlator A Novel Programmable Device," *Proc. IEEE*, Vol. 64, 1976, pp. 764–769.
[35] Defranould, P., and C. Maerfeld, "A SAW Planar Piezoelectric Convolver," *Proc. IEEE*, Vol. 64, No. 5, May 1976, pp. 748–751.
[36] Smythe, D. L., R. W. Ralston, B. E. Burke, and E. Stern, "An Acoustic SAW/CCD Buffer Memory Device," *Appl. Phys. Lett.*, Vol. 33, Dec. 15, 1978, pp. 1025–1027.
[37] Smythe, D. L., and R. W. Ralston, "Improved SAW Time Integrating Correlator with CCD Readout," *Proc. IEEE Ultrasonic Symp.*, 1980, pp. 14–17.
[38] Panasik, C. M., and J. R. Toplicar, "Adaptive Interference Suppression Using SAW Hybrid Programmable Transversal Filter," *Proc. IEEE Ultrasonic Symp.*, 1983, pp. 170–173.
[39] Panasik, C. M., and D. E. Zimmerman, "A 16 Tap Hybrid Transversal Filter Using Monolithic GaAs Dual-Gate FET Array," *Proc. IEEE Ultrasonic Symp.*, 1985, pp. 130–133.
[40] Cowan, C. F. N, and P. M. Grant, *Adaptive Filters*, Englewood Cliffs, NJ: Prentice-Hall, 1985.
[41] Grant, P. M., and G. S. Kino, "Adaptive Filter Based on SAW Monolithic Storage Correlator," *Electronics Lett.*, Vol. 14, August 17, 1978, pp. 562–564.
[42] Bowers, J. E., G. S. Kino, D. Behar, and H. Olaisen, "Adaptive Deconvolution Using a SAW Storage Correlator," *IEEE Trans. on Microwave Theory Technology*, Vol. MTT-29, No. 5, May 1981, pp. 491–498.
[43] Jack, M. A., P. M. Grant, and J. H. Collins, "The Theory Design and Applications of SAW Fourier Transform Processors," *Proc. IEEE*, Vol. 68, No. 4, April 1980, pp. 450–468.
[44] Williamson, R. C., V. S. Dolat, R. R. Rhodes, and D. M. Boroson, "A Satellite-Borne SAW Chirp-Transform System for Uplink Demodulation of FSK," *Proc. IEEE Ultrasonic Symp.*, 1979, pp. 741–747.
[45] Arsenault, D. R., "Wideband Chirp-Transform Adaptive Filter," *Proc. IEEE Ultrasonic Symp.*, 1985, pp. 102–107.
[46] Morgul, A., P. M. Grant, and C. F. N. Cowan, "Wideband Hybrid Analog/Digital Frequency Domain Adaptive Filter," *IEEE Trans. on Acoustics, Speech, and Signal Processing*, Vol. ASSP-32, No. 4, August 1984, pp. 762–769.
[47] Withers, R. S., "Electron Devices on Piezoelectric Semiconductors: Advice Model," *IEEE Trans. on Sonics and Ultrasonics*, Vol. SU-31, No. 2, March 1984, pp. 117–123.
[48] Wagers, R. S., and M. R. Melloch, "GaAs Strip Coupled Memory Correlator," *Proc. IEEE Ultrasonics Symp.*, 1983, pp. 377–385.

[49] Oates, D. E., P. G. Gottschalk, and P. V. Wright, "Holographic-Grating Acoustic Devices," *Appl. Phys. Lett.*, Vol. 46, No. 12, June 15, 1985, pp. 1125–1127.

Selected Bibliography

Buss, D. D., and W. H. Bailey, "Application of Charge Transfer Devices to Communication," *Tech. Digest, 1973 CCD Applications Conf.*, San Diego, CA, Sept. 1973, pp. 83.

Chen, P. C., R. H. Dyck, and E. G. Magill, "Design, Operation, and Application of a High-Speed Charge-Coupled Programmable Transversal Filter," *Proc. 1978 Int. Symp. Applications of Charge-Coupled Devices*, San Diego, CA, Oct. 25–27, 1978.

Gandolfo, D. A., J. R. Tower, J. I. Pridgen, and S. C. Munroe, "Analog-Binary CCD Correlator: A VLSI Signal Processor," *IEEE Trans. Electron Devices*, Vol. ED-26, April 1979.

Grieco, D. M., "Inherent Signal to Noise Ratio Limitations of Charge Coupled Devices Pseudonoise Matched Filters," *IEEE Trans. on Communications*, Vol. COM-28, No. 5, May 1980, pp. 729–732.

Grieco, D. M., "The Application of Charge Coupled Devices to Spread-Spectrum Systems," *IEEE Trans. on Communications*, Vol. Com-28, No. 9, Sept. 1980, pp. 1693–1705.

Magill, E. G., D. M. Grieco, R. H. Dyck, and P. C. Chen, "Charge-Coupled Device Pseudonoise Matched Filter Design," *Proc. IEEE (Special Issue on Miniaturized Filters)*, Jan. 1979.

Masenlen, W. K., "Adaptive Signal Processing," *Case Studies in Advanced Signal Processing*, P. M. Grant (ed.), *IEEE Conf. Publication*, Vol. 180, 1979, pp. 168–177.

Melen, R., and D. Buss (Eds.), *Charge-Coupled Devices:Technology and Applications*, New York: IEEE Press, 1977.

Sequin, C. H., and M. F. Tompsett, *Charge Transfer Devices*, New York: Academic Press, 1975.

CHAPTER 5

CDMA Systems

In this chapter we present an overview of those aspects of spread spectrum techniques that are specific to providing multiple access in cellular mobile communication networks and principal issues that arise in the application of spread spectrum techniques in the multiuser cellular environment. In particular, we will consider the concept of CDMA based on spread spectrum transmission and design and analysis of CDMA communication systems for cellular mobile communications. The user capacity is studied for voice communications over large areas and power control schemes aimed at optimizing the capacity.

5.1 A CELLULAR CDMA SYSTEM

An important issue when designing mobile radio systems is the selection of the multiple access scheme that enables sharing of the common radio spectrum. Multiple access can be organized according to the major multiple access principles:

- *Frequency division multiple access* (FDMA);
- TDMA;
- CDMA.

In FDMA schemes a unique frequency slot is assigned to each user who retains that particular slot during the period of the call. In TDMA each user is allocated a unique time slot during the period of the call. With ideal separation between individual channels, the number of users is determined by the number of distinct time or frequency slots available. In a CDMA scheme, on the other hand, all users transmit simultaneously and at the same frequency. The transmitted signals occupy the entire system bandwidth. Code sequences are used to separate one user from another.

The three multiple-access schemes have equivalent capacity on Gaussian channels assuming that the code sequences in CDMA are orthogonal. If the channel is subject to time-varying amplitudes and frequency selectivity, which are typical for mobile radio, CDMA is superior to other multiple-access techniques. It can average out the variations in the channel transfer function caused by frequency selectivity. CDMA receivers can be designed to take advantage of the multipath characteristics associated with frequency-selective fading and to minimize their effect on the system capacity.

The main capacity advantage of CDMA is achieved in a multicell radio environment. In early mobile communication a single high-powered base station was used to provide large-area coverage. The system was severely bandlimited and could not meet demands for mobile services. In a cellular system, the single base station transmitter is replaced by many low-power transmitters, each providing coverage in a small honeycomb-shaped cell. In FM/FDMA or TDMA/FDMA, each cell is allocated a portion of the available bandwidth. The bandwidth used in a given cell can be reused in another cell sufficiently far away so that interference is not significant. The L cells that collectively use the whole available spectrum are called a *cluster*. The clusters can be placed in a geometric pattern, as in Figure 5.1. The system can grow over an arbitrary large area.

Hexagonal shapes have been universally adopted to simplify the plan and design of a cellular system. In practical systems such a shape cannot be achieved. The real cell shapes are typically amorphous.

Some other forms of multiple access are discussed in Chapter 8.

In a cellular FM/FDMA or TDMA/FDMA system the number of channels per cell is reduced by the reuse factor, L, relative to the total number of channels for the given spectrum, since each cell within a cluster is only assigned $1/L$ of the total available spectrum in the system. The reuse factor in Figure 5.1 is 7.

On the other hand, CDMA can reuse the same entire spectrum for all cells. The reuse factor in a CDMA cellular system is thus equal to 1. This results in the improved capacity of the overall system. Note that capacity is measured as the maximum number of active users in an overall multicell network rather than the maximum number of active users per bandwidth or per isolated cell. The improvement in the overall capacity, as defined, of CDMA over digital TDMA or FDMA is on the order of 4 to 6 and over analog FM/FDMA is almost a factor of 20 [1].

Other user signals active at the same time in the same frequency band as a user of interest cause cochannel interference. Cochannel interference is the limiting factor in a cellular mobile radio system. The frequency reuse method in TDMA/

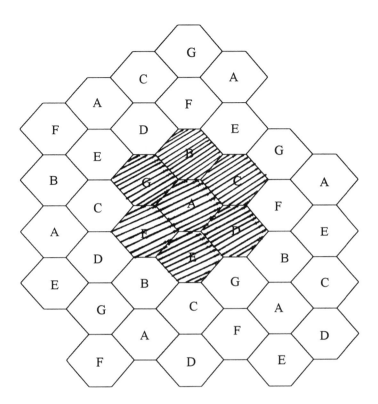

Figure 5.1 Basic structure of a cellular system.

FDMA and FM/FDMA results in cochannel interference because the same frequency band is used repeatedly in different cells. In most mobile radio systems, the use of a seven-cell cluster is not sufficient to avoid cochannel interference. Increasing $L > 7$ would reduce the number of channels per cell, which would reduce the system capacity. Alternatively, the reuse factor of seven might be retained, and the area of the cell could be sectored radially. Each cell is divided into three or six sectors and uses three or six directional antennas at a base station both for transmitting and receiving. Each sector utilizes a different frequency band than the other sectors of the same base station. For example, if a cell is divided into three sectors, the interference received by any antenna is approximately one-third of those received by one omnidirectional antenna. Using three sectors, the number of users per cell can be increased by a factor of three, for the same cluster size.

Another important consideration in increasing the system capacity is voice activity monitoring. In a two-person conversation, each speaker is active only 35% to 40% of the time and listens the rest of the time [2]. In CDMA all the users share one radio channel. When users assigned to the channel do not talk, all other active users experience lower interference. Thus, the voice activity monitoring reduces multiple-access interference by 65%. This translates into an increase of the system capacity by a factor of 2.5.

The other two multiple-access schemes, FDMA and TDMA, which allocate the frequency or time slots for the period of the call and reallocation of these two resources to other users for a brief period of time when a given channel is silent, are not practical, as this requires very fast switching between various users.

In FDMA and TDMA, frequency management is a critical task, since it controls cochannel interference. Since there is only one common channel in CDMA, there is no need for frequency management.

In FDMA and TDMA, when the mobile unit moves out of the coverage area of a cell during the call, the received signal becomes weak and the present cell site requires a *handoff*. The system switches to a new channel while the call continues. In CDMA different cells use the same frequency bandwidth and differ only in assigned code sequences. Therefore, no handoff from one frequency to another is needed. This is called a *soft handoff*.

In CDMA, there is no distinct limit on the number of users as in TDMA and FDMA. Rather, the system performance for all users degrades gracefully as the number of active users increases. Naturally, when the number of users exceeds a certain level, interference would cause unintelligible speech and system instability. However, in CDMA we are concerned with the "soft blocking" condition, which can be relaxed, as contrasted to the "hard blocking" condition in FDMA and TDMA where all channels are occupied.

There are some disadvantages associated with CDMA. The two most prominent are *self-jamming* and the *near-far* effect. Self-jamming manifests in multiple-access interference caused by nonorthogonal code sequences. In cellular mobile systems, mobile units transmit independently from each other and their signals arrive asynchronously at the base station. Since their relative time delays are randomly distributed, the crosscorrelation between the received signals coming from different users is nonzero. To achieve a low level of interference, the assigned signatures need to have low crosscorrelations for all relative time delays. A low crosscorrelation between signatures is obtained by designing a set of orthogonal sequences. However, there is no known set of code sequences that is completely orthogonal when used in an asynchronous system. The nonorthogonal components of signals of other users will appear in the demodulated desired signal as interference. If the matched filter receiver is used, in such a system, the number of users is limited by interference coming from other users. This is distinct from TDMA or FDMA where orthogonality of the received signals is preserved by accurate synchronization and filtering.

The major drawback associated with CDMA is the near-far effect. It manifests when a weak signal received at the base station from a distant mobile unit is overpowered by a strong signal coming from a nearby interferer. An interfering signal with a power n times higher than that of the desired signal will approximately have the same effect on system capacity as n interferes of the same power as the desired signal. To combat the near-far effect *power control* is used in most CDMA applications. In a cellular system, power control is carried out by each base station, which periodically instructs the mobile units to adjust their transmitter powers so that all signals are received at the base station at roughly the same level. In the next sections we will examine all these phenomena in more detail and quantify their effects on system performance.

5.2 A SINGLE-CELL CDMA DEMODULATOR IN A GAUSSIAN CHANNEL

We examine an asynchronous one-cell CDMA cellular mobile radio system, consisting of mobile units as transmitters and the base station as the receiver. The block diagram of the transmitter is shown in Figure 5.2.

We will refer to the *subscribers* as the authorized users of the CDMA system. Each subscriber is assigned a unique code sequence. We assume that the number of subscribers is equal to the number of code sequences for a given code sequence of length N. We will refer to the *users* as those subscribers actively engaged in transmission.

As Figure 5.2 shows, K independent users use the same carrier frequency and may transmit simultaneously. The transmitter for the kth user, denoted by T_{xk} in Figure 5.2, is shown in Figure 5.3.

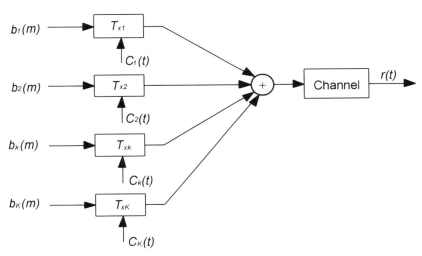

Figure 5.2 Transmitter model of an asynchronous DS/CDMA system.

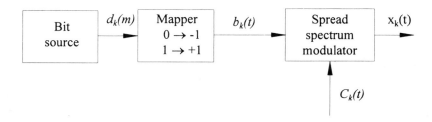

Figure 5.3 Block diagram of a single-user DS/CDMA transmitter.

The kth binary source generates a binary sequence $d_k(m)$, where m is the time instant. The mapper maps binary symbols 1 and 0 into bipolar symbols +1 and −1, respectively. The bipolar symbol $b_k(m)$ is given by

$$b_k(m) = 2d_k(m) - 1 \tag{5.1}$$

The kth bipolar signal, $b_k(t)$, is defined as

$$b_k(t) = \sum_{m=-M}^{M} b_k(m) U_T(t - mT) \tag{5.2}$$

where $2M + 1$ is the number of transmitted symbols, T is the duration of a rectangular pulse corresponding to the transmission time of an information bit, and $U_T(t)$ is the unit rectangular pulse of T-sec duration defined as

$$U_T(t) = \begin{cases} 1 & t \in [0, T] \\ 0 & t \notin [0, T] \end{cases} \tag{5.3}$$

Each transmitter is assigned a pseudorandom code sequence C_k of length N that is approximately orthogonal to all other code sequences. The symbol of the kth spreading sequence, denoted by C_k^j, assigned to the kth user at instant jT_c, where T_c is the chip interval, takes values from the set $\{+1, -1\}$.

In the spread spectrum modulator, the bipolar signal is multiplied by the code waveform, denoted by $C_k(t)$, given by

$$C_k(t) = \begin{cases} \sqrt{2}\cos[\omega_c t + \varphi_k] \sum_{j=0}^{N-1} C_k^j \psi_{T_c}(t - jT_c) & t \in [0, T] \\ 0 & t \notin [0, T] \end{cases} \tag{5.4}$$

where $\psi_{T_c}(t)$ is the normalized chip waveform of T_c sec, $\omega_c = 2\pi f_c$, where f_c is the carrier frequency and φ_k is the random phase of the kth carrier.

As the expression for $C_k(t)$ shows, the spread spectrum modulator includes frequency shifting by the carrier frequency f_c, multiplication by the code sequence C_k, and waveshaping filtering. The transmitted signal of the kth user is given by

$$x_k(t) = \sum_{m=-M}^{M} \sqrt{E_{ck}} b_k(m) C_k(t - mT - \tau_k) \qquad (5.5)$$

where τ_k is the time shift of the kth user and E_{ck} is the kth transmitted energy per chip. In an asynchronous system, the transmitted signals have different time shifts, but the symbol intervals T for all transmitters are assumed to be equal. The transmit filter suppresses the signal spectrum outside the bandwidth $B_c = 1/T_c$. We assume that the symbol interval T is a multiple of the chip interval T_c. The spreading gain is defined as

$$G_p = \frac{T}{T_c} = \frac{B_c}{B} \qquad (5.6)$$

where B_c is the spread spectrum signal bandwidth and B is the baseband signal bandwidth.

For simplicity we assume that the sequence length N is equal to the processing gain G_p. The model could be generalized to include nonbinary modulation, nonbinary sequences, and error control coding. For conceptual clarity we will focus on the simplest model. The channel is modeled by the zero-mean *additive white Gaussian noise* (AWGN) $n(t)$ with variance σ_n^2, and there is no other distortion in the channel apart from constant linear scaling of signal amplitudes and multiple-access interference caused by the presence of other active users.

The received waveform, denoted by $r(t)$, can be modeled as consisting of K superimposed modulated signals observed in AWGN

$$r(t) = \sum_{k=1}^{K} \sqrt{E_{ck}} A_k \sum_{m=-M}^{M} b_k(m) C_k(t - mT - \tau_k) + n(t) \qquad (5.7)$$

where A_k is the attenuation of the kth signal, due to propagation.

In a synchronous DS/CDMA system, individual user signals are symbol synchronized. An example of this system is the forward link in a cellular radio network. The received signal in a synchronous DS/CDMA system is obtained as a special case of (5.7), when the time shifts, τ_k, $k = 1, 2, \ldots, K$, are the same and equal to zero

$$r(t) = \sum_{k=1}^{K} \sqrt{E_{ck}} A_k b_k(t) C_k(t) + n(t) \qquad (5.8)$$

The conventional matched filter receiver for simultaneous demodulation of K user signals is shown in Figure 5.4. This receiver consists of a bank of matched filters followed by a decision device. The kth filter is matched to the corresponding incoming signal $\psi_{T_c}(t)$. The output of the kth filter is sampled at the end of the mth symbol interval. It can be written as

$$y_k(m) = \frac{1}{T} \int_{\tau_k+mT}^{\tau_k+(m+1)T} r(t) C_k(t - mT - \tau_k) dt \quad -M \leq m \leq M \quad (5.9)$$

We assume ideal carrier phase tracking and that the filter has unity power gain.

The final processing operation in the demodulator adds the received samples $y_k(m)$ for all sampling instants within one bit and forms the decision variable Y_k defined as

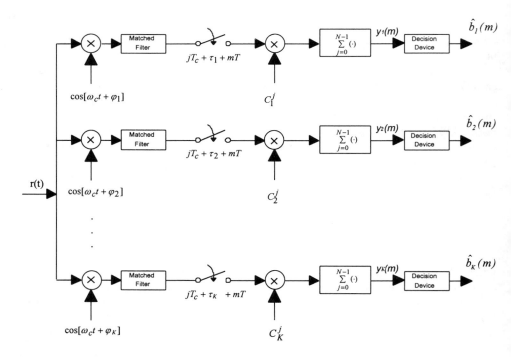

Figure 5.4 Conventional matched filter receiver for simultaneous demodulation of K asynchronous DS/CDMA user signals.

$$Y_k = \sum_{m=1}^{G_p} y_k(m) \qquad (5.10)$$

where G_p is the number of chips per bit, which is assumed to be equal to the code sequence length N.

The kth decision device estimates the mth symbol of the kth user by examining the sign of the decision variable Y_k

$$b_k(m) = \text{sgn}[\,Y_k\,] = \begin{cases} +1 & \text{if } Y_k \geq 0 \\ -1 & \text{if } Y_k < 0 \end{cases} \qquad (5.11)$$

where m is the sampling instant, $-M \leq m \leq M$.

Substituting $r(t)$ from (5.7) into (5.9) we get for $y_k(m)$

$$y_k(m) = A_k\sqrt{E_{ck}}\,b_k(m) + \frac{1}{T}\int_{\tau_k+mT}^{\tau_k+(m+1)T} \sum_{\substack{j=1 \\ j \neq k}}^{K} A_j\sqrt{E_{cj}} \qquad (5.12)$$

$$\sum_{i=-M}^{M} b_j(i) C_j(t - iT - \tau_j) C_k(t - mT - \tau_k)\,dt + n_k(m)$$

where $n_k(m)$ is the Gaussian noise sample at the sampling instant m. The first term in (5.12) represents the scaled BPSK modulated desired signal, while the second term represents multiple access interference. Expression (5.12) can be further written as

$$y_k(m) = A_k\sqrt{E_{ck}}\,b_k(m) + \sum_{\substack{i=-M \\ j \neq k}}^{M}\sum_{}^{K} \sqrt{E_{cj}}\,A_j b_j(i)\rho_{jk}(l) + n_k(m) \qquad (5.13)$$

where ρ_{jk} is the crosscorrelation of code sequences defined as

$$\rho_{jk}(l) = \frac{1}{T}\int_{\tau_k+mT}^{\tau_k+(m+1)T} C_j(t - iT - \tau_j) C_k(t - mT - \tau_k)\,dt \qquad (5.14)$$

where $l = i - m$. As this expression shows, the interference term is linear in the amplitude of each of the interfering users. If some interferers are strong, they cause high interference and can mask weak signals. This represents the main drawback of the matched filter receiver. Even if the powers of all users are equal, the interference term might be significant since the cross-correlation between the desired and interfering signature waveforms can be high due to different time delays.

The multiple-access interference term in (5.13) can be reduced either by maintaining low powers of interfering signals or by selecting a set of code sequences with low crosscorrelations for all time delays between them.

Thus, the performance of the matched filter receiver is nonoptimum in the presence of interfering signals. To maintain interference at an acceptable level, the number of users must be reduced from its theoretical limit of the spreading gain N to only about 10% of N. That is, the bandwidth efficiency of a single-cell DS/CDMA system with the conventional matched filter receiver is reduced by a factor of 10 relative to traditional methods of multiplexing such as FDMA and TDMA.

5.3 BIT ERROR PROBABILITY

The filter output in (5.13) can be represented as

$$y_k(m) = y_{k1}(m) + y_{k2}(m) + y_{k3}(m) \tag{5.15}$$

where $y_{k1}(m)$ represents the desired signal, $y_{k2}(m)$ is the multiple-access interference, and $y_{k3}(m)$ is the Gaussian noise term.

The decision variable, Y_k, can be then written as

$$Y_k = \sum_{m=1}^{G_p} [y_{k1}(m) + y_{k2}(m) + y_{k3}(m)] \tag{5.16}$$

The average probability of bit error, P_b, can be expressed as

$$P_b = \frac{1}{2} P_r \{Y_k > 0 | b_k = -1\} + \frac{1}{2} P_r \{Y_k < 0 | b_k = +1\} \tag{5.17}$$

assuming that the probabilities of transmitting symbols -1 and $+1$ are equal.

The bit probability can then be written as

$$P_b = P_r \{Y_k < 0 | b_k = +1\} \tag{5.18}$$

since

$$P_r \{Y_k > 0 | b_k = -1\} = P_r \{Y_k < 0 | b_k = +1\} \tag{5.19}$$

Assuming that the number of chips per bit, G_p, is large, the decision variable Y_k can be approximated according to the central limit theorem by a Gaussian random variable. The bit error probability is then given as

$$P_b = Q\sqrt{\frac{[E(Y_k)]^2}{\text{Var}(Y_k)}} \quad (5.20)$$

where $Q(-)$ is the complementary error function given by

$$Q(x) = \frac{1}{\sqrt{2\pi}} \int_0^\infty e^{-t^2/2} dt$$

and $E[Y_k]$ is the mean and $\text{var}[Y_k]$ is the variance of the decision variable Y_k. The mean value of Y_k is given by

$$E[Y_k] = E\left[\sum_{m=1}^{G_p} y_{k1}(m)|b_k = 1 + y_{k2}(m) + y_{k3}(m)\right]$$

$$E[Y_k] = E\left[\sum_{m=1}^{G_p} y_{k1}(m)|b_k = 1\right] + E\left[\sum_{m=1}^{G_p} y_{k2}(m)\right] + \left[\sum_{m=1}^{G_p} y_{k3}(m)\right] \quad (5.21)$$

Since

$$E\left[\sum_{m=1}^{G_p} y_{k2}(m)\right] = 0 \quad \text{and} \quad E\left[\sum_{m=1}^{G_p} y_{k3}(m)\right] = 0$$

$$E[Y_k] = E\left[\sum_{m=1}^{G_p} y_{k1}(m)|b_k = 1\right] = G_p\sqrt{E_{ck}} \quad (5.22)$$

the variance $\text{var}(Y_k)$ is given by

$$\text{var}(Y_k) = G_p \text{var}[y_{k1}(m)] + \text{var}[y_{k2}(m)] + \text{var}[y_{k3}(m)] \quad (5.23)$$

The desired signal variance is

$$\text{var}[y_{k1}(m)] = 0 \quad (5.24)$$

The variance due to thermal Gaussian noise is

$$\text{var}[y_{k1}(m)] = \frac{N_o}{2} \quad (5.25)$$

where N_o is the one-sided thermal noise power spectral density. The variance of the interfering signals can be computed assuming that the interfering signal is modeled as white noise with the two-sided power spectral density of E_c/T_c.

Taking into account the relative phase differences between the desired signal and interfering signals and averaging over them, it is possible to show that for bandlimited systems with constant transfer function of the matched filter [3] the variance of $Y_{k3}(m)$ can be power bounded by

$$\text{var}[y_{k3}(m)] \geq \sum_{\substack{j=1 \\ j \neq k}}^{K} \frac{E_{ck}(j)}{2} \qquad (5.26)$$

Then the bit error probability is lower bounded by

$$P_b \geq Q\left[\sqrt{\frac{2E_{ck}G_p}{N_o + \sum_{\substack{j=1 \\ j \neq k}}^{K} E_{cj}}}\right] \qquad (5.27)$$

The term $2E_{ck}G_p$ in (5.27) is the double bit energy $2E_b$, and the denominator represents the total power spectral density coming for the thermal noise and multiple-access interference. If we denote this power spectral density by I_o, it is given by

$$I_o = N_o + \sum_{\substack{j=1 \\ j \neq k}}^{K} E_{cj} \qquad (5.28)$$

The bit error probability lower bound can then be written as

$$P_b \geq Q\left(\sqrt{\frac{2E_b}{I_o}}\right) \qquad (5.29)$$

The bit error probability P_b on a Gaussian channel can be approximated by [4]

$$P_b \approx Q\left\{\left[\frac{K-1}{3G_p} + \frac{N_o}{3E_b}\right]^{-1/2}\right\} \qquad (5.29a)$$

where N_o is the Gaussian noise one-sided power spectral density.

This expression for the bit error probability is obtained assuming perfect power control. The bit error probability curves from (5.29a) for various number of active users K and processing gain of 31 are shown in Figure 5.5.

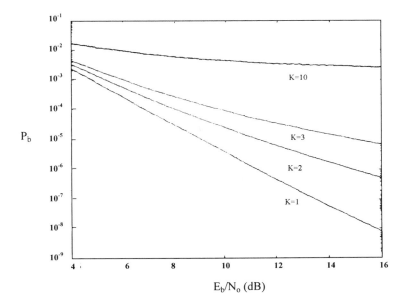

Figure 5.5 The bit error probability for a CDMA system with K active users and perfect power control.

5.4 CAPACITY OF A SINGLE-CELL CDMA SYSTEM

The capacity of CDMA systems is limited by multiple-access interference. In order to derive the expression for the CDMA system capacity, let us consider first a single-cell system. We will assume that as a result of perfect power control the energy per chip of all users is the same and equal to E_c. If we ignore the contribution of the thermal noise power spectral density to the effective noise power spectral density in (5.28), the effective noise power spectral density, I_o, becomes

$$I_o = (K - 1)E_c \qquad (5.30)$$

The bit energy-to-effective noise ratio, E_b/I_o is then

$$\frac{E_b}{I_o} = \frac{E_b}{(K-1)E_c} = \frac{G_p \cdot E_c}{(K-1)E_c} = \frac{G_p}{K-1} \qquad (5.31)$$

If we take into account the background thermal noise with the variance σ_n^2, E_b/I_o becomes

$$\frac{E_b}{I_o} = \frac{G_p}{K - 1 + N_o/E_c} = \frac{G_p}{K - 1 + \sigma_n^2/S} \qquad (5.32)$$

where S is the signal power and N_o is the one-sided noise power spectral density.

For a given E_b/I_o level, specified by the required bit error probability, the number of users that can access the system is given by

$$K \leq 1 + \frac{G_p}{E_b/I_o} - \frac{\sigma_n^2}{S} \qquad (5.33)$$

If we ignore the effect of thermal noise, the number of users is given by

$$K \leq 1 + \frac{G_p}{E_b/I_o} \qquad (5.34)$$

The capacity of a CDMA system can be increased by reducing the interference from other users, given by (5.28) or (5.30). One way of reducing interference is using antenna sectorization. In a cell with three sector antennas the capacity is increased roughly by a factor of 3. Another way to increase the capacity is by monitoring the voice activity of each user. When there is no voice activity the transmitter is switched off. The voice activity factor, denoted by α, is defined as the ratio of the time that a speaker is active to the total transmission time [1]. The effective noise power spectral density in (5.30) becomes $\alpha(K_s - 1)E_c$, where K_s is the number of users per sector.

If both antenna sectorization and voice activity monitoring are used, the bit energy-per-effective-noise power spectral density per sector is given by

$$\frac{E_b}{I_o} = \frac{G_p}{(K_s - 1)\alpha + \sigma_n^2/S} \qquad (5.35)$$

Ignoring σ_n^2/S, the number of users per sector is given by

$$K_s \leq 1 + \frac{1}{\alpha} \frac{G_p}{E_b/I_o} \qquad (5.36)$$

The total number of users in a cell is

$$K = 3K_s \qquad (5.37)$$

If the voice activity factor is 3/8, the total number of users is increased by a factor of 8 relative to a system with an omnidirectional antenna with no voice activity monitoring.

Note that sectored antennas can be also used to increase capacity in other multiple access schemes.

5.5 MULTIPATH PROPAGATION

In a cellular radio environment the mobile unit is surrounded by various objects, such as cars, buildings, and trees, that act as scatterers of radio waves. These obstacles produce reflected waves with the same frequency as the incident radio wave, attenuated amplitude and phase dependent on the incident angle. For a single transmitted wave, the mobile unit will receive one direct wave and many reflected waves arriving from different angles with a spread of arrival times. Due to the different arrival angles and times, the reflected waves at the receiver site will have different phases. When they are collected by the mobile unit antenna, their sum results in constructive or destructive interference. With relation to a stationary user, the sum of reflected waves forms a spatially varying standing wave field. The motion of the receiver through this pattern leads to a signal with fluctuating amplitudes and phases. Phases and amplitudes can change drastically within only a fraction of a wavelength. The depth of fades in moving just a short distance may typically be up to 20 dB and occasionally greater than 30 dB.

The receiver motion introduces a frequency shift in every wave called the *Doppler shift*. A single tone transmitted gives rise to a received signal with a spectrum of nonzero width.

The path lengths of various reflected waves at the receiver site are different. The different path lengths result in different propagation time delays. If, for example, a short pulse were transmitted, the received signal may appear as a train of pulses. The phenomenon is called *time dispersion*, and it causes intersymbol interference.

Because of multiplicity of factors involved in propagation in a cellular mobile environment, it is necessary to apply statistical techniques to describe signal variations. In the next section we will introduce the statistical channel models used to describe signal variations in a multipath environment.

5.5.1 Fading and Doppler Shifts for Single Tone Transmission

In a common land mobile radio situation, as depicted in Figure 5.6, the direct wave is obscured and the mobile unit receives only reflected waves.

Consider a situation when only one tone of frequency f_c is transmitted and the mobile receives N reflected signals.

The ith reflected signal, coming in at an incident angle θ_i, with respect to the direction of travel of the mobile, will experience a Doppler shift given by

$$f_{di} = \frac{vf_c}{c}\cos\theta_i \qquad (5.38)$$

where v is the vehicle speed and c is the speed of light. The ith received ray is given by

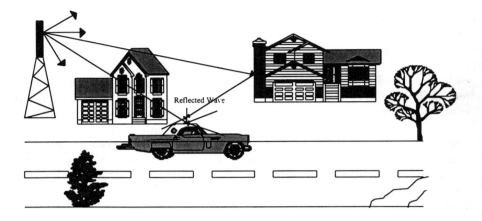

Figure 5.6 A typical mobile propagation environment.

$$r_i(t) = R_i \cos\left(2\pi f_c t - \frac{2\pi f_c vt}{c}\cos\theta_i - \phi_i\right) \quad (5.39)$$

where R_i is the random amplitude of the ith wave and ϕ_i is its random uniformly distributed phase. The frequency of the ith wave can be represented as

$$f_i(\theta) = f_c - \frac{v f_c}{c}\cos\theta_i \quad (5.40)$$

This shift from the carrier frequency is known as the Doppler shift of the ith component. The ith received signal can be written as

$$r_i(t) = R_{Ii}(t)\cos 2\pi f_c t + R_{Qi}(t)\sin 2\pi f_c t \quad (5.41)$$

where $R_{Ii}(t)$ is the in-phase signal component given by

$$R_{Ii}(t) = R_i \cos\left(\frac{2\pi f_c vt}{c}\cos\theta_i + \phi_i\right) \quad (5.42)$$

and $R_{Qi}(t)$ is the quadrature signal component given by

$$R_{Qi}(t) = -R_i \sin\left(\frac{2\pi f_c vt}{c}\cos\theta_i + \phi_i\right) \quad (5.43)$$

The resulting signal at the receiver is given by the sum of these N independent waves

$$r(t) = \sum_{i=1}^{N} R_i \cos[2\pi f_i(\theta) - \phi_i] \tag{5.44}$$

One can also express the resulting signal in terms of its in-phase and quadrature components as

$$r(t) = R_I(t)\cos 2\pi f_c t + R_Q(t)\sin 2\pi f_c t \tag{5.45}$$

The resulting $R_I(t)$ and $R_Q(t)$ components can be expressed as the sum of the corresponding individual components as

$$R_I(t) = \sum_{i=1}^{N} R_{Ii}(t) \tag{5.46}$$

$$R_Q(t) = \sum_{i=1}^{N} R_{Qi}(t) \tag{5.47}$$

As expressions (5.42) and (5.43) indicate, $R_{Ii}(t)$ and $R_{Qi}(t)$ are independent random variables. As a consequence of the Central Limit Theorem, when N gets large $R_I(t)$ and $R_Q(t)$ each can be approximated by uncorrelated Gaussian random processes with zero mean and variance σ_s^2.

The signal envelope is given by

$$R(t) = \sqrt{R_I^2(t) + R_Q(t)^2} \tag{5.48}$$

Its probability density function is Rayleigh given by

$$p(R) = \begin{cases} \dfrac{R}{\sigma_s^2} \cdot e^{-R^2/2\sigma_s^2} & \text{for } R \geq 0 \\ 0 & \text{otherwise} \end{cases} \tag{5.49}$$

The phase of the received signal is uniformly distributed between $-\pi/2$ and $\pi/2$.

If an unmodulated carrier frequency f_c is transmitted and noting that $-1 \leq \cos \theta_i \leq 1$, then the received signal components, according to (5.40) will be spread into the range $f_c \pm v f_c/c$. The maximum Doppler shift $v f_c/c$ is also referred as the maximum *fade rate*.

Consider a mobile unit with an antenna characterized by the gain $G(\theta)$. If the distribution of θ is $p(\theta)$, then within a differential angle $d\theta$, the normalized power is $G(\theta)p(\theta)d\theta$. This power is equal to the differential variation of the received power with frequency, $S(f)df$. Noting that $f(\theta) = f(-\theta)$, from (5.40), we have

$$S(f)|dt| = \{G(\theta)p(\theta) + G(-\theta)p(-\theta)\}|d\theta| \tag{5.50}$$

Also from (5.40) we get

$$|df| = f_d|\sin\Theta d\Theta| = \sqrt{f_d^2 - (f - f_c)^2}|d\Theta| \tag{5.51}$$

Substituting (5.51) into (5.50) and assuming an omnidirectional antenna (the $\lambda/4$ monopole with $G(\theta) = 1.5$, for example) we get for the power spectral density

$$S(f) = \begin{cases} \dfrac{3\{p(\theta) + p(-\theta)\}}{2f_d\sqrt{1 - \left(\dfrac{f - f_c}{f_d}\right)^2}} & f_c - f_d \le |f| \le f_c + f_d \\ 0 & \text{otherwise} \end{cases} \tag{5.52}$$

where $f_d = \dfrac{vf_c}{c}$ is the maximum Doppler shift. Using the assumption that the power received in an incremental angle $d\theta$ is uniform, $p(\theta)$ can be represented by

$$p(\theta) = \frac{2\sigma_s^2}{2\pi f_d}$$

and assuming $G(\theta) = 1$, the resultant received power spectrum for the transmitted unmodulated carrier is then often quoted as [5]

$$S(f) = \begin{cases} \dfrac{2\sigma_s^2}{2\pi f_d}\left[1 - \left(\dfrac{f - f_c}{f_d}\right)^2\right]^{-1/2} & f_c - f_d \le |f| \le f_c + f_d \\ 0 & \text{otherwise} \end{cases} \tag{5.53}$$

where $2\sigma_s^2$ is the power of the received signal. This spectrum is sketched in Figure 5.7.

Note that this spectrum contains the unrealistic result of infinite power at $f_c \pm f_d$, as a consequence of the simplistic original assumption of uniform power distribution.

If there is a direct path present in the received signal, the composite signal is the sum of the constant direct signal and a Rayleigh distributed resulting reflected signal. The resulting component has a Rician distribution given by

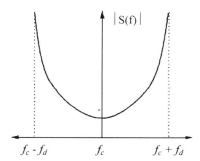

Figure 5.7 The ideal Doppler spectrum for an unmodulated carrier f_c.

$$p(R) = \begin{cases} \dfrac{R}{\sigma_s^2} \cdot e^{-R^2+D^2/2\sigma_s^2} I_0\left(\dfrac{RD}{\sigma_s^2}\right) & R \geq 0 \\ 0 & R < 0 \end{cases} \quad (5.54)$$

where D^2 is the direct signal power and $I_0(-)$ is the modified Bessel function of order zero.

5.5.2 Frequency Selective Fading Channel

Let us now consider transmitting a signal that occupies a finite bandwidth for the same propagation scenario shown in Figure 5.6. Assume that a rectangular pulse of duration T is transmitted and that there are three reflected paths. The path lengths of the reflected path are different and arrive with different propagation delays. The received signal consists of three resolvable waves, as shown in Figure 5.8(a). However, due to the motion of the mobile unit, if the same pulse is transmitted at another time, the amplitudes, phases, delays, and number of resolvable paths will change, as in Figure 5.8(b). These channel characteristics can be represented by time-varying random variables.

If a sequence of rectangular pulses, each of duration T, is transmitted over a multipath channel, each of the transmitted pulses will be replicated at the receiver several times. If the delay of reflected waves becomes comparable to the pulse duration, T, consecutive received pulses will overlap, resulting in significant intersymbol interference. The channels that introduce this type of distortion are called *time-dispersive channels*.

This phenomenon can be considered in the frequency domain as well. Each radiated tone in the signal spectrum will have a different propagation path. If the

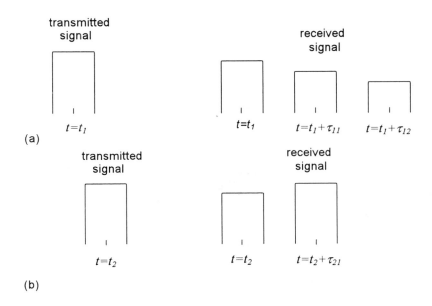

Figure 5.8 Multipath channel response.

frequency separation between two tones is small, their fades will approximately coincide in time, so that at each time instant they will have the same attenuation. As the frequency separation between the two tones increases, they tend to fade independently. That means the attenuation of two tones will be different at a particular time. The frequency separation above which spectral components start to fade independently is called the *coherence bandwidth*.

Signals that occupy a bandwidth that is greater than the coherence bandwidth will become distorted since the relationship between various components in the received signal is not the same as in the transmitted signal. As different frequencies in the spectrum are attenuated differently by the multiphase channel, this distortion is also known as *frequency-selective fading*. In the channels where the signal bandwidth is smaller than the coherence bandwidth, all spectral components are attenuated equally. Such channels are called *flat fading* channels.

5.5.3 Wideband Channel Models

In order to determine the impulse response of a multipath channel, let us consider the transmission of a single impulse with amplitude a at time t_1. The received signal consists of a direct path impulse followed by a number of delayed echoes, as shown in Figure 5.9(a). If we transmit the same single impulse at time t_2, the channel

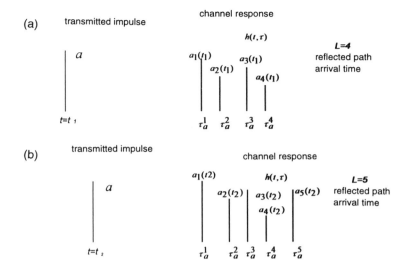

Figure 5.9 Multipath channel impulse response, time (a) t_1 and (b) t_2.

response will change, as shown in Figure 5.9(b). The channel impulse response $h(t, \tau)$ is a random variable, which is a function of time t and reflected path arrival time τ.

From Figure 5.9, the channel impulse response can be represented as

$$h(t, \tau) = \sum_{l=1}^{L} |a_l| \delta[\tau - \tau_{al}] e^{j\phi_l} \qquad (5.55)$$

where τ is the time delay, $|a_l|$, τ_{al}, and ϕ_l are the amplitude, arrival delay, and phase of the lth reflected wave, respectively; L is the number of reflected waves; and $\delta(\ )$ is the delta function. The output of the channel $r(t)$ for a given input $x(t)$ is given by the complex convolution

$$r(t) = x(t) * h(t, \tau) \qquad (5.56)$$

where $*$ stands for convolution. In a cellular mobile environment with an obscured direct path, a_l is a Rayleigh distributed random variable, subject to Doppler spread. The Doppler frequency information can be directly obtained from the function $S(\tau, v)$, which is defined as the inverse transform of $h(t, \tau)$

$$h(t, \tau) = \int_{-\infty}^{\infty} S(v, \tau) e^{j 2\pi v t} dv \qquad (5.57)$$

The channel output is related to this function and the input as

$$r(t) = \int_{-\infty}^{\infty}\int_{-\infty}^{\infty} x(t - \tau)S(\tau, \nu)e^{j2\pi\nu t}\,d\nu\,d\tau \tag{5.58}$$

Another way of computing the multipath spread is based on the autocorrelation function of $h(t, \tau)$, denoted by $\phi_c(\tau_1, \tau_2, \Delta t)$ and defined as

$$\phi_c(\tau_1, \tau_2; \Delta t) = \frac{1}{2}E[h^*(\tau_1; t)h(\tau_2; t + \Delta t)] \tag{5.59}$$

where $E[-]$ denotes expectation and $h^*(t, \tau)$ is the complex conjugate of $h(t, \tau)$. If we assume uncorrellated scatterers and use no time offset ($\Delta t = 0$), then $\phi_c(\tau)$ gives the average power output of the channel as a function of delay τ. Typically, $\phi_c(\tau)$ has a shape as shown in Figure 5.10. $\phi_c(\tau)$ is called the *multipath intensity profile*.

The range of values over which $\phi_c(\tau)$ is essentially nonzero is called the *multipath spread* of the channel and is denoted by Δ.

In real systems, the delay spread varies from fractions of a microsecond to many microseconds. In urban areas it is typically longer than 3 μs, while in suburban and open areas it is shorter and typically less than 0.5 μs and 0.2 μs, respectively.

A similar characterization of the multipath channel in the frequency domain can be obtained by defining the autocorrelation function of $S(f, t)$ as

$$\Phi_c(f_1, f_2, \Delta t) = \frac{1}{2}E[s^*(f_1; t)S(f_2; t + \Delta t)] \tag{5.60}$$

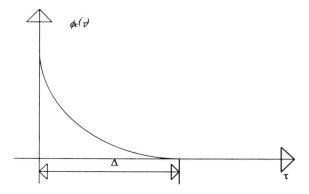

Figure 5.10 Multipath intensity profile.

$\Phi_c(f_1, f_2, \Delta t)$ is also the Fourier transform of the multipath intensity profile $\phi_c(\tau, \Delta t)$. It is called the *spaced-frequency spaced-time correlation function* of the multipath channel.

If we set $\Delta t = 0$, we get

$$\Phi_c(\Delta f) = \int_{-\infty}^{\infty} \phi_c(\tau) e^{-j2\pi \Delta f \tau} \, d\tau \tag{5.61}$$

A typical $\Phi_c(\Delta f)$ is shown in Figure 5.11. Since $\Phi_c(\Delta f)$ is an autocorrelation function in the frequency domain, it provides the information about the coherence bandwidth. The range of values over which $\Phi_c(\Delta f)$ is essentially nonzero is defined as the coherence bandwidth. Since $\phi_c(\tau)$ and $\Phi_c(\Delta f)$ are related by the Fourier transform, the reciprocal of the multipath spread gives a measure of the coherence bandwidth, denoted by $(\Delta f)_c$ as

$$(\Delta f)_c \approx \frac{1}{\Delta} \tag{5.62}$$

If the signal bandwidth is higher than $(\Delta f)_c$ various spectral components are attenuated differently by multipath fading and the channel is frequency selective.

A more practical way of computing the multipath delay spread is based on the probability density function of the impulse arrival time random variable τ, denoted $p(\tau)$. The mean delay is computed as

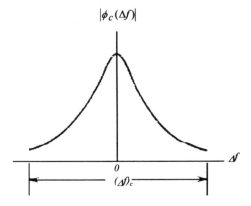

Figure 5.11 An example of $\Phi_c(\Delta f)$.

$$\bar{\tau} = \int_0^\infty \tau p(\tau) d\tau \tag{5.63}$$

The multipath delay spread is obtained as

$$\Delta = \int_0^\infty (\tau - \bar{\tau})^2 p(\tau) d\tau = \int_0^\infty \tau^2 p(\tau) d\tau - \bar{\tau}^2 \tag{5.64}$$

In general, a negative exponential distribution is used to describe the impulse arrival time density function. This function can be written as

$$p(\tau) = \frac{1}{\bar{\tau}} e^{-\tau/\bar{\tau}} \tag{5.65}$$

For this distribution both the mean time delay and the delay spread are equal to $\bar{\tau}$.

Note that when the channel impulse response $h(t,\tau)$ and its Fourier transform $S(\tau, \nu)$ are known, the received signal can be computed using (5.57) and (5.58) as

$$r(t) = \int_{-\infty}^\infty x(t - \tau) h(t, \tau) d\tau \tag{5.66}$$

Using relationship (5.58) a physical implementation of the channel model is apparent. It could be represented by a tapped delay line, with a specific $S(\tau, \nu)$ associated with each tap. The model is shown in Figure 5.12, where F represents the Fourier transform and F^{-1} the inverse Fourier transform.

Another approach is based on relationship (5.57) and is shown in Figure 5.13. In this model, the kth tap is characterized by the random amplitude a_l with a white spectrum and random phase and the spectral spreading Ψ_l, which is modeled by a filter with the transfer function equal to the Doppler spectrum of the kth component.

In CDMA systems with code sequences with chip duration T_c, the individual paths are resolvable as long as their relative delay is larger than T_c. Therefore, the tapped delay line model in Figure 5.13 can be truncated to L taps, where L is given by

$$L = \left\lfloor \frac{\Delta}{T_c} \right\rfloor + 1 \tag{5.67}$$

CDMA Systems 235

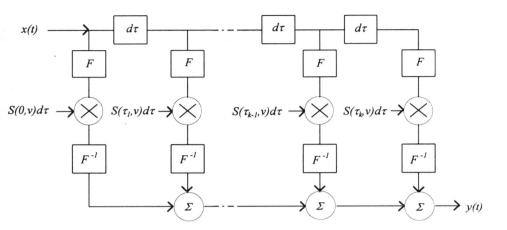

Figure 5.12 Tapped delay line model of the multipath channel based on $S(\tau, v)$.

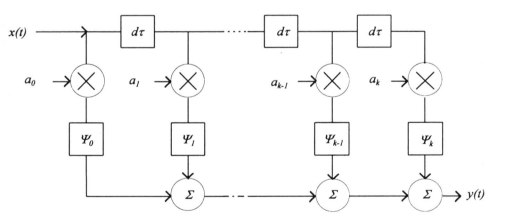

Figure 5.13 Tapped delay line model based on impulse response $h(t, \tau)$.

where Δ is the delay spread. The time delay $\Delta \tau$ in the model shown in Figure 5.12 is replaced by T_c in CDMA systems.

Thus, the received DS/CDMA signal in a multipath environment can be represented as

$$r(t) = \sum_{k=1}^{K} \sqrt{E_{ck}} \sum_{m=-M}^{M} b_k(m) \sum_{l=1}^{L} a_l C_k(t - mT_b - lT_c - \tau_k) + n(t) \qquad (5.68)$$

Each resolvable path component is characterized by its random amplitude a_l. Each of these resolvable paths can be represented as the sum of a number of unresolvable paths, with random amplitudes and phases. Thus, the amplitude of their sum, a_l, is Rayleigh distributed. In an environment where the direct path is obscured by obstacles, each tap coefficient amplitude in the model in (5.55) is Rayleigh distributed. This model is typical for urban areas or for large cells. However, if there is a direct path, the first tap coefficient in the model will be the sum of a constant amplitude component and a Rayleigh distributed reflected component. The amplitude of the resulting component has a Rician distribution. This model is representitive of small cells or satellite mobile systems, where there is a line of sight path between the base station and the mobile unit.

5.6 RAKE RECEIVER

In the previous section we showed that wideband signals in a multipath environment are frequency selective. Such channels are described by a tapped delay line model. Since CDMA spreading codes are designed to have very low crosscorrelation between successive chips, multipath components delayed by more than one chip duration are uncorrelated and appear as resolvable paths in the model. Typically, CDMA systems are designed to have several resolvable paths within the multipath delay spread. At the same time, the delay spread is chosen to be lower than the bit duration T. If the delay spread is greater than the bit duration T, then the data rate is higher than the coherence bandwidth, which results in intersymbol interference. To avoid intersymbol interference, the data rate should be maintained below the coherence bandwidth.

When the delay spread is lower than T and there are several delayed versions of the transmitted code sequence with delay differences greater than T_c, they will have a low correlation with the original code sequence. Thus, each of these delayed signals will appear at the receiver as another uncorrelated user and will be ignored by the matched filter receiver of the desired signal.

However, spread spectrum signals are inherently resistant to multipath fading since multipath components carry the information about the transmitted signal and they are independent. Thus, if one of the multipath components is attenuated by fading, the other may not be and the receiver could use unfaded components to make the decision. The CDMA receiver that takes advantage of the multiple paths to provide diversity is called a Rake receiver.

The Rake receiver, shown in Figure 5.14, consists of a bank of correlators. Each of them is used to detect separately one of the L strongest multipath components. This receiver is basically a diversity receiver based on the fact that the multipath components in a CDMA system are uncorrelated if the relative delays are larger than the chip period.

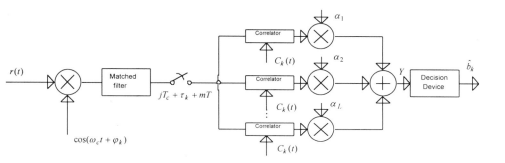

Figure 5.14 A Rake receiver with L branches.

As in other diversity receivers, the outputs from the correlators are weighted and added to compute the estimate for the transmitted signal. If the maximal ratio combining technique, which gives the highest reduction of fading, is used, the weighting coefficient is the complex conjugate of the corresponding channel tap coefficient $\alpha_k = a_k^*$.

Each multipath demodulator in the Rake receiver is called a *finger*. In the original Rake receiver [6] the delay between consecutive taps and the number of taps was fixed. These receivers required a large number of taps in order to capture major multipath components. Modern receivers have only a few Rake fingers and are capable of adjusting the tap positions.

For the operation of the Rake receiver it is necessary to identify and track major multipath components as well as to estimate their relative delays, amplitudes, and phases. The estimation of these parameters is best performed by transmitting unmodulated signals in the form of periodic preambles or pilot tones [1].

5.6.1 Performance of Rake Receiver

Let us consider a Rake receiver shown in Figure 5.14. Assume we transmit only one DS spread spectrum signal of the form as given in (5.12) and that the demodulator provides perfect estimates of the channel so that $\alpha_l = |a_l|$. The decision variable Y at the input of the decision device is given by

$$Y = \sum_{l=1}^{L} \sum_{m=1}^{G_p} \sqrt{E_c} b(m) a_l^2 + \sum_{l=1}^{L} \sum_{m=1}^{G_p} n_m a_l \qquad (5.69)$$

where n_m is Gaussian noise. For a fixed set of values a_l, $l = 1, \ldots, L$, Y has the mean value

$$E[Y] = G_p\sqrt{E_c}\sum_{l=1}^{L} a_l^2 \tag{5.70}$$

and the variance

$$\text{var}[Y] = G_p \cdot \frac{N_o}{2}\sum_{l=1}^{L} a_l^2 \tag{5.71}$$

The probability of error for constant a_l is the probability that the decision variable Y is smaller than zero, which is given by

$$P_b(\gamma_b) = Q\left(\sqrt{\frac{(E[Y])^2}{\text{var}(Y)}}\right) \tag{5.72}$$

or

$$P_b(\gamma_b) = Q\left(\sqrt{\frac{2G_p E_c}{N_o}\sum_{l=1}^{L} a_l^2}\right) \tag{5.73}$$

$$P_b(\gamma_b) = Q\left(\sqrt{\frac{2E_b}{N_o}\sum_{l=1}^{L} a_l^2}\right) \tag{5.74}$$

If we denote by γ_k the instantaneous bit-energy-to-noise spectral density ratio in one Rake receiver finger, it can be represented as

$$\gamma_k = \frac{E_b}{N_o} a_l^2 \tag{5.75}$$

Then the total instantaneous bit energy-to-noise spectral density ratio, denoted by γ_b, is given by

$$\gamma_b = \frac{E_b}{N_o}\sum_{l=1}^{L} a_l^2 = \sum_{l=1}^{L} \gamma_l \tag{5.76}$$

If a_l is Rayleigh distributed, a_l^2 has a chi-square distribution with two degrees of freedom. Therefore, γ_l is also chi-square distributed. Accordingly

$$p(\gamma_l) = \frac{1}{\overline{\gamma_l}} e^{-\gamma_l/\overline{\gamma_l}} \qquad \gamma_l \geq 0 \tag{5.77}$$

where $\bar{\gamma}_l$ is the average bit-to-noise spectral density ratio defined as

$$\bar{\gamma}_l = \frac{E_b}{N_0} E[a_l^2] \tag{5.78}$$

where $E[a_l^2]$ is the average value of a_l^2. The characteristic function of γ_l is

$$\Psi_{\gamma_l}(jv) = E[e^{jv\gamma_l}] = \frac{1}{1 - jv\bar{\gamma}_l} \tag{5.79}$$

Let us assume that the mean square values $E[a_l^2]$ are identical for all tap values. The average bit-energy-per-noise spectral density ratio per finger is given by

$$\bar{\gamma}_l = \frac{E_b}{N_0} E[\alpha_l^2] = \bar{\gamma}_c \tag{5.80}$$

Since the fading of the L channels is statistically independent, γ_l are statistically independent. Thus the characteristic function of γ_b is

$$\Psi_{\gamma_b}(jv) = \frac{1}{(1 - jv\bar{\gamma}_c)^L} \tag{5.81}$$

The probability density function of γ_b is obtained as the inverse Fourier transform of the characteristic function $\Psi_{\gamma_b}(jv)$

$$p(\gamma_b) = \frac{1}{(L-1)!\bar{\gamma}_c^L} \gamma_b^{L-1} e^{-\gamma_b/\bar{\gamma}_c} \tag{5.82}$$

The average bit error probability is obtained by averaging the conditional error probability in (5.74) over all γ_b

$$P_b = \int_0^\infty P_b(\gamma_b) \cdot p(\gamma_b) \cdot d\gamma_b \quad \text{or} \quad P_b = \left(\frac{1-\mu}{2}\right)^L \sum_{l=0}^{L-1} \binom{L-1}{l} \left(\frac{1+\mu}{2}\right)^l \tag{5.83}$$

where $\mu = \sqrt{\bar{\gamma}_c/(1 + \bar{\gamma}_c)}$

For large average bit-to-noise spectral density ratio per finger $\bar{\gamma}_c$, we have

$$\frac{1+\mu}{2} \approx 1 \quad \text{and} \quad \frac{1-\mu}{2} \approx \frac{1}{4\bar{\gamma}_c} \tag{5.84}$$

Furthermore,

$$\sum_{l=0}^{L-1} \binom{L+l-1}{l} = \binom{2L-1}{L} \quad (5.84a)$$

Then the bit error probability can be approximated as

$$P_b \approx \left(\frac{1}{4\bar{\gamma}_c}\right)^L \binom{2L-1}{L} \quad (5.85)$$

If the mean square values $\bar{\gamma}_l$ are distinct, the bit error probability can be approximated for large average ratio $\bar{\gamma}_c$

$$P_b \approx \binom{2L-1}{L} \prod_{l=1}^{L} \frac{1}{2\bar{\gamma}_l} \quad (5.86)$$

The bit error probabilities for equal $\bar{\gamma}_l = \bar{\gamma}_c$ are plotted versus the average bit-energy-per-noise spectral density $\bar{\gamma}_b$ and shown in Figure 5.15.

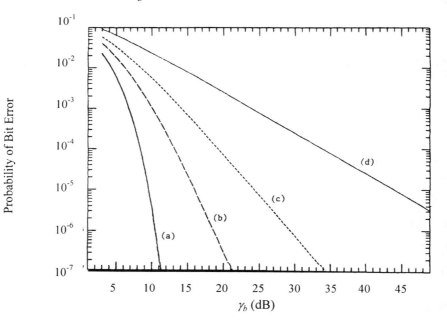

Figure 5.15 Probability of bit error for BPSK modulation for a spread spectrum system over a Rayleigh fading channel for a Rake receiver: (a) no fading, (b) $L = 4$ in fading, (c) $L = 2$ in fading, and (d) $L = 1$ in fading.

5.7 CAPACITY OF A MULTIPLE-CELL CDMA SYSTEM

In Section 5.4 we estimated the capacity of a single-cell CDMA system with ideal power control. CDMA has a real capacity advantage over other multiple-access techniques in a high-density multicell network. This is a direct consequence of (1) spatial isolation due to high propagation losses in the UHF band, typically proportional to the fourth distance power; (2) spread spectrum immunity to interference; and (3) monitoring of voice activity. As a result, in a CDMA multiple-cell network it is possible to use the same frequency band in all cells as opposed to narrowband multiple access techniques in which the bandwidth used in a given cell can be reused only in sufficiently far away cells so that the interference is not significant. In order to compare CDMA with other multiple-access schemes, capacity is measured as the total number of users in a multiple-cell network rather than the number of users per bandwidth or per isolated cell. Based on this criterion, capacity of a CDMA high-density multiple-cell network is much higher than with narrowband multiple-access techniques. The key difference is that CDMA allows 100% frequency of reuse among all cells in a network.

Another important point is that the capacity of practical CDMA systems is interference limited. Thus, the key issue in CDMA network design is minimization of multiple access interference. Power control is critical to multiple-access interference. Each base station controls the transmit power of its own users. However, a given base station is unable to control the power of users in neighboring cells; and these users introduce additional interference, thereby reducing the capacity of the reverse link. The interference coming from other cells determines the real reuse factor of a CDMA cellular system.

CDMA has an additional advantage in voice transmission where interference can be further reduced with the use of voice activation.

In narrowband systems, additional capacity is needed to maintain low cochannel interference. In TDMA and FDMA guard times and guard bands take up to 20% of the overall capacity. CDMA networks are designed to tolerate a certain level of interference and thus have a capacity advantage in this respect compared to narrowband techniques.

In the next section we consider the main factors affecting the capacity of multiple-cell CDMA networks and derive the expressions for reverse and forward link interference, signal-to-interference ratio, number of users, probability of error, and Erlang capacity.

5.7.1 Propagation Loss Model

In Section 5.5 we considered multipath propagation and presented multipath wideband channel models. In this analysis we concentrated on signal amplitude and phase variations and assumed that the mean of the signal amplitude is constant.

Due to shadowing, the mean of the received signal fluctuates with a log-normal distribution. Shadowing is caused by terrain features such as buildings and hills. These large objects cause diffraction and scattering losses. The effect is a very slow change in the local mean value.

Shadowing is typically modeled by a log-normal random variable. If S is the signal power, its probability density function is then given by

$$f(S) = \frac{1}{\sqrt{2\pi}\sigma S} e^{-(\ln S - \bar{S}_d)^2 / 2\sigma^2} \qquad (5.87)$$

The power expressed in decibels is given by $S_d = 10 \log_{10} S$. It is normally distributed with $\bar{S}_d$ being the mean and σ^2 the variance. Typically, σ ranges from 6 dB to 12 dB with $\sigma = 8$ dB most often observed in field measurements [1,6,7,8,9]

The random variable power S can be expressed as

$$S = 10^{S_d/10} \qquad (5.88)$$

where S_d is a Gaussian variable.

The received signal power in a given position in a cell falls off with distance according to a power law. That is, the path loss between the user and the base station is proportional to r^μ, where r is the distance between the user and the base station and μ is a constant. Taking into account both the shadowing and propagation power law, the simplest theoretical model for the propagation signal power loss variation, denoted by L_p, is

$$L_p = r^\mu 10^{L_d/10} \qquad (5.89)$$

where L_d is the power loss variation Gaussian random variable in decibels due to shadowing and μ is a constant that ranges from 2 in the free space to 5.5 in a very dense urban environment [1,7], with a typical value of 4.

5.7.2 Power Control

In Section 5.4 we derived the capacity formula for a single-cell CDMA system under the assumption of perfect power control. On a reverse link, each signal is received by the base station with a different path loss due to (1) differing distances between each user and the base station and (2) statistical variations in radio propagation paths. That is, if the all mobile units transmit at the same power, their signals will arrive at the base station with vastly different power levels.

If a particular signal received at the base station is too weak, it will be masked by other stronger signals. On the other hand, if it has too high power, its performance will be satisfactory but will introduce interference to other user signals in the same cell and degrade their performance. For example, if the signal power is ten times greater than the power of other user signals, the interference caused by this signal is equivalent to the interference of ten other users. Thus, the capacity of the system is reduced by nine. Received signal powers can vary over a range as wide as 80 dB.

The capacity of a conventional CDMA system is maximized if each mobile-transmitted power is controlled so that at the base station the same power level is received from each mobile. Power control is used on the reverse link to combat the near-far problem and thus minimize the effect of interference on system capacity.

For the forward link, no power control is needed in a single-cell system, since for any user interference caused by other user signals remains at the same level as the desired signal. All signals are transmitted together, and there is no difference in propagation path losses as on the reverse link. In addition to mitigating the near-far problem, power control is also used to reduce the effects of shadowing and to keep the transmitted energy per user, required to keep the bit error rate at a certain level, at a minimum. Thus, power control has another benefit of prolonging battery life in portable handsets.

One way of controlling the mobile transmitter power is by applying *automatic gain control* (AGC) measurements in the mobile receiver. Prior to any transmission, the mobile monitors the total power received by the base station. The measured power provides an indication of the propagation loss to each user. The mobile adjusts its transmitter power so that it is inversely proportional to the total signal power it receives. It may be necessary to control power over a dynamic range on the order of 80 dB. This is called *open-loop power control*.

A difficult problem in using AGC measurements for power control is a wide variation in forward and reverse link propagation losses. Forward and reverse link center frequencies are typically in different frequency bands. In this case, reciprocity does not hold. As a result, propagation losses for these two links are different. For example, in the IS-95 cellular system, in which the two center frequencies differ by 45 MHz, the two respective propagation losses may differ by several decibels.

More effective average or long-term power control is achieved by a *closed-loop power control* scheme. This requires that the base station provides continuous feedback to each mobile so that the mobile varies its power accordingly. The base station estimates the user signal power on the reverse link and compares it with the desired nominal threshold. Based on whether the measured received power is higher or lower than the desired level, the base station transmits a one-bit command to the mobile to lower or raise its transmitter power by a fixed step, expressed in decibels. This is called a *bang-bang* control loop [10].

The loop introduces a delay that is the sum of the transmission time and the time needed to execute the command in the transmitter.

In a real system a combination of open- and closed-loop control can be used [3]. The nominal desired power can be fixed to a level required to achieve a given bit error rate. However, due to shadowing, this power level might vary. Thus, an additional power control loop, called *outer loop*, might be implemented to adjust the desired nominal power level according to the specified bit error rate.

The speed at which power level control is performed depends upon whether the control function is designed to follow slow log-normal shadowing and propagation loss variations only, or it is required to mitigate variations caused by fast fading as well. If the power adjustments are to follow the fast multipath fading, the rate of power control must be higher than ten times the maximum fade rate [11]. If the power control mechanism is designed to follow only the slow log-normal variations due to shadowing, the received signal will exhibit fast Rayleigh fading after power control.

Figure 5.16(a) shows sample waveforms of power-controlled signals using the fixed step closed loop power control scheme. The power control sampling rate $1/T_p$ is assumed to be ten times the maximum fade rate f_d. For perfect power control, the transmitter signal power should be identical to the inverse channel variation, shown in the same figure. However, due to fast signal fading variations in the channel, the transmit signal power does not follow the inverse channel variations. As a result, there is significant residual fading in the received signal after power control (Figure 5.16(b)).

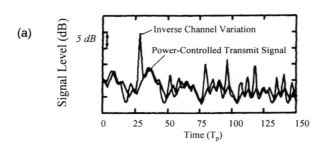

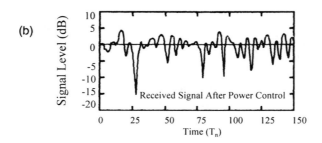

Figure 5.16 An example of power controlled transmit and received signals. (© 1993 IEEE. *From:* [11].)

The commands for power control are transmitted over the forward link without error control coding in order to minimize the loop delay.

In real systems, power control is not ideal. The received signal power variations in the base station after power control are approximately log-normally distributed, as evidenced by experimental data [12]. The standard deviation of the received signal-to-noise ratio is typically between 1 dB and 2 dB.

Figure 5.17 shows variations of E_b/I_o obtained from experimental data. E_b/I_o in a power-controlled system with all cells fully loaded is varied to maintain frame error rate below 1%. The histogram shown by the solid line is closely approximated by a log-normal probability density, represented by the dotted line, with the mean of 7 dB and the standard deviation of 2.4 dB for the normal exponent. The standard deviation is generally higher when E_b/I_o is varied, compared to the case when it is maintained constant.

The received power after power control, S_p, can be expressed as

$$S_p = 10^{S_{pd}/10} \tag{5.90}$$

where S_{pd}, which is a Gaussian random variable, is S_p expressed in decibels.

In multiple-cell CDMA systems the mobile receiver in the forward link receives interference from other cells. In these systems power control on the forward link becomes necessary to allocate the power to individual users according to their actual needs determined by the level of interference to which they are exposed.

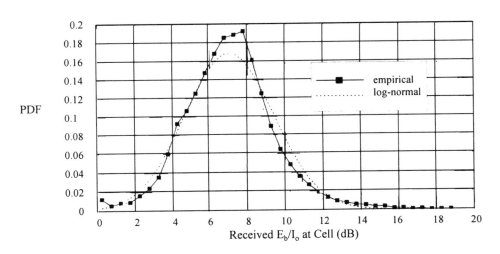

Figure 5.17 E_b/I_o histogram and log-normal approximation in a power control scheme. (© 1993 IEEE. From: [12].)

5.7.3 Reverse Link Capacity in Cellular CDMA

A CDMA system with multiple base stations is generally referred to as a *cellular system*. Each base station in a cellular system receives interference from the users in the same cell, called *intracell interference*, and interference coming from mobiles in neighboring cells, called *intercell interference*. In order to compute the capacity of the reverse link in a cellular system we compute the intercell and intracell interference powers received by the base station. The CDMA concept allows each cell to use the same broadband frequency channel, eliminating the need for a mobile to change its frequency when moving into another cell. This process is called soft handoff. With soft handoff a mobile user can communicate with two different base stations at the same time. The two base station signals can be combined on the forward link to improve the performance. On the reverse link, the two different stations typically detect the signals independently. If the two received signals are different, the switching center can select the better one.

In establishing the connection, the mobile detects and tracks the base station with the highest power. At the same time the mobile monitors the powers of neighboring base stations. The second base station is introduced when its power becomes significant relative to the first base station power. When the first base station's signal becomes too weak relative to the second one, it will be abandoned.

Each mobile user is power-controlled by the base station whose total signal power is maximum to that mobile. That is, the mobile will be controlled by the base station to which the propagation loss is minimum.

We consider a cluster of hexagonal cells, as shown in Figure 5.18.

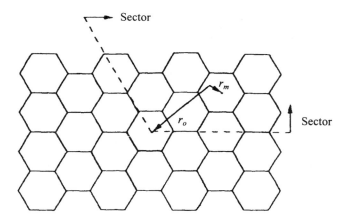

Figure 5.18 Hexagonal cellular structure cells.

We assume ideal power control at the base station. All user signals within a given cell are power-controlled by the same base station and, so, ideally are received with the same power S. For K active users per cell, the total intracell interference is not greater than $(K-1)S$. In voice transmission this power is reduced by the voice activity factor α.

The mobiles in other cells are controlled by other base stations. We consider a mobile in another cell at a distance r_m from its base station and r_o from the given base station. Due to the perfect power control assumption, the transmit power of this mobile is

$$S_T(r_m) = \frac{Sr_m^\mu}{10^{L_{dm}/10}} \tag{5.91}$$

where μ is the power loss exponent and L_{dm} is a Gaussian variable representing power variation due to shadowing.

This implies that the power received by its base station is

$$S_R(r_m) = \frac{Sr_m^\mu}{10^{L_{dm}/10}} \cdot \frac{10^{L_{dm}/10}}{r_m^\mu} = S \tag{5.92}$$

The interference power from the mobile in another cell at a distance r_o from the given base station depends on (1) the propagation loss from that mobile to the given base station and (2) propagation loss from the interfering mobile to its own base station. This interference power can be written as

$$I(r_o, r_m) = S \frac{10^{L_{do}/10}}{r_o^\mu} \cdot \frac{r_m^\mu}{10^{L_{dm}/10}} \tag{5.93}$$

where L_{do} is power variation due to log-normal shadowing on the propagation path from the mobile at a distance r_o to the given base station.

A generally accepted propagation model, supported by experimental data, has the power exponent μ of 4. L_{dm} and L_{do} are independent Gaussian random variables each with a standard deviation σ of 8 dB. We assume that fast fading does not affect the average power level.

The intercell interference power can be written as

$$I(r_o, r_m) = S\left(\frac{r_m}{r_o}\right)^\mu 10^{\frac{L_{do} - L_{dm}}{10}} \tag{5.94}$$

$I(r_o, r_m)/S$ is assumed to be equal or less than 1, otherwise the mobile would be controlled by another base station, for which it is equal to or less than one. As L_{do}

and L_{dm} are independent Gaussian random variables with the same mean and each of them has the variance σ^2, their difference $L_{do} - L_{dm}$ is also a Gaussian random variable with a zero mean and a variance $2\sigma^2$.

We assume the cell radius is normalized to unity and that users are uniformly distributed. In a system with three sector antennas the user density is given by

$$\delta = \frac{2K}{3\sqrt{3}} = \frac{2K_s}{\sqrt{3}} \tag{5.95}$$

where K is the total number of users per cell and K_S is the number of users per sector.

If the propagation loss was only a function of the distance from the base station, the mobile would be controlled by the nearest base station. To simplify the analysis we will assume that the nearest base station exercises power control. By doing this, the derived expression for intercell interference power will be an upperbound.

The total normalized intercell interference power received by the given base station can be obtained by integrating over the two-dimensional area A comprising all cells in the system (Figure 5.18)

$$\frac{I}{S} = \iint_A \psi \cdot \left(\frac{r_m}{r_o}\right)^\mu 10^{\frac{L_{do} - L_{dm}}{10}} \Theta\left(L_{do} - L_{dm}, \frac{r_m}{r_o}\right) \delta\, dA \tag{5.96}$$

where m is the cell index for which $r_m = \min_{k \neq 0} r_k$ and

$$\Theta\left(L_{do} - L_{dm}, \frac{r_m}{r_o}\right) = \begin{cases} 1 & \text{if } \left(\frac{r_m}{r_o}\right)^\mu 10^{(L_{do} - L_{dm})/10} \leq 1 \\ 0 & \text{otherwise} \end{cases} \tag{5.97}$$

ψ is the voice activity random variable defined as

$$\psi = \begin{cases} 1 & \text{with probability } \alpha \\ 0 & \text{with probability } (1 - \alpha) \end{cases} \tag{5.98}$$

For $\mu = 4$, the mean value of the random variable I/S is given by [1]

$$E\left(\frac{I}{S}\right) = \alpha \iint_A \left(\frac{r_m}{r_o}\right)^4 f\left(\frac{r_m}{r_o}\right) \delta\, dA \tag{5.99}$$

where

$$f\left(\frac{r_m}{r_o}\right) = e^{[\sigma(\ln 10)/10]^2}\left\{1 - Q\left[\frac{40}{\sqrt{2\sigma^2}} \cdot \log_{10}\left(\frac{r_o}{r_m}\right) - \sqrt{2\sigma^2}\frac{\ln 10}{10}\right]\right\} \quad (5.100)$$

Assuming the spatial autocorrelation of L_{do} and L_{dm} to be very narrow in all directions, which is equivalent to a two-dimensional white random process, the variance of the normalized intercell interference power is given by [1]

$$\text{var}\left(\frac{I}{S}\right) \leq \iint_A \left(\frac{r_m}{r_o}\right)^4 \left[\alpha g\left(\frac{r_m}{r_o}\right) - \alpha^2 f\left(\frac{r_m}{r_o}\right)\right] \delta \, dA \quad (5.101)$$

where

$$g\left(\frac{r_m}{r_o}\right) = e^{((\sigma \ln 10)/5)^2}\left[\left\{1 - Q\left[\frac{40}{\sqrt{2\sigma^2}}\log_{10}\frac{r_o}{r_m} - \sqrt{2\sigma^2}\left(\frac{\ln 10}{5}\right)\right]\right\}\right] \quad (5.102)$$

For a typical value of the power variance due to shadowing of $\sigma = 8$ dB, the mean and the variance of I/S obtained by numerical integration are

$$E\left(\frac{I}{S}\right) \leq 0.247 K_S \quad (5.103)$$

$$\text{var}\left(\frac{I}{S}\right) \leq 0.078 K_S \quad (5.104)$$

The results of an improved upper bound on the mean normalized intercell interference divided by the number of users per cell are derived in [13]. The intercell interference factor, denoted f, is defined as

$$f = \frac{E(I/S)}{K_S} \quad (5.105)$$

For $\mu = 4$, $\sigma = 8$ dB, the upper bound on the inter-cell interference factor is 0.55 [13].

The random variable interference power I/S, given by integral (5.96), can be approximated by a Gaussian random variable following the assumption of spatial whiteness of L_{do} and L_{dm}.

The received effective bit energy-to-noise density ratio on the reverse link, following the same reasoning as in derivation of (5.35), is computed as

$$\frac{E_b}{I_o} = \frac{G_p}{\sum_{i=1}^{K_s-1} \psi_i + \frac{I}{S} + \frac{\sigma_n^2}{S}} \qquad (5.106)$$

where I/S is computed in (5.96), σ_n^2 is the thermal noise power, and ψ_i are binomial random variables that model the voice activity of users within the given cell.
They are given by

$$\psi_i = \begin{cases} 1 & \text{with probability } \alpha \\ 0 & \text{with probability } 1 - \alpha \end{cases} \qquad (5.107)$$

The performance of a link can be measured by the probability that it has the bit error rate higher than a specified value BER_S. This is called the *outage probability* and is denoted by P_{out}. It is given by

$$P_{out} = P_r\{BER > BER_S\} \qquad (5.108)$$

or

$$P_{out} = P_r\left\{\sum_{i=1}^{K_s-1} \psi_i + \frac{I}{S} > \Delta_r\right\} \qquad (5.109)$$

where

$$\Delta_r = \frac{G_p}{(E_b/I_o)_s} - \frac{\sigma_n^2}{S} \qquad (5.110)$$

and $(E_b/I_o)_s$ is the value of E_b/I_o for which the specified value of bit error rate BER_S is achieved with the given modulation, coding, demodulation, and decoding schemes.
For the binomial random variables ψ_i defined in (5.107) and assuming that I/S is a Gaussian random variable with the mean and variance defined by (5.99) and (5.101), respectively, the upper bound on the outage probability can be calculated as

$$P_{out} < \sum_{K=0}^{K_s-1} P_r\left(\frac{I}{S} > \Delta_r - k | \Sigma\psi_i = k\right) P_r(\Sigma\psi_i = k)$$

$$= \sum_{k=0}^{K_s-1} \binom{K_s-1}{k} \alpha^k (1-\alpha)^{K_s-1-k} Q\left[\frac{\Delta_r - k - E(I/S)}{\sqrt{\text{var}(I/S)}}\right] \qquad (5.111)$$

The curves for the outage probability for the propogation model with $\mu = 4$ and $\sigma = 8$ dB and the system parameters of the IS-95 system are shown in Figure 5.19. For this system the spreading gain is $G_p = W/R = \dfrac{1228800}{9600} = 128$, the voice activity factor $\alpha = 3/8$, and the received signal-to-thermal noise ratio $S/\sigma_n^2 = -1$ dB. The specified bit error rate is 10^{-3}; and the required E_b/I_o with the 1/3 rate convolutional code, two-antenna diversity, and differential detection at the reverse link is 5 or 7 dB.

The outage probability is plotted in Figure 5.19 for various values of the mean and variance of I/S. The leftmost curve is obtained for the full load of neighboring cells corresponding to the mean and variance of I/S given by (5.103) and (5.104), respectively. The other two curves are obtained for lower loads corresponding to the mean of 1/2 and 1/4 of the given user cell, and the rightmost curve is obtained to a single cell with no interference from other cells.

As Figure 5.19 shows the reverse link can support 36 users per sector or 108 users per cell with the BER performance lower than 10^{-3} for 99% of the time, under the full load. If the number of users in the neighboring cells is halved, the total number of users per cell increases to 144. If the total spectral allocation is 12.5 MHz and CDMA is employed across the entire cellular system, the total system capacity is increased ten times, which for the full load is 1,080 users.

In an FM/FDMA analog cellular system the channel allocation is 30 KHz, so the number of users in a cell with 1.25-MHz band is about 42. With a frequency reuse factor of 7, the total number of users that the system can support with the same spectral allocation of 12.5 MHz as for CDMA is 60. Thus, CDMA has a higher capacity on the reverse link by a factor of 18 relative to an FM/FDMA system.

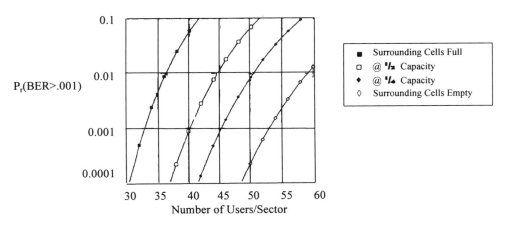

Figure 5.19 Outage probability of the cellular CDMA system reverse link. (© 1991 IEEE. *From:* [1].)

In the US IS-54 digital standard, the channel allocation is 30 KHz and three users share one frequency channel on the basis of TDMA. This increases its capacity by a factor of three compared to FM/FDMA and it is still lower than CDMA capacity by a factor of 6.

5.7.4 Erlang Capacity of the Reverse Link in a Single-Cell CDMA

In telephone networks the quality of service is measured by the *blocking probability*, which is defined as the probability that a user attempting a cell finds all channels busy. The service is considered to be adequate if the blocking probability is less than 2%. The average number of users requesting service that give a specified blocking probability is called the *Erlang capacity* of the system.

In narrowband FDMA and TDMA systems the number of users is limited by the number of frequency and time slots, respectively. Blocking occurs when the number of users exceeds the number of available frequency or time slots. In CDMA systems the number of users is not strictly limited, since the system is designed to tolerate a certain level of interference. However, in order to ensure good quality service the total interference-to-thermal noise ratio in CDMA systems is maintained above a certain level. Blocking in CDMA occurs when this level is exceeded.

It should be pointed out, though, that this is a soft blocking condition that can be relaxed by increasing I_o/N_o as opposed to hard blocking in narrowband multiple-access schemes when all channels are busy. Though this ratio I_o/N_o could in principle increase arbitrarily, when it exceeds a certain level, the increased interference would cause unintelligible voice, loss of synchronization, and system instability.

In a single cell the total normalized noise plus interference power, assuming perfect power control, is given by

$$\sum_{i=2}^{K} \psi_i + \frac{\sigma_n^2}{S} \qquad (5.112)$$

where ψ_i is a binary variable used to represent voice activity given by (5.107) and σ_n^2 is the thermal noise power.

The first term in (5.112) is the interference normalized power, and the second term is the thermal noise power. To ensure good service quality and limit the dynamic range of the receiver, in CDMA systems it is required that the total interference plus thermal noise power-to-thermal noise power ratio be lower than a prescribed value $1/\eta$

$$\frac{\sum_{i=2}^{K} \psi_i + \sigma_n^2/S}{\sigma_n^2/S} < \frac{1}{\eta} \qquad (5.113)$$

where $1/\eta$ is in the range between 10 to 40, or 10 dB to 16 dB.

Since the bit energy-to-interference plus thermal noise ratio for a single-cell system can be represented as

$$\frac{E_b}{I_o} = \frac{G_p}{\sum_{i=2}^{K} \psi_i + \sigma_n^2/S} \tag{5.114}$$

condition (5.113) can be expressed as

$$\sum_{i=2}^{K} \psi_i < \frac{G_p}{E_b/I_o}(1 - \eta) \tag{5.115}$$

If we introduce a new parameter defined as

$$\Delta_r' = \frac{G_p}{E_b/I_o}(1 - \eta) \tag{5.116}$$

the good service condition becomes

$$\sum_{i=2}^{K} \psi_i < \Delta_r' \tag{5.117}$$

Since a CDMA system is interference limited, when the condition (5.117) is not met, a blocking event will occur, meaning that the user call will be rejected. Thus, the blockage probability can be expressed as

$$P_{bl} = P_r\left(\sum_{i=2}^{K} \psi_i > \Delta_r'\right) < P_r\left(\sum_{i=1}^{K} \psi_i > \Delta_r'\right) \tag{5.118}$$

In order to compute the Erlang capacity of a single-cell CDMA system we assume that the number of active users k is modeled by a Poisson distribution

$$P_k = \frac{(\lambda/\mu)^k}{k!} e^{-\lambda/\mu} \tag{5.119}$$

where λ/μ is the offered average traffic measured in Erlangs, called the *occupancy parameter*, λ is the total average arrival rate from the entire population of users, and $1/\mu$ is the average time per call.

The call service time τ per user is assumed to be exponentially distributed, so the probability that τ exceeds T is given by

$$P_r(\tau > T) = e^{-\mu T} \quad T > 0 \tag{5.120}$$

Thus, the number of active users in a CDMA system in (5.118), denoted by K, is a Poisson random variable, with the distribution given by (5.119).

In order to compute the blockage probability bound in (5.118) we need to find the distribution of the random variable

$$z = \sum_{i=1}^{K} \psi_i \tag{5.121}$$

It can be obtained by computing the generating function

$$\Psi(jv) = E(e^{jvz}) = E\left(e^{jv\sum_{i=1}^{K}\psi_i}\right) = E_K \cdot \prod_{i=1}^{K} E(e^{jv\Psi_i})$$

$$= \sum_{K=0}^{\infty} \frac{(\lambda/\mu)^K}{K!} e^{-\lambda/\mu} \left[\alpha(e^{jv} - 1) + 1\right]^K = \exp\left[\frac{\lambda\alpha}{\mu}(e^{jv} - 1)\right] \tag{5.122}$$

where α is the voice activity factor. This is a generating function of a Poisson random variable with the occupancy parameter $\alpha\mu/\lambda$. The blocking probability P_{bl} can then be expressed as

$$P_{bl} = e^{-\alpha\lambda/\mu} \sum_{K=\lceil \Delta'_r \rceil}^{\infty} \left(\frac{\alpha\lambda}{\mu}\right)^K \frac{1}{K!} \tag{5.123}$$

which is the sum of the tail of the Poisson distribution with the occupancy parameter $\alpha\mu/\lambda$. For large Δ'_r, the Poisson distribution can be approximated by a Gaussian distribution with same mean and variance, so that the blockage probability can be written as

$$P_{bl} \approx Q\left(\frac{\Delta'_r - \alpha\lambda/\mu}{\sqrt{\alpha\lambda/\mu}}\right) \tag{5.124}$$

The blocking probability for a system with $\eta = 0.1$, $E_b/I_o = 5$ or 7 dB, $G_p = 1280$, voice activity $\alpha = 0.4$ is shown in Figure 5.20 as a function of the Poisson parameter $\alpha\lambda/\mu$. Note that in computing the processing gain for this example, it is assumed that the total frequency bandwidth B_w allocation is equal to the total frequency allocation of 12.5 MHz, so the effective processing gain is

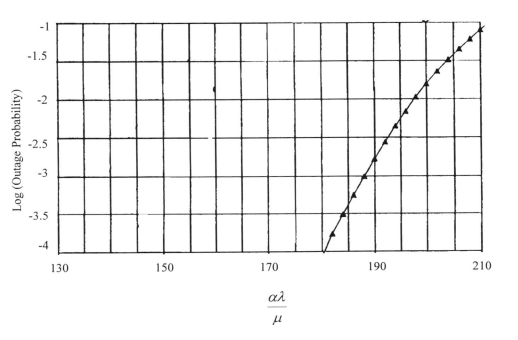

Figure 5.20 Single-cell CDMA reverse link outage probability with perfect power control [3].

$$G_p = \frac{B_w}{R_b} = \frac{12.5 \text{ MHz}}{9.76 \text{ KHz}} = 1280$$

where R_b is the average data rate of 9.76 KHz.

5.7.5 Erlang Capacity of the Reverse Link in a Cellular CDMA

In a cellular CDMA system the other cells introduce interference into the given cell. We have shown in Section 5.7.3 that the effect of users in other cells is to introduce interference with the normalized power of

$$E\left(\frac{I}{S}\right) = f \cdot K \tag{5.125}$$

where f is the intercell interference factor. The values of factor f for various values of the variance of log-normal shadowing σ are shown in Table 5.1 for the propagation path loss exponent $\mu = 4$ [13].

Table 5.1
Factor f for Various Values of σ

σ (dB)	f
0	0.44
2	0.43
4	0.45
6	0.49
8	0.55
10	0.66
12	0.91

The blocking probability for the cellular system can then be represented as

$$P_{bl} \leq P_r\left[\sum_{i=1}^{K(1+f)} \psi_i > \Delta_r'\right] \quad (5.126)$$

The mobiles in neighboring cells are assumed to have the same activity factor α as the users in the given cell and the same occupancy distribution given by the Poisson formula (5.119) with the parameter $\alpha\frac{\lambda}{\mu}(1+f)$. Then the blockage probability for a cellular CDMA system can be obtained from (5.124) by replacing $\alpha\lambda/\mu$ by $\alpha\frac{\lambda}{\mu}(1+f)$

$$P_{bl} \approx Q\left(\frac{\Delta_r' - \frac{\alpha\lambda}{\mu}(f+1)}{\sqrt{\frac{\alpha\lambda}{\mu}(f+1)}}\right) \quad (5.127)$$

5.7.6 Erlang Capacity of the Reverse Link in a Cellular CDMA System With Imperfect Power Control

The performance of a power control loop in a CDMA system depends on the power control algorithm, speed of the adaptive power control system, dynamic range of the transmitter, spatial distribution of users, and propagation statistics. Due to a multiplicity of factors involved, the variations of the signal power due to the combined effect of these factors can be described by a log-normal random variable. Experimental results confirm this assumption, as shown in Figure 5.17. Typically, the standard deviation of the log-normal distribution varies between 1.5 dB to 2.5 dB.

To obtain the blockage probability in a cellular system with imperfect power control, in the expression for the blockage probability (5.126), the binomial random variable ψ_i is replaced by a log-normal random variable γ_i, as

$$P_{bl} < P_r\left[\sum_{i=1}^{K(1+f)} \gamma_i > \Delta'_r\right] \tag{5.128}$$

This expression can be approximated as

$$P_{bl} \approx Q\left[\frac{\Delta'_r - \alpha(\lambda/\mu)(1+f)e^{(c\sigma)^2/2}}{\sqrt{\alpha(\lambda/\mu)(1+f)e^{(c\sigma)^2}}}\right] \tag{5.129}$$

where $c = (\ln 10)/10$. The blockage probability curve given by (5.129) is shown in Figure 5.21 along with the simulation result represented by the leftmost curve, for this same system parameters as for the single-cell CDMA system and the variance of the log-normal shadowing of 2.5 dB [12].

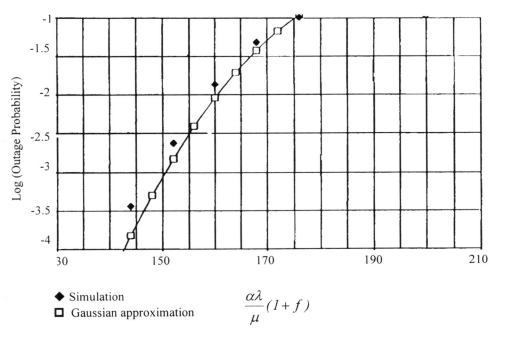

Figure 5.21 Outage probabilities for the reverse link in a cellular communication system with power control [3].

For a given outage probability, the Erlang capacity is equal to the average traffic load λ/μ. For the system parameters previously noted and the blocking probability of 1%, the Erlang capacity of the CDMA reverse link can be determined from Figure 5.21. From Figure 5.21, the value of $\alpha\frac{\lambda}{\mu}(1 + f)$ is 160. Thus, for $f = 0.55$, $\alpha = 0.4$, we get for λ/μ the value of 250.

For the FM/FDMA analog AMPS system with 30-KHz channel allocations, the frequency reuse factor of seven and three sectors, there are 19 channels per sector, within the total frequency allocation of 12.5 MHz

$$N_{FDMA} = \frac{12.5 \text{ MHz}}{30 \text{ KHz} \cdot 7 \cdot 3} = 19 \text{ channels/sector}$$

In a TDMA system, similarity, with three time slots per channel there are three times more channels per sector, denoted by N_{TDMA}, is $N_{TDMA} = 57$ channels/sector. In FDMA and TDMA systems, the probability that a new user enters the system and finds all servers busy is given by

$$P_K = \frac{(\lambda/\mu)^K/K!}{\sum_{j=0}^{K}(\lambda/\mu)^j/j!} \quad (5.130)$$

where the parameters are defined as before. This is called the Erlang-B formula.

Using the Erlang-B formula with $P_K = 10^{-2}$, for example, to compute the offered traffic Erlang capacity, we get

$$(\lambda/\mu)_{FDMA} = 12.34 \text{ Erlangs/sector}$$

$$(\lambda/\mu)_{TDMA} = 46.8 \text{ Erlangs/sector}$$

Compared to the Erlang capacity of an FDMA system, the CDMA system achieves an increase by a factor of 20

$$\frac{(\lambda/\mu)_{CDMA}}{(\lambda/\mu)_{FDMA}} = \frac{250}{12.34} \approx 20$$

When the fading rate is high and closed-loop power control is not able to mitigate fast Raleigh fading, the Erlang capacity of the reverse link is 3 to 4 Erlangs less than that determined from Figure 5.21 [10].

5.7.7 Forward Link Capacity in a Cellular CDMA System

The forward link does not suffer from the near-far effect, and in a single-cell CDMA system in the absence of log-normal shadowing there is no need for power control.

In a cellular system, however, the mobile receives interference from neighboring base stations. In order to maximize the capacity of the forward link it is essential to control the power of the base station as to allocate the power to individual users according to their needs. More power should be provided to those mobiles that receive highest interference from neighboring base stations.

Power control on the forward link is accomplished by measuring the mobile power received from its own base station and the total received power. The information about these two power values is transmitted to the base station.

The ith mobile monitors the powers received from the neighboring base stations and finds the maximum one, denoted $(S_{T1})_i$. The base station from which power $(S_{T1})_i$ is received is designated as its own base station. The powers from other neighboring base stations are denoted $(S_{T2})_i$, $(S_{T3})_i$, ..., $(S_{TQ})_i$. The notation is chosen so that

$$(S_{T1})_i > (S_{T2})_i > (S_{T3})_i > \cdots > (S_{TQ})_i > 0 \tag{5.131}$$

where it is assumed that the total Q powers are significant and all other base station powers are negligible. The received bit energy-to-interference plus thermal noise ratio is bounded by

$$\left(\frac{E_b}{I_o}\right)_i \geq \frac{\beta w_i (S_{T1})_i / R_b}{\sum_{j=1}^{Q(S_{T_j})_i} S_{Tj}/B_w + N_o} \tag{5.132}$$

N_o is the thermal noise power spectral density; S_{Tj} are defined in (5.131); R_b is the data rate; B_w is the spread spectrum bandwidth; β is the fraction of the total base station power assigned to the subscriber, while the rest of $(1 - \beta)$ is assigned to the transmission of a pilot tone or preamble necessary for acquisition and tracking (for example, in the IS-95, a pilot tone that takes 20% of the total power is transmitted on the forward link); and w_i is the fraction of the total power allocated to the ith user, where

$$\sum_{i=1}^{K} w_i \leq 1 \tag{5.133}$$

the weighting factor w_i is proportional to the total sum of other base station powers, S_{T2}, S_{T3}, ..., S_{TQ}, relative to the mobile's own base station power $(S_{T1})_i$. From (5.132), the weighting factor w_i can be expressed as

$$w_i \leq \frac{E_b/I_o}{\beta G_p}\left[1 + \left(\frac{\sum_{j=2}^{Q} S_{Tj}}{S_{T1}}\right)_i + \frac{\sigma_n^2}{(S_{T1})_i}\right] \quad (5.134)$$

where G_p is the processing gain ($G_p = B_w/R_b$) and σ_n^2 is the thermal noise power. If we define

$$f_i \equiv 1 + \left(\frac{\sum_{j=2}^{Q} S_{Tj}}{S_{T1}}\right)_i \quad (5.135)$$

then combining (5.133) and (5.134) we get

$$\sum_{i=1}^{K} f_i \leq \frac{G_p \beta}{E_b/I_o} - \sum_{i=1}^{k} \frac{\sigma_n^2}{(S_{T1})_i} \equiv \Delta_f \quad (5.136)$$

The second term in (5.136) typically can be neglected relative to the first term. The capacity can be estimated from the outage probability, defined as

$$P_{out} = P_r[BER > (BER)_s] \quad (5.137)$$

where $(BER)_S$ corresponds to a specified bit error rate for which E_b/I_o is

$$\frac{E_b}{I_o} = \left(\frac{E_b}{I_o}\right)_s \quad (5.138)$$

If we compute Δ_f' for $\left(\frac{E_b}{I_o}\right)_s$, the outage probability can be written as

$$P_{out} = P_r\left[\sum_{i=1}^{K} f_i > \Delta_f'\right] \quad (5.139)$$

The distribution of $\sum_{i=1}^{K} f$, cannot be expressed in a closed form. The simulation results for the outage probability for the forward link of the IS-95 CDMA system are shown in Figure 5.21 [1], with the system parameters of $G_p = 1280$, $E_b/I_o = 5$ dB, and $\beta = 0.8$. In the simulation, the powers $(S_{T1})_i$, $(S_{T2})_i$, ..., $(S_{TQ})_i$ in (5.136) are represented as the product of the fourth-order power of distance and a log-normally distributed attenuation. With these parameters, as Figure 5.22 shows, the forward link can support the bit error rate of 10^{-3} for more than 99% of the time for 38 users per sector or 114 users per cell.

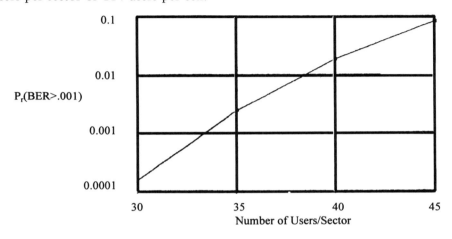

Figure 5.22 Forward link capacity of a cellular CDMA system. (© 1991 IEEE. *From:* [1].)

References

[1] Gilhousen, K., I. Jacobs, R. Padovani, A. Viterbi, L. Weaver, and C. Wheatley, III, "On the Capacity of a Cellular CDMA System," *IEEE Trans. on Vehicular Technology*, Vol. 40, No. 2, May 1991, pp. 303–312.
[2] Rappaport, T., *Wireless Communications, Principles & Practice*, Englewood Cliffs, NJ: Prentice Hall, 1996.
[3] Viterbi, A., "Principles of Spread Spectrum Multiple Access Communications," *Qualcomm*, 1993.
[4] Pursley, M., "Performance Evaluation for Phase Coded Spread Spectrum Multiple Access Communications—Part I," *IEEE Trans. on Communications*, Vol. 25, Aug. 1977, pp. 795–799.
[5] Miller, M., B. Vucetic, and L. Berry, *Satellite Mobile Communications*, Kluwer, 1993.
[6] Price, R., and P. E. Green "A Communication Technique for Multipath Channels," *Proc. IRE*, Vol. 46, 1958, pp. 555–570.
[7] Lee, W. C. Y., *Mobile Communications Design Fundamentals*, New York: Howard W. Sams, 1986.
[8] Lee, W. C. Y., *Mobile Communications Engineering*, New York: McGraw-Hill, 1982.
[9] French, R. C., "The Effect of Fading and Shadowing on Channel Reuse in Mobile Radio," *IEEE Trans. on Vehicular Technology*, Vol. VT-28, Aug. 1979, pp. 171–181.
[10] Viterbi, A. J., A. M. Viterbi, and E. Zehavi, "Performance of Power Controlled Wideband Terrestrial Digital Communication," *IEEE Trans. on Communications*, Vol. 41, April 1993, pp. 559–569.

[11] Ariyavisitakul, S., and L. F. Chang, "Signal and Interference Statistics of a CDMA System with Feedback Power Control," *IEEE Trans. on Communications*, Vol 41, No. 11, Nov. 1993, pp. 1626–1634.
[12] Viterbi, A. M., and A. J. Viterbi, "Erlang Capacity of a Power Controlled CDMA System," *IEEE JSAC*, Vol. 11, Aug. 1993, pp. 892–899.
[13] Viterbi, A. J., A. M. Viterbi, and E. Zehavi, "Other-Cell Interference in Cellular Power-Controlled CDMA," *IEEE Trans. on Communications*, Vol. 42, No. 2/3/4, Feb./March/April, 1994, pp. 1501–1504.

Selected Bibliography

Lee, W., *Mobile Cellular Telecommunications Systems*, New York: McGraw-Hill, 1989.
Shen, Q., and W. Krsymich, "The Effect of Fading on the Erlang Capacity of the IS-95 CDMA Cellular System," *ICC'96*, Dallas, June 1996, pp. 1829–1833.

Appendix A5 Interim Standard 95

In this appendix we briefly describe an example of the standardized CDMA system adopted by the TIA in 1993. A block diagram of the reverse CDMA channel structure is shown in Figure A5.1. Depending on the voice activity, one of four different

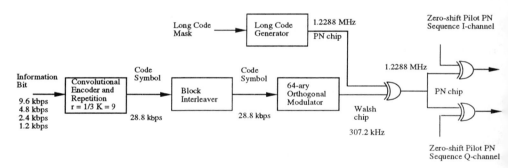

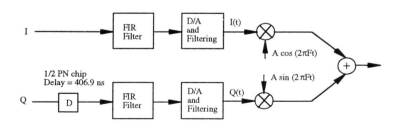

Figure A5.1 Reverse CDMA channel data path example.

information rates is selected. These rates are 9.6, 4.8, 2.4, and 1.2 Kbps. Convolutional coder with $r = 1/3$ and $K = 9$ and block interleaver are used. Code symbols are of constant rate 28.8 Kbps. For maximum bit rate $R_b = 9.6$ Kbps, this rate is obtained directly after convolutional coding as $9.6 \times 3 = 28.8$ Kbps. For all other, lower values of $R_{bin} = 9.6/m$ ($m = 0, 2, 4, 8$), the constant code rate is achieved by repeating the interleaver block m times. In the next step 64-ary orthogonal modulation is used utilizing a set of 64 Walsh functions. Using such a highly efficient modulation scheme, the required SNR for a given quality of transmission is minimized and capacity $C \propto G/Y_b$ is maximized. Every combination of six bits is represented by one out of $2^6 = 64$ different Walsh functions, hence the rate of $28.8 \cdot 64/6 = 307.2$ KHz. This signal is then combined with long code generated with four times higher clock 1.2288 MHz. The rest of the transmitter is shown in the figure. For this signal a frequency band of 1.23 MHz is allocated, and in the same band up to 63 traffic channels and up to 32 access channels can be situated. The forward link channel structure is shown in Figure A5.2. The traffic channel structure is very much similar to the structure of the reverse link except that the Walsh functions are now not used for orthogonal modulation but for channelization purposes. Different applications of the Walsh functions in the reverse and forward links come from the fact that the orthogonality of this functions can be achieved only in a synchronous mode. In addition to the traffic and paging channels, in the forward link two additional channels are used. First, a pilot channel with no modulation is spread only by the short code. After the synchronization of the short code in a mobile unit the level of the signal in this channel is used for open-loop power control. The lower level of this signal is an indication that the attenuation losses are higher, so the mobile unit should increase the transmitted signal level accordingly. Once the short code is synchronized, the information about the phase of the long code is extracted from the synchro-channel so that the long code is synchronized too. Thereafter, the despreading in the traffic channel is possible. Due to different characteristics of the forward channel, the convolutional coder used in this channel has parameters $S = 1/2$ and $K = 9$, resulting in the constant rate at the output of the interleaver of 19.2 Kbps. In addition to the open-loop power control, the base station measures the level of the received signal of each mobile unit and periodically sends the information to the mobile to increase or decrease the level of the signal for a discrete step. The correction step ranges from 0.5 to 1.75 dB depending on the adaptation rate.

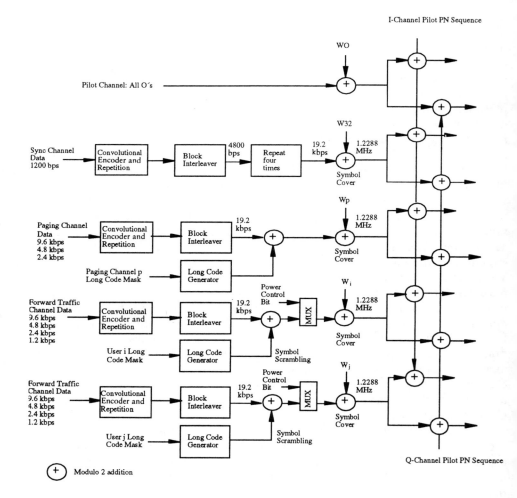

Figure A5.2 Forward CDMA channel structure.

CHAPTER 6

Multiuser CDMA Receivers

6.1 INTRODUCTION

Until now we assumed that detection in direct sequence CDMA systems is performed by matched filter receivers. The operation of the conventional matched filter receiver after down conversion may be viewed as the correlation of the received signal with a specific user signature sequence followed by a hard decision of the correlator output. For a system with orthogonal signatures, in the presence of additive white Gaussian noise, this receiver is optimum, whereby the criterion for optimality is minimization of the average symbol error. In a practical CDMA system user signatures are nonorthogonal, causing *multiple-access interference* (MAI), which is highly structured and essentially non-Gaussian. The matched filter receiver enhances the desired signal while suppressing other interfering signals viewed as Gaussian noise. Since the interfering signals are non-Gaussian, the performance of the conventional matched filter receiver is far from optimum. Furthermore, the performance is particularly sensitive to the unbalanced power of individual signals. For example, in a cellular radio system strong interfering signals coming from mobile users close to the base station might swamp out a weak signal from a distant user. In the previous chapter we showed how to counteract the near-far effect using tight power control. Transmitters that adjust the power at a moderate rate of a few updates per second can only account for power variations due to propagation loss and shadowing.

Multipath power variations will remain and will be the cause of a residual near-far effect. In order to eliminate the effect of multipath on the near-far effect, power control must be performed at a rate that is one hundred times higher than the maximum fade rate.

Even under ideal power control the performance of the conventional matched filter receiver is not optimum and essentially limited by MAI. That is, when a number of users share the common channel, increasing SNR of each transmitted signal will not improve the received SNR.

In a system with the conventional matched filter receiver the immunity against MAI depends on the selection of the signature sequences. A low crosscorrelation between signatures is obtained by designing a set of mutually orthogonal signatures. As explained in previous chapters, using sequences that are orthogonal or almost orthogonal yields better interference rejection, calls for user synchronization, and limits the number of allowable simultaneous users. Most CDMA systems of practical importance are asynchronous. That is, the relative delays of signals coming from different users are arbitrary, resulting in nonzero crosscorrelation between the received signals. To achieve a low level of interference in an asynchronous system the assigned signatures need to have low crosscorrelation for all relative time delays.

Even with orthogonal sequences and synchronous systems, shadowing and multipath propagation distort received signatures, hence reducing their orthogonality.

Another approach in demodulation of a number of CDMA signals sharing a multiple-access channel is to track and demodulate all user waveforms simultaneously. This strategy is referred to as *multiuser demodulation*. The receiver makes a symbol decision based on the observation of the whole received waveforms for all users.

Optimum demodulators are a class of multiuser demodulators with the decision criteria based on minimizing the sequence or bit error probability. The receivers minimizing the sequence error probability can be implemented by a forward or backward dynamic programming algorithm [1], where the forward algorithm is equivalent to the Viterbi detection algorithm [2]. If the objective is to minimize the bit error probability, the optimum demodulator can be implemented as a backward-forward dynamic programming algorithm [3].

The computational complexity of optimum receivers increases exponentially with the number of users. A number of suboptimum multiuser receivers with the computational complexity linear in the number of users have been developed [4–6]. These receivers also simultaneous detect all user signals. However, instead of maximum likelihood detection, they perform a set of linear transformations on the outputs of a matched filter bank.

In other approach, referred to as *adaptive cancellation* [7], the receiver estimates the transmitted sequence for each user by the conventional matched filters. Then the strongest estimated sequence is subtracted from the received waveform, resulting in a signal clean from interference from the strongest interferer, assuming that the previous decision was correct. The cancellation operation is repeated with other interferers in the order of decreasing power.

Adaptive *minimum mean square error* (MMSE) receivers [8,9] take advantage of the cyclostationary structure of interference to improve capacity and error performance. An adaptive MMSE receiver consists of an adaptive fractionally spaced filter. The receiver requires a training sequence for initial estimation of filter parameters. This imposes additional overheads in data transmission but enables easy synchronization unlike other receiver structures. The receiver can be easily implemented as a single-user structure retaining most of the capacity gain as in multiuser structure.

We begin the study of multiuser detectors by describing a model of an asynchronous *direct sequence* (DS) CDMA system, which provides a framework for the development of the theory of multiuser detection. Later in the chapter we consider various receiver structures and their performances.

6.2 THE SYSTEM MODEL FOR MULTIUSER DEMODULATION

Detecting DS/CDMA signals is most difficult in an asynchronous system in which the signals arrive at the receiver site with different delay times, phases, frequency shifts, and amplitudes. Examples of such systems are reverse links in cellular mobile and satellite mobile networks. The transmitters are mobile phones, and the receivers are the base stations. In such systems the cost of implementing a centralized multiuser receiver in a base station is justified as the number of base stations is small. Synchronous CDMA systems, such as forward links in cellular and satellite mobile systems, will be addressed as a special case of the asynchronous system.

The transmitter model is the same as shown in Figures 5.2 and 5.3. There are K users sharing the same CDMA channel. User k is assigned a signature waveform $C_k(t)$, given by (5.4).

Each user transmits digital information at the same data rate $1/T$. We assume that the modulation is bipolar, so the mth symbol of the kth user $b_k(m)$ in Figure 5.3 can have values $+1$ or -1. The bipolar data stream of the kth user, denoted by $\mathbf{b}_k$, is given by

$$\mathbf{b}_k^T = \{b_k(-M), \cdots, b_k(0), \cdots, b_k(M)\}$$

where $(2M + 1)$ is the number of transmitted data symbols and T denotes the transpose. The transmitted signal of the kth user is given by

$$x_k(t) = \sum_{m=-M}^{M} b_k(m) C_k(t - mT - \tau_k) \qquad (6.1)$$

where τ_k is the delay time of the kth user.

For convenience, we further assume that the sequence length N is equal to the spreading gain $G_p = T/T_c$, where T_c is the chip interval. Moreover, we assume without loss of generality that the signature waveforms have unit power

$$\frac{1}{T}\int_0^T C_k^2(t)\,dt = 1 \qquad k = 1, \cdots, K \tag{6.2}$$

The received energy per bit is given by

$$E_{ck}\int_0^T b_k^2(m)C_k^2(t)\,dt = E_{ck} \tag{6.3}$$

where $\sqrt{E_{ck}}$ is the received amplitude of the kth user. The received waveform, denoted by $r(t)$, consists of the sum of K transmitted waveforms embedded in additive white Gaussian noise. It can be represented as

$$r(t) = \sum_{k=1}^{K}\sqrt{E_{ck}}\sum_{m=-M}^{M} b_k(m)C_k(t - mT - \tau_k) + n(t) \tag{6.4}$$

where $n(t)$ is white Gaussian noise with one-sided power spectral density N_o.

We assume that the delays τ_k are smaller than the bit period time T, which implies frame-synchronous operation. In a synchronous system, individual users maintain synchronism between transmitted signals, so the received signal is given by

$$r(t) = \sum_{k=1}^{K}\sqrt{E_{ck}}\sum_{m=-M}^{M} b_k(m)C_k(t - mT) + n(t) \tag{6.5}$$

Since there is not offset between symbols, it is enough to consider the one-shot version of the signal, corresponding to any arbitrary transmitted bit, which for $m = 0$ is given by

$$r(t) = \sum_{k=1}^{K}\sqrt{E_{ck}}\,b_k(0)C_k(t) + n(t) \tag{6.6}$$

The performance of an asynchronous CDMA system depends on the crosscorrelations between every pair of signature waveforms. In asynchronous systems every symbol of a given user overlaps with the two consecutive symbols of each of the interferers, introducing the memory in the channel.

The crosscorrelations of signature waveforms are defined as

$$\rho_{ln}(j) = \frac{1}{T} \int_{\tau_l+mT}^{\tau_l+(m+1)T} C_l(t - \tau_l) C_n(t + jT - t_n) \, dt$$

$$l = 1, 2, \cdots, K, \quad n = 1, 2, \cdots, K \quad j = -M, \cdots, 0, \cdots, M \quad (6.7)$$

For synchronous systems time offset are equal to zero, so all crosscorrelations are equal to zero

$$\rho_{ln}(j) = 0 \quad \text{for } l \neq n \tag{6.8}$$

6.3 OPTIMUM RECEIVER

As (5.13) shows, the output of the matched filter consists of a term that is proportional to the transmitted symbol in that interval and the interference term computed as the sum of the crosscorrelations of the signature for the user of interest and all other interferers. The interference term depends on the sequences of transmitted symbols for all interferers before and after the symbol of interest.

In the conventional matched filter receiver, the decision device, which follows the matched filter, makes one-shot decisions. That is, it estimates the transmitted symbol on the basis of the received signal only in the interval corresponding to that symbol. In this approach, detection is not optimum since the information on interference coming from overlapping symbols from other users is ignored. In optimum detection of asynchronous DS/CDMA signals, observation of the whole received waveform for all users at the output of the matched filter is necessary.

For optimum demodulation we assume that the receiver has the information about the signature waveforms for each user as well as time delays, phase shifts, and amplitudes of each user signal.

Optimum receivers could be designed to select the bit sequence

$$\hat{\mathbf{b}} = \begin{bmatrix} \hat{b}_1(-M) & \cdots & \hat{b}_1(M) \\ \vdots & & \\ \hat{b}_K(-M) & \cdots & \hat{b}_K(M) \end{bmatrix} \tag{6.9}$$

which maximizes the conditional probability

$$P[\hat{\mathbf{b}}|r(t)] \quad t \in (-MT, MT + 2T) \tag{6.10}$$

This decision rule results in the minimum sequence probability.

Assuming that the transmitted bits are independent and equip probable, maximizing the probability in (6.10) is equivalent to maximizing the likelihood

$$P[r(t)|\hat{\mathbf{b}}] = Ce^{-\frac{1}{2\sigma_n^2}\int_0^T \left[y_k(t) - \sum_{k=1}^{K} \hat{b}_k(0)\sqrt{E_{ck}}C_k(t)\right]^2 dt} \quad \text{for } t \in [0, T] \quad (6.11)$$

where C is a constant, σ_n^2 is the noise power, and $y_k(t)$ is the output of the kth matched filter.

The optimization of (6.11) can be achieved by dynamic programming. The Viterbi algorithm is one example of a forward dynamic programming algorithm that selects a sequence $\mathbf{b}$ as to maximize the likelihood in (6.11). The input to the Viterbi algorithm are output samples of the matched filters. The Viterbi algorithm operates on a trellis with the number of states proportional to 2^{K-1}. Therefore, the complexity of the algorithm is an exponential function of the number of users.

Optimum receivers can make decisions by selecting the transmitted sequence to minimize either the sequence error probability or the symbol error probability. In both cases the optimum decision algorithm can be implemented as a dynamic program with complexity per binary decision exponential in the number of users.

We will focus on the first decision type called *maximum likelihood sequence detection* proposed by Verdu [1].

The maximum likelihood detector for an asynchronous DS/CDMA consists of a front-end of a matched filter bank followed by the maximum likelihood Viterbi decision algorithm. The block diagram of the optimum receiver is shown in Figure 6.1.

In order to illustrate the operation of the optimum receiver let us consider the simplest asynchronous DS/CDMA system consisting only of two users and the Gaussian channel model. The front end of the optimum receiver is the bank of two matched filters. Therefore, the outputs of the matched filters can be computed from (5.11). Without the loss of generality we can assume that the first timing offset τ_1 is zero and $\tau_2 \leq T$. The matched filter outputs at the mth sampling instant are given by

$$y_1(m) = \frac{1}{T} \int_{mT}^{(m+1)T} r(t)C_1(t)\,dt \quad (6.12)$$

$$y_2(m) = \frac{1}{T} \int_{\tau_2+mT}^{\tau_2+(m+1)T} r(t)C_2(t-\tau_2)\,dt \quad (6.13)$$

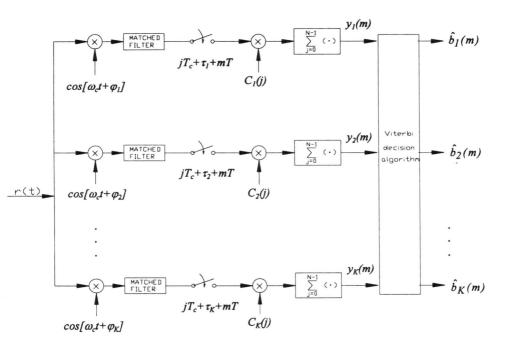

Figure 6.1 Optimum receiver for asynchronous DS/CDMA systems.

where the received signal $r(t)$ is given by

$$r(t) = \sqrt{E_{c1}} \sum_{i=-M}^{M} b_1(i) C_1(t - iT_b) + \sqrt{E_{c2}} \sum_{i=-M}^{M} b_2(i) C_2(t - iT_b - \tau_2) \quad (6.14)$$

After substituting (6.14) into (6.12) and (6.13) and performing integration we get

$$y_1(m) = \sqrt{E_{c1}} b_1(m) + \sqrt{E_{c2}} b_2(m - 1)\rho_{12}(1) \quad (6.15)$$
$$+ \sqrt{E_{c2}} b_2(m)\rho_{12}(0) + \sqrt{E_{c2}} b_2(m + 1)\rho_{12}(-1) + n_1(m)$$

$$y_2(m) = \sqrt{E_{c2}} b_2(m) + \sqrt{E_{c1}} b_1(m - 1)\rho_{21}(1) \quad (6.16)$$
$$+ \sqrt{E_{c1}} b_1(m)\rho_{21}(0) + \sqrt{E_{c1}} b_1(m + 1)\rho_{21}(-1) + n_2(m)$$

where $n_1(m)$ and $n_2(m)$ are the noise components and $\rho_{12}(j)$ and $\rho_{21}(j)$, $j = -1, 0, 1$, are crosscorrelations of signature waveforms.

The crosscorrelations of signature waveforms are defined as

$$\rho_{ln}(j) = \frac{1}{T} \int_{\tau_l+nT}^{\tau_l+(n+1)T} C_l(t - \tau_l) C_n(t + jT - \tau_n) \, dt \qquad (6.17)$$

$$l = 1, 2 \quad n = 1, 2 \quad j = -M, \cdots, 0, \cdots, M$$

As (6.15) and (6.16) show, the current matched filter output $y_k(0)$, $k = 1, 2$, can be represented by the sum of the desired signal given by the first term, the multiple access interference given by the sum of the next three terms and the noise component. The sum of the desired signal and multiple access interference represents the noiseless signal, denoted by $z_k(\mathbf{b}_1, \mathbf{b}_2)$, $k = 1, 2$, given by

$$z_1(\mathbf{b}_1, \mathbf{b}_2) = \sqrt{E_{c1}} b_1(m) + \sqrt{E_{c2}} b_2(m - 1) \rho_{12}(1) \qquad (6.18)$$
$$+ \sqrt{E_{c2}} b_2(m) \rho_{12}(0) + \sqrt{E_{c2}} b_2(m + 1) \rho_{12}(-1)$$

$$z_2(\mathbf{b}_1, \mathbf{b}_2) = \sqrt{E_{c2}} b_2(m) + \sqrt{E_{c1}} b_1(m - 1) \rho_{21}(1) \qquad (6.19)$$
$$+ \sqrt{E_{c1}} b_1(m) \rho_{21}(0) + \sqrt{E_{c1}} b_1(m + 1) \rho_{21}(-1)$$

where $\mathbf{b}_1 = \{b_1(m - 1), b_1(m), b_1(m + 1)\}$ and $\mathbf{b}_2 = \{b_2(m - 1), b_2(m), b_2(m + 1)\}$. The filter outputs then are

$$y_1(m) = z_1(\mathbf{b}_1, \mathbf{b}_2) + n_1(m)$$
$$y_2(m) = z_2(\mathbf{b}_1, \mathbf{b}_2) + n_2(m) \qquad (6.20)$$

As (6.18) and (6.19) show, the current noiseless filter outputs depend not only on the current system input $\{b_1(m), b_2(m)\}$ but also on the previous and subsequent pairs of inputs $\{b_1(m - 1), b_2(m - 1)\}$ and $\{b_1(m + 1), b_2(m + 1)\}$. The system can be described by a tapped delay line model shown in Figure 6.2. If we assume that $\tau_2 \leq T$, we get for the crosscorrelations

$$\rho_{12}(j) = \frac{1}{T} \int_0^T C_1(t) C_2(t + jT - \tau_2) \, dt = 0 \quad \text{for } j \geq 2 \qquad (6.21)$$

and

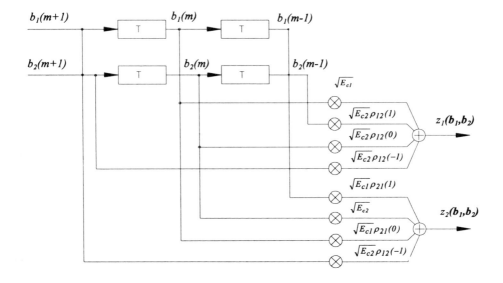

Figure 6.2 A tapped delay line model for a synchronous DS/CDMA system with two users.

$$\rho_{21}(j) = \frac{1}{T} \int_{\tau_2}^{T+\tau_2} C_2(t - \tau_2) C_1(t + jT) \, dt = 0 \quad \text{for } j \geq 2 \tag{6.22}$$

Note that we assumed

$$\rho_{kk}(0) = \frac{1}{T} \int_{\tau_k}^{\tau_k+T} C_k^2(t - \tau_k) \, dt = 1 \tag{6.23}$$

$$\rho_{kk}(j) = \frac{1}{T} \int_{\tau_k}^{\tau_k+T} C_k(t - \tau_k) C_k(t + jT - \tau_k) = 0 \tag{6.24}$$

$$j = \pm 1, \pm 2, \cdots, \pm M \quad k = \pm 1, \pm 2, \cdots, \pm M$$

The tapped delay line in Figure 6.2 can be considered as a two-dimensional finite state machine, and its operation is described by a state diagram. The matched filter outputs can be considered as the finite state machine outputs corrupted by noise.

A state of the state diagram is defined as the content of the tapped delay line

$$S_m = \{b_1(m), b_1(m-1), b_2(m), b_2(m-1)\} \quad (6.25)$$

Since a state contains four symbols and each of them can take values ±1, the total number of states is 16.

The trellis diagram is obtained by expanding the state diagram in time. The trellis diagram for the tapped delay line from Figure 6.2 is shown in Figure 6.3. The states are represented by circles and state transitions by directed lines or branches. Each branch is assigned a pair of input symbols $\{b_1(m), b_2(m)\}$ and a corresponding pair of the tapped delay line outputs $\{z_1(m), z_2(m)\}$. They are not shown on the graph to avoid clutter.

The Viterbi algorithm searches the trellis diagram to find the sequence of the finite machine output symbols at the minimum Euclidean distance from the sequence of the matched filter outputs. The sequence of the finite machine outputs is then mapped, on the basis of the trellis diagram, into an estimated sequence of transmitted symbols

$$\hat{\mathbf{b}} = \{\hat{b}_1(m), \hat{b}_2(m), \cdots, \hat{b}_K(m)\} \quad m = -M, \cdots, 0, \cdots, M$$

The algorithm makes decisions over a finite window of sampling instants rather than waiting for all the data to be received.

The algorithm can be easily generalized for any number of users K. The number of operations in the Viterbi algorithm per binary decision is proportional to the number of states, and the number of states is exponential in the number of users $(\sim 2^K)$.

It can be shown that the multiuser algorithm that minimizes the probability of error has the same structure, but it uses a backward-forward dynamic programming algorithm instead [10].

As a special case, consider a two-user synchronous DS/CDMA system. The received waveform for a synchronous system is given by (6.5) The matched filter outputs are

$$y_1(m) = \sqrt{E_{c1}} b_1(m) + \sqrt{E_{c2}} b_2(m) \rho_{12}(m) + n_1(m)$$
$$y_1(m) = z_1\{b_1(m), b_2(m)\} + n_1(m) \quad (6.26)$$
$$y_2(m) = \sqrt{E_{c2}} b_2(m) + \sqrt{E_{c1}} b_1(m) \rho_{21}(m) + n_2(m)$$
$$y_2(m) = z_2\{b_1(m), b_2(m)\} + n_2(m) \quad (6.27)$$

since $\rho_{12}(-1) = \rho_{12}(1) = \rho_{21}(-1) = \rho_{21}(1) = 0$ in a synchronous system.

Multiuser CDMA Receivers 275

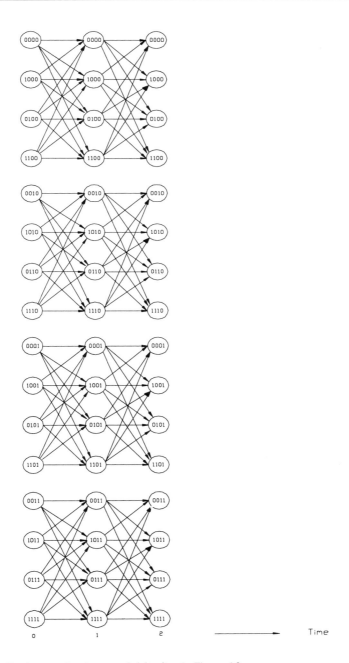

Figure 6.3 Trellis diagram for the tapped delay line in Figure 6.2.

As (6.26) and (6.27) show, the filter outputs depend only on the current system inputs $b_1(m)$ and $b_2(m)$. The estimates of the transmitted symbols $b_1(m)$ and $b_2(m)$ thus can be obtained by a minimum distance one-shot decision device. The decision device computes the Euclidean distance from the matched filter output $\{y_1(m), y_2(m)\}$ and all possible $\{z_1\{b_1(m), b_2(m)\}, z_2, \{b_1(m), b_2(m)\}\}$ values computed for all possible inputs $b_1(m)$ and $b_2(m)$. It then selects a pair $z_1\{\hat{b}_1(m), \hat{b}_2(m)\}, z_2\{\hat{b}_1(m), \hat{b}_2(m)\}$ that has the minimum Euclidean distance from $\{y_1(m), y_2(m)\}$. The values $\hat{b}_1(m)$ and $\hat{b}_2(m)$ are the estimates of the transmitted symbols. In a synchronous system with K users the decision operation consists of computing and comparing 2^K Euclidean distances. The optimum receiver for a synchronous DS/CDMA system is shown in Figure 6.4.

The bit error probabilities of user 1 in a two equal energy user system for the conventional and optimum receiver are shown in Figure 6.5. For SNR higher than 6 dB the upper bounds on the worst and best cases of the optimum receiver are indistinguishable from each other and from the single user lower bound.

6.4 SUBOPTIMUM DS/CDMA RECEIVERS

The performance gains obtained by optimum receivers are achieved by centralized implementation, which involves a high degree of complexity. The optimum receiver

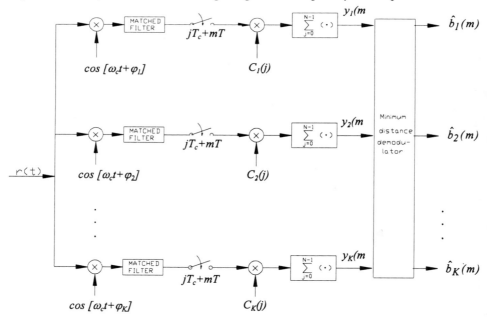

Figure 6.4 Optimum receiver for a synchronous DS/CDMA system.

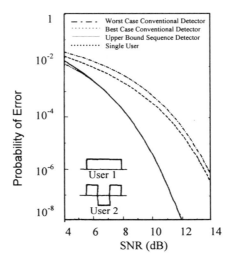

Figure 6.5 Best and worst cases of error probability of user 1 achieved by conventional and optimum receivers. (© 1986 IEEE. *From:* [1].)

complexity is exponential in the number of users. For example, in a system with 50 users, the number of computations per received symbol is proportional to 2^{50}, which is an extremely high number. For practical systems, implementation complexity needs to be reduced to a reasonable level even if the performance is somewhat degraded from the optimum one. Following this line of thought, a number of suboptimum receivers linear in the number of users have been proposed.

6.4.1 Decorrelating Receiver

The decorrelating receiver [11,12] is a multiuser detector that linearly transforms each vector of matched filter outputs. It is near-far resistant, has suboptimum performance, and its complexity increases linearly with the number of users. For the sake of conceptual clarity let us first consider a synchronous DS/CDMA system.

The matched filter outputs for a two-user synchronous system are given by (6.26) and (6.27). These equations can be easily generalized for a K-user system. They can be written in a matrix form as

$$y = RAb + n \qquad (6.28)$$

where R is the crosscorrelation matrix whose ρ_{ij} coefficient is given by (6.17) and

$$A = \text{diag}\{\sqrt{E_{c1}}, \cdots, \sqrt{E_{cK}}\}$$

is a diagonal matrix of the signal amplitudes,

$$\mathbf{b} = \begin{bmatrix} b_1(-M) & \cdots & b_1(0) & \cdots & b(M) \\ \vdots & & & & \\ b_K(-M) & \cdots & b_K(0) & \cdots & b_K(M) \end{bmatrix}$$

and $\mathbf{n}$ is a zero mean Gaussian K vector.

Multiplying the filter output vector $\mathbf{y}$ with the inverse of $\mathbf{R}$, $\mathbf{y} = \{y_1, \ldots, y_i, \ldots, y_K\}$ denoted by $\mathbf{R}^{-1}$, from the left-hand side, we get

$$\mathbf{R}^{-1}\mathbf{y} = \mathbf{A}\mathbf{b} + \mathbf{R}^{-1}\mathbf{n} \qquad (6.29)$$

In the absence of noise the first term on the right-hand side will give a scaled transmitted sequence. The decorrelating receiver is based on the prior linear transformation. It consists of the matched filter bank as the front-end processor producing the vector $\mathbf{y}$. It is followed by a decorrelator computing the crosscorrelation matrix inverse $\mathbf{R}^{-1}$ and multiplying it by $\mathbf{y}$. The estimates of the transmitted sequences are obtained by taking the sign of $\mathbf{R}^{-1}\mathbf{y}$. The decorrelating receiver is shown in Figure 6.6. Note that in a synchronous DS/CDMA system with an AWGN channel the entries of the crosscorrelation matrix ρ_{ij} remain constant in time.

The most significant feature of this receiver is its relatively low complexity compared to the optimum receiver. Another important attribute is that the receiver does not require knowledge of the received amplitudes, making it near-far resistant. However, the receiver needs the information on the signature waveforms, carrier phases, and timing of all the users.

In an asynchronous DS/CDMA system, the crosscorrelation coefficients ρ_{ij} depend on time. Let us again consider a two-user system example. Assuming that the maximum timing offset is equal to the symbol period, the only nonzero crosscorrelation coefficients are $\rho_{ij}(-1)$, $\rho_{ij}(0)$, and $\rho_{ij}(1)$.

The matched filter outputs for a two-user asynchronous CDMA system are given by (6.15) and (6.16). In vector form, these equations can be written as

$$\mathbf{y}(m) = \mathbf{R}(-1)\mathbf{A}(m+1)\mathbf{b}(m+1) + \mathbf{R}(0)\mathbf{A}(m)\mathbf{b}(m) + \mathbf{R}(1)\mathbf{A}(m-1)\mathbf{b}(m-1) + \mathbf{n}(m) \qquad (6.30)$$

where

$$\mathbf{y}(m) = \{y_1(m), \cdots, y_K(m)\}^T$$
$$\mathbf{b}(m) = \{b_1(m), \cdots, b_K(m)\}^T$$

and $\mathbf{n}(m)$ is a Gaussian K-vector with a zero mean.

Multiuser CDMA Receivers 279

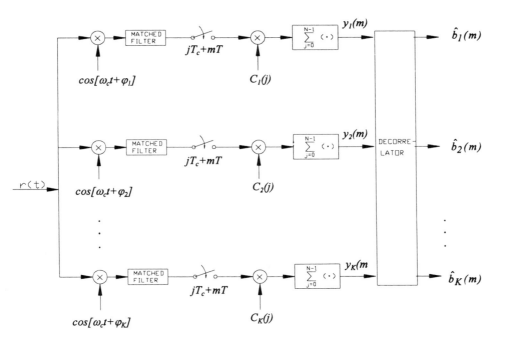

Figure 6.6 Decorrelating receiver for a synchronous DS/CDMA system.

We assume that $\mathbf{b}(-M - 1) = \mathbf{b}(M + 1) = 0$. The entries of the matrix $\mathbf{R}(i)$, $i = -1, 0, 1$, are obtained from (6.17). It is possible to make this notation more compact by introducing a $(2M + 1)K \times (2M + 1)K$ block-Toeplitz matrix $\mathbf{R}'$ and a $(2M + 1)K \times (2M + 1)K$ diagonal matrix $\mathbf{A}'$ as

$$\mathbf{R}' = \begin{bmatrix} \mathbf{R}(0) & \mathbf{R}(-1) & 0 & \cdot & \cdot & 0 \\ \mathbf{R}(1) & \mathbf{R}(0) & \mathbf{R}(-1) & \cdot & \cdot & \cdot \\ 0 & \mathbf{R}(1) & \mathbf{R}(0) & \cdot & \cdot & \cdot \\ \cdot & \cdot & \cdot & & & \cdot \\ \cdot & \cdot & \cdot & & & \mathbf{R}(-1) \\ 0 & \cdot & \cdot & & \mathbf{R}(1) & \mathbf{R}(0) \end{bmatrix}$$

$$\mathbf{A}' = \text{diag}[\sqrt{E_{c1}}(-M), \cdots, \sqrt{E_{cK}}(-M), \cdots, \sqrt{E_{c1}}(0),$$
$$\cdots, \sqrt{E_{cK}}(0), \cdots, \sqrt{E_{c1}}(M), \cdots, \sqrt{E_{cK}}(M)]$$

With this notation the matched filter output vector $\mathbf{y}$ can be represented as

$$y = R'A'b + n \qquad (6.31)$$

The matched filter output vector has the same form as the corresponding vector for a synchronous system given by (6.19). Then the estimate of the transmitted matrix **b** can be obtained by a similar equation as for a decorrelating receiver for a synchronous system by multiplying the matched filter output matrix with the matrix R^{-1}

$$R'^{-1}y = A'b + R'^{-1}y \qquad (6.32)$$

Figure 6.7 shows the average probability of error for a decorrelating receiver versus the SNR compared to the probability of error for the conventional matched filter receiver for a system with maximum length sequences [13] of length 31 with three users whose delay relative to each other is fixed. The error probability of user 1 is computed for different values of the energy ratio, denoted by E_j/E_1 averaged over the bit sequences of the two interferers and over the delay of user 1. Also shown is the error probability of the single-user channel. As Figure 6.7 shows, the decorrelating receiver is insensitive to variations of the energy of interfering users. On the other hand, the performance of the conventional receiver deteriorates considerably with increasing interference. At the energy ratio of 5 dB the conventional receiver becomes

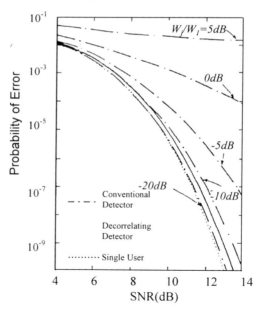

Figure 6.7 Error probability of user 1 with two active equal energy interferers, each of energy E_j. (© 1990 IEEE. *From:* [12].)

multiple-access limited. The decorrelating receiver has superior performance to the conventional receiver for all significant interference levels. Only when interference drops much beyond the noise level, the conventional receiver outperforms the decorrelating detector since the conventional receiver is optimum under Gaussian noise.

6.4.2 Interference Canceler

The operation of the interference canceler is based on successive cancelations of interference from the received waveform. It is important to cancel the strongest detected signal before detection of the other signals because it has the most severe effect in producing interference and it is easier to achieve acquisition and demodulation of strong signals. Thus, the operation of the interference canceler consists of

1. Ranking the user signals so that

$$\sqrt{E_{c1}} > \sqrt{E_{c2}} > \sqrt{E_{c3}} > \cdots > \sqrt{E_{cK}}$$

2. Detection of the strongest user by the conventional receiver;
3. Regeneration of the strongest user spread spectrum signal $\hat{x}_k(t)$ using its chip sequence and the estimate of its amplitude, $\hat{x}_k(t) = \sqrt{E_{ck}}\hat{b}_k(t)C_k(t)$;
4. Canceling the strongest user interferer;
5. Repeating until all users signals are detected.

The block diagram of an interference canceler is shown in Figure 6.8.

Since all user signals must be detected independently by the conventional receiver, the signatures, timing, and phases of all users must be known. In order to rank the user signals, very accurate amplitude estimation is required. The complexity of the receiver is linear in the number of users.

The receiver can be simplified by canceling only a number of strongest signals instead of canceling all interferers. There is a number of various implementations of interference cancelers. In the multistage detector proposed by Varanasi and Aazhang the front end is either a conventional matched filter receiver [5] or a decorrelating detector [14]. The subsequent stage subtracts the MAI estimated in the previous stage, modulated by the corresponding signatures, from the received waveform. The block diagram of the multistage receiver with the matched filter in the first stage is shown in Figure 6.9.

Note that in this receiver all interferers are canceled from every user signal unlike the receiver in Figure 6.8, where only strong signals are canceled from each signal being detected. A multistage receiver for two users is shown in Figure 6.10.

The conventional receiver in Figure 6.10 contains the matched filter and the multiplier, which multiplies the received signal with the code sequence. The procedure shown in the first stage of Figure 6.10 can be repeated as many times as desired.

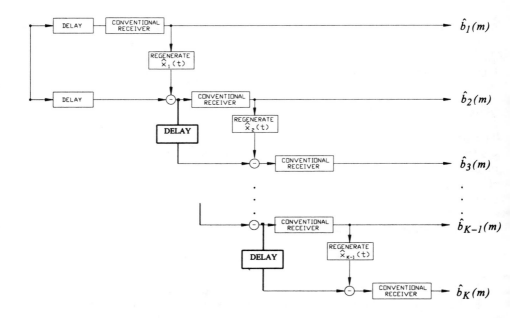

Figure 6.8 Interference canceler.

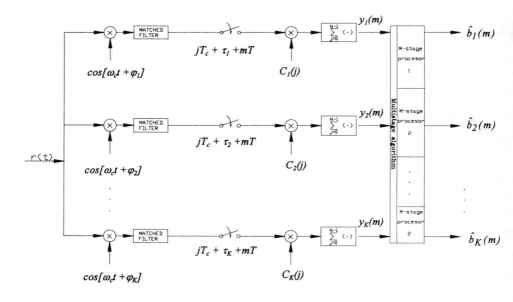

Figure 6.9 Multistage receiver.

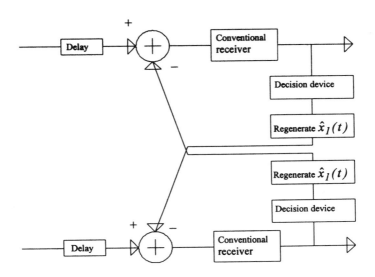

Figure 6.10 Multistage receiver for two users.

The interference canceler proposed by Kohno in [7] also consists of the conventional matched filter receiver as the first stage and a decision feedback equalizer as the second stage. The decorrelating decision-feedback receiver of Duel-Hallen [6] has a decorrelating receiver as the first stage and decision feedback filters to eliminate multiuser interference. Decisions on user signal values are made in the order of decreasing received amplitudes. Thus, the receiver for the strongest user signal does not involve feedback and is equivalent to the decorrelator. On the other hand, the receiver for the nearest signal combines decisions on all other user signals. If there are many users with almost equal powers, the performance of the interference canceler deteriorates since no user signal can be accurately detected.

In cellular radio systems the performance improvement relative to the conventional receiver is limited due to interference coming from other cells. These interferers cannot be controlled by the base station and hence cannot be canceled.

6.4.3 Adaptive MMSE Receivers

In detecting DS/CDMA signals by the suboptimum demodulators considered until now, the essential assumption was that signatures, timing, and carrier phases are known; while in the case of the interference canceller very accurate estimates of signal amplitudes are further required. In addition, the receiver structure is centralized, demodulating jointly all user signals. In asynchronous CDMA systems affected by multipath propagation and shadowing, it is difficult to accurately estimate timing and carrier phase of received user signals. Furthermore, a centralized receiver structure is not desirable in terminals for detection of one user signal, such as in mobile phones.

Adaptive *minimum mean square error* (MMSE) receivers do not require signature, timing, and carrier phase information about the interferers, but only the timing and carrier phase of the desired user. They can be implemented as both single- and multiuser receivers, with a high degree of flexibility in implementation complexity, depending on required application and performance.

They are simple to implement even in asynchronous and multipath channels. The implementation complexity is independent of the number of users for single-user implementation and linear in the number of users for centralized multiuser receivers. The block diagram of a single-user DS/CDMA adaptive MMSE receiver for simultaneous detection of all user signals is shown in Figure 6.11.

The adaptive single-user receiver consists of an adaptive fractionally spaced digital filter followed by a decision device. A training sequence is used to estimate the initial filter parameters. In the training mode, the transmitter generates a data symbol sequence known to the receiver. This sequence is used to establish synchronization and to estimate filter parameters. In the transmission mode, the adaptation of the receiver is decision directed. This means that the receiver decisions are used instead of the training signal to adjust the filter coefficients. Of course, the decision device makes occasional errors, but due to the long averaging time of the filter coefficient adjustment algorithm, these errors have no significant effect.

The timing recovery of the receiver is greatly simplified due to the fractional filter structure, which makes the detected desired signal insensitive to time differences in the signal arrival times of various users and also simplifies the timing recovery of the useful signal itself.

After being shifted to the baseband by a down-converter and filtered by an antialiazing filter with the bandwidth B_c, the signal is sampled every T_f seconds, where $T_f = 1/2B_c$. In practical realizations it is convenient to adopt $T_f = T_c/2$.

The adaptive filter consists of a tapped delay line with $2L + 1$ taps. The tap coefficients, denoted by a_l, where $-L \leq l \leq L$, are determined by minimizing the *mean square error* (MSE) between the filter output and a known sequence. This sequence is agreed upon between the transmitter and receiver in the training mode. During transmission, the sequence is obtained from the output of the decision device.

The sampled filter input is

$$r(mT_f) = \sum_{i=-M}^{M} \sum_{k=1}^{K} \sqrt{E_{ck}} b_k C_k(mT_f - iT - \tau_k) + n(mT_f) \qquad (6.33)$$

The filter output, which is only calculated at the instants T sec apart, is given by

$$y(mT) = \sum_{l=-L}^{L} a_l r(mT - lT_f) \qquad (6.34)$$

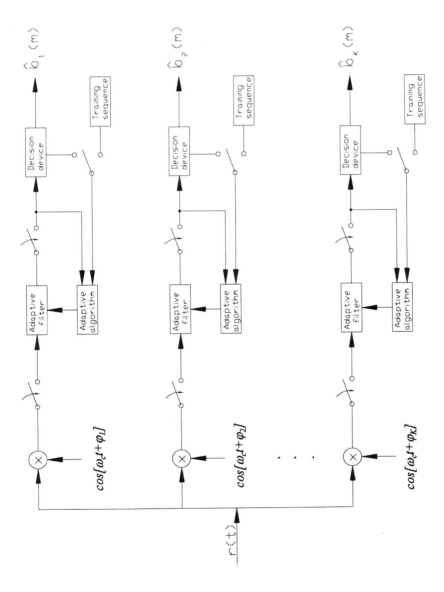

Figure 6.11 Adaptive single-user DS/CDMA receiver for simultaneous detection of all user signals.

Without the loss of generality we assume that the user of interest is the user number 1, that is, $k = 1$, and the subscript for the user 1 is omitted. For the finite-length filter the MSE is given by

$$\epsilon = E(e_m^2) = E(|y(mT) - b(m)|^2) \tag{6.35}$$

or after substitution

$$\epsilon = E(|\mathbf{a}^T \mathbf{r} - b(m)|^2) \tag{6.36}$$

where the tap vector and the delay-line sample vector are given by

$$\mathbf{a} = (a_{-L}, \cdots, a_L)^T$$
$$\mathbf{r} = [r(mT + LT_f), \cdots, r(mT), \cdots, r(mT - LT_f)]^T$$

After the indicated expectation is performed, the MSE is

$$\epsilon = \mathbf{a}^* \mathbf{D} \mathbf{a} - (\mathbf{a}^* \mathbf{f} + \mathbf{f}^* \mathbf{a}) - \sigma_b^2 \tag{6.37}$$

where the asterisk denotes the transpose of the complex conjugate matrix. $\mathbf{D}$ is the $(2L + 1) \times (2L + 1)$ Hermitian multiple-access channel correlation matrix given by $\mathbf{D} = E(\mathbf{r} \mathbf{r}^*)$. The $(2L + 1) \times 1$ single-user channel vector $\mathbf{f}$ is given by $E[b^*(m)\mathbf{r}]$. The data power σ_b^2 is defined by $\sigma_b^2 = E[|b(m)|^2]$.

A direct calculation of the lth element, f_l, of the single-user channel vector $\mathbf{f}$ gives

$$f_l = E[b(m)r(mT + lT_s)] = E[|b(m)|^2]C(lT_s)$$

The optimum tap setting is obtained as

$$\mathbf{a}_{\text{opt}} = 1 - \mathbf{f}^* \mathbf{D}^{-1} \mathbf{f} \tag{6.38}$$

However, in practice this cannot be computed due to nonavailability of $\mathbf{f}$ and $\mathbf{D}$. One of the advantages of the linear MMSE receiver is the ease with which it lends itself to adaptive implementation with training sequences.

6.4.4 The Adaptive Algorithm

The simplest adaptive control algorithm makes use of the gradient of the squared error with respect to the tap weights. Taking these derivatives and writing the result in complex notation produce the adjustment algorithm

$$a_{n+1} = a_n - \alpha \epsilon_n r_n \quad n = 0, 1, 2, \cdots \quad (6.39)$$

where a_n is the complex tap vector at the nth iteration; ϵ_n is a single error sample; and α is a small positive number, usually called the step size, which affects the algorithm's convergence rate and fluctuations about the minimum attainable MSE. The algorithm is updated once every symbol interval. The adaptive algorithms are discussed in more detail in Chapter 7.

The performance of single-user adaptive algorithms can be improved by centralized structures. A multiuser adaptive receiver is shown in Figure 6.12.

The receiver is near-far resistant and does not require strict power control. The single-user receiver can increase the number of users to 70% to 80% of the spreading gain. It is also able to remove the effect of multipath propagation providing that the multipath parameter variations are slower than the adaptive algorithm convergence speed.

The simplest adaptive algorithm is the *least mean square* (LMS) algorithm [15]. The algorithm is relatively slow and cannot be used on fast-varying fading channels. The convergence speed can be significantly improved by *recursive least square* (RLS) algorithms or in lattice structures [15]. The need for training sequences could be sidestepped by blind algorithm [16,17].

Figure 6.13 shows the BER performance for a single-user adaptive receiver compared to the optimum receiver and the conventional matched filter receiver for 16 QAM, modulation in a CDMA system with the spreading gain of 31, 10 users, and the signal-to-interference ratio of 0 dB. As the figure indicates, the performance of the adaptive receiver is only slightly degraded relative to the optimum receiver, while the MFR has the bit error rate of approximately 0.5.

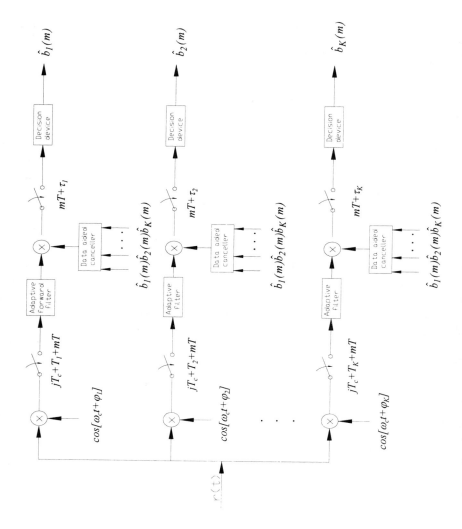

Figure 6.12 Multiuser adaptive receiver for asynchronous DS/CDMA systems.

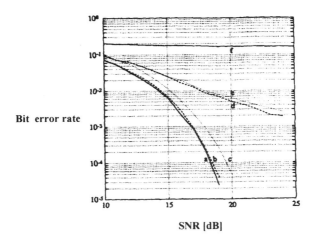

Figure 6.13 BER of a single-user adaptive receiver. (© 1994 IEEE. *From:* [8].)

References

[1] Verdu, S., "Minimum Probability of Error for Asynchronous Gaussian Multiple-Access Channels," *IEEE Trans. on Information Theory*, Vol. IT-32, No. 1, Jan. 1986, pp. 85–96.
[2] Forney, G. D., "The Vitervi Algorithm," *Proc. IEEE*, March 1973, pp. 268–278.
[3] Verdu, S., and H. V. Poor, "Abstract Dynamic Programming Models under Commutativity Conditions," *SIAM J. Control Optimization*, Vol. 24, July 1987, pp. 990–1006.
[4] Lupas, R., and S. Verdu, "Linear Multiuser Detectors for Synchronous Code—Division Multiple Access Channels," *IEEE Trans. on Information Theory*, Vol. 35, No. 1, Jan. 1989, pp. 123–136.
[5] Varanasi, M. K., and B. Aazhang, "Multistage Detection in Asynchronous Code Division Multiple Access Communications," *IEEE Trans. on Communications*, Vol. 38, April 1990, pp. 509–519.
[6] Duel-Hallen, A., "Decorrelating Decision-Feedback Multiuser Detector for Synchronous Code-Division Multiple Access Channel," *IEEE Trans. on Communications*, Vol. COM-41, Feb. 1993, pp. 285–290.
[7] Kohno, R., H. Imai, and M. Hatori, "Cancellation Techniques of Co-Channel Interference in Asynchronous Spread Spectrum Multiple Access Systems," *Trans. IECE (Electronics and Communications in Japan)*, Vol. 66, May 1983, pp. 416–423.
[8] Rapajic, P., and B. Vucetic, "Adaptive Receiver Structures for Asynchronous CDMA Systems," *IEEE J. Selected Areas in Communications*, Vol. 12, No. 4, May 1994, pp. 685–697.
[9] Abdulrahman, M., D. D. Falconer, and A. U. H. Sheinh, "Equalisation for Interference Cancellation in Spread Spectrum Multiple Access Systems," *Proc. IEEE on Vehicular Technology Conf.*, Denver, CO, May 1992.
[10] Verdu, S., and H. V. Poor, "Backward, Forward and Backward-Forward Dynamic Programming Models under Commutativity Conditions," *Proc. 23rd IEEE Cont. Decision Contr.*, Dec. 1984, pp. 1081–1086.
[11] Lupas, R., and S. Verdu, "Linear Multiuser Detectors for Synchronous Code—Division Multiple Access Channels," *IEEE Trans. on Information Theory*, Vol. 35, No. 1, Jan. 1989, pp. 123–136.
[12] Lupas, R., and S. Verdu, "Near-Far Resistance of Multiuser Detectors in Asynchronous Channels," *IEEE Trans. on Communications*, Vol. 38, No. 4, April 1990, pp. 496–508.

[13] Garber, F. D., and M. B. Pursley, "Optimal Phases of Maximal-Length Sequences for Asynchronous Spread-Spectrum Multiplexing," *Electron. Lett.*, Vol. 16, Sept. 1980, pp. 756–757.
[14] Varanasi, M. K., and B. Aazhang, "Near-Optimum Detection in Synchronous Code Division Multiple Access Systems," *IEEE Trans. on Communications*, Vol. 39, May 1990, pp. 725–736.
[15] Haykin, S., *Adaptive Filter Theory*, Englewood Cliffs, NJ: Prentice-Hall, 1986.
[16] Sato, Y., "A Method of Self-Recovering Equalisation for Multilevel Amplitude Modulation Systems," *IEEE Trans. on Communications*, Vol. COM-23, No. 6, 1975, pp. 679–682.
[17] Honig, M., V. Madhow, and S. Verdu, "Blind Adaptive Multiuser Detection," *IEEE Trans. on Information Theory*, Vol. 41, No. 4, July 1995, pp. 944–960.

Selected Bibliography

Madhow, U., and M. Honig, "MMSE Interference Suppression for Direct Sequence Spread Spectrum CDMA," *IEEE Trans. on Communications*, Vol. 42, Dec. 1994, pp. 3178–3188.

Miller, S. L., "An Adaptive Direct Sequence Code Division Multiple Access Receiver for Multiuser Interference Rejection," *IEEE Trans. on Communications*, Vol. 43, Nos. 2/3/4, Feb/Mar/Apr 1995, pp. 1746–1755.

Proakis, J., *Digital Communications*, New York: McGraw Hill, 1989.

Pursley, M. B., "Performance Evaluation for Phase-Coded Spread-Spectrum Multiple-Access Communication—Part I: System Analysis," *IEEE Trans. on Communications*, Vol. COM-25, No. 8, Aug. 1977, pp. 795–799.

CHAPTER 7

ADAPTIVE INTERFERENCE SUPPRESSION AND CDMA OVERLAY SYSTEMS

In the previous chapter we showed how to use signal processing to deal with MAI in demodulating the signal in a spread spectrum network. The MAI caused by other spread spectrum signals represents a *wideband* type of interference.

It has been shown that DS spread spectrum systems can share common spectrum with the currently operating cellular or fixed microwave systems [1–3] in order to achieve efficient bandwidth utilization. The signals of coexisting users would appear as narrowband interference in the spectrum of a wideband signal. As noted in Chapter 1, spread spectrum modulation has an inherent ability to reject narrowband interference. The despreading operation in the receiver spreads the narrowband interference across the frequency band occupied by the spreading sequence and thus significantly reduces the level of interference. That is, the interference is transformed into a lower level noise with a relatively flat spectrum. At the same time, the spectrum of the desired signal is converted back into one occupied by the information signal prior to spreading.

As shown in Chapter 1, the effectiveness of narrowband interference rejection in spread spectrum systems depends on the processing gain. When the processing gain is limited due to bandwidth constraints, interference suppression can be substantially enhanced by filtering the received signal prior to despreading. This operation is motivated by the great disparity in the bandwidths of the desired and narrowband interfering signals. Hence, the interference can be successfully suppressed by notch filters while only slightly reducing the desired signal power. This filtering operation can effectively complement spread spectrum modulation in eliminating narrowband interference.

The techniques for suppressing narrowband interference are classified into three main categories. The algorithms in the first category are based on estimation processing [4–6]. The interference is modeled as white Gaussian noise passed through an all-pole filter. Linear prediction is used to estimate the coefficients of an all-pole model. The estimated coefficients specify an all-zero transversal filter. The received signal is passed through this filter, which suppresses the narrowband interference prior to detection.

The algorithms in the second category are based on transform domain filtering [7–9]. The receiver computes the real-time fast Fourier transform of the received signal. On the basis of the signal spectrum, a transversal notch filter is designed for suppressing interference. The transform is multiplied by the filter transfer function and the inverse transform of the product is computed prior to detection by a matched filter.

The algorithms in the third category are based on adaptive CDMA receivers. An adaptive CDMA linear receiver [10,11] consists of a *least mean square error* (LMSE) transversal digital filter and a decision device. The receiver estimates the optimum initial filter coefficients in a training mode and adjusts them continuously in the transmission mode using the detected data symbols as reference. The receiver thus does not require a separate despreading device in addition to a filter for suppression of narrowband interference.

In all these techniques, interference suppression is conceptually accomplished in two steps. In the first step the interfering signals are estimated and a set of coefficients are computed for use in the transversal filter. In the second step the actual filtering of the received signal is carried out. For interference that is stationary over a specified interval, the filter coefficients can be used for the entire interval. As the interference varies with time, the filter coefficients must be adaptively tracked.

All these filters can be implemented by SAW technology, CDD, or most frequently by *digital signal processing* (DSP) technology.

In this chapter signal processing techniques for narrowband interference suppression over and above the performance improvement that comes automatically by employing spread spectrum are considered. In particular, estimation and transform domain filters are presented and a brief overview of other techniques is given. We also examine the effectiveness of the interference suppression algorithms by estimating the SNR improvements and probability of error. An example of network implementation based on CDMA overlay of a narrowband cellular system is presented. The example can model a local wireless loop or a microcell within a macrocell.

7.1 ESTIMATION FILTERS

7.1.1 DS Spread Spectrum Receiver and Estimation Filter Structures

As we noted, estimation filters can be used in conventional DS spread spectrum systems to enhance their inherent ability to suppress narrowband interference [4-6,12-14]. Figure 7.1 shows how the estimation filter fits into the structure of a DS spread spectrum receiver.

The transmitted signal in a DS spread spectrum system is obtained by modulating a random information signal with a PN sequence. As a result this signal tends to have a flat spectrum that is much wider than the spectrum of the original information signal.

Assuming that the channel introduces only Gaussian noise $n(t)$ and narrowband interference $I(t)$, the received signal $r(t)$ consists of the sum of the transmitted signal $s(t)$, narrowband interference, and thermal noise. The thermal noise is also assumed to have a flat spectrum. The interference is generally assumed to be nonwhite with the bandwidth much smaller than the bandwidth of the transmitted signal.

The filtering operation is performed at baseband, and perfect carrier and chip timing synchronization is assumed.

The estimation filters exploit a large disparity in the transmitted spread spectrum and narrowband interference bandwidths. Since the spectrum of the spread spectrum signal is wide and flat, its samples are uncorrelated and cannot be predicted by the narrowband estimation filter.

On the other hand, a strong narrowband interference will have highly correlated samples that can be easily predicted from their past values by linear prediction algorithms. Hence, a prediction of the *received signal* based on its previous samples will, in effect, be an estimate of the narrowband interference signal. Once this estimate is obtained, it is subtracted from the received signal. Thus, the effect of narrowband interference on the received signal is canceled if its estimate was correct. The received signal now consists of the sum of the transmitted signal, thermal noise, and residual interference, which causes a small degradation of the desired signal.

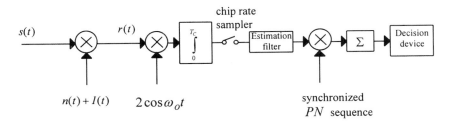

Figure 7.1 Receiver block diagram.

Two-sided and one-sided filter structures, shown in Figures 7.2(a,b), respectively, can be used for interference suppression.

The one-sided filter is also called the prediction error filter or Wiener filter. It uses past samples of the input signal only to produce an estimate of the current sample of the input signal [5].

The effectiveness of the interference suppression can be measured by the SNR or error probability improvements.

7.1.2 Narrowband Interference Suppression by Estimation Filters

The SNR improvement is defined as the ratio of the SNR in a system with a suppression filter to the SNR of the system operating without a suppression filter [4–6].

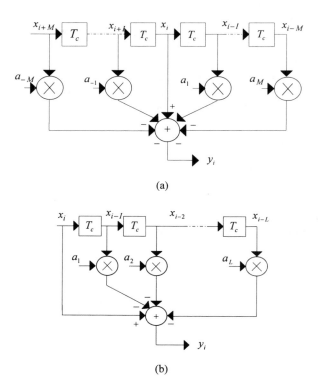

Figure 7.2 Transversal filters: (a) two-sided transversal filters (2SF) and (b) one-sided transversal filter (1SF) (linear prediction filter).

In order to illustrate the SNR improvement in a DS spread spectrum system consider the receiver shown in Figure 7.1 with a two-sided suppression filter as in Figure 7.2(a). This filter uses both past and future samples to estimate interference. To simplify the analysis we will assume that narrowband interference is modeled by a single tone signal.

The received signal, $r(t)$, consists of a sum of a binary PSK direct sequence spread spectrum signal, $s(t)$, a single tone interferer, $I(t)$, and thermal noise, $n(t)$

$$r(t) = s(t) + I(t) + n(t) \tag{7.1}$$

where

$$s(t) = Ac(t)b(t)\cos \omega_o t \tag{7.2}$$

$$I(t) = a \cos[(\omega_o + \Omega)t + \Theta] \tag{7.3}$$

and $n(t)$ is AWGN with the two-sided spectral density of $N_0/2$. In (7.2) and (7.3) A and α are constant amplitudes; Θ is a random phase uniformly distributed in $[0, 2\pi]$; $b(t)$ is a random sequence of binary symbols of duration T_b sec, taking on values ± 1 with equal probability; Ω is the frequency offset of the interference tone relative to the carrier frequency ω_0; and $c(t)$ is the spreading sequence taking on values ± 1, which last for T_c sec, where $T_c \ll T_b$.

The received signal $r(t)$ is coherently demodulated by multiplying it with the reference carrier of $2 \cos \omega_0 t$. We assume that the carrier phase and timing of the chip rate sampler are ideally estimated. The sample on the central tap of the estimation filter in Figure 7.2(a) at time iT_c is given by

$$x_i = d_i + V \cos(\Omega i T_c + \phi) + n_i \tag{7.4}$$

where

$$d_i = \pm A T_c \tag{7.5}$$

$$V = 2\alpha \frac{\sin(\Omega T_c/2)}{\Omega} \tag{7.6}$$

$$\phi = \Theta - \frac{\Omega T_c}{2} \tag{7.7}$$

To simplify notation we let $AT_c = \sqrt{S}$, so that $d_i = \pm\sqrt{S}$, where $\pm$ is determined by the combination of algebraic signs of the data symbol and the symbol of the spreading chip sequence at the current sampling instant. The current filter central tap sample is predicted from the past and future samples as

$$\hat{x}_i = \sum_{\substack{l=-M \\ l \neq i}}^{M} a_l x_{i-l} \tag{7.8}$$

where a_l, $l = -M, \ldots, M$, are the filter tap coefficients and $(2M + 1)$ is the number of the filter taps.

The filter output, denoted by y_i, represents the prediction error, expressed as the difference between the current sample x_i and its predicted value $\hat{x}_i$.

$$y_i = x_i - \hat{x}_i = x_i - \sum_{\substack{l=-M \\ l \neq i}}^{M} a_l x_{i-l} \tag{7.9}$$

Define the $2M$-dimensional vector of the off-center filter tap samples, at time iT_c, denoted by $\mathbf{X}_i$, as

$$\mathbf{X}_i \equiv \{x_{i+M}, x_{i+M-1}, \cdots, x_{i+1}, x_{i-1}, \cdots, x_{i-M}\}^T \tag{7.10}$$

where T denotes transpose.

Let us also define the $2M$-dimensional vector of adjustable filter tap coefficients, denoted by $\mathbf{W}$, as

$$\mathbf{W} \equiv \{a_{-M}, a_{-M+1}, \cdots, a_{-1}, a_1, \cdots, a_M\}^T \tag{7.11}$$

The filter output sample at time iT_c is given by

$$y_i = x_i - \mathbf{W}^T \mathbf{X}_i \tag{7.12}$$

The filter coefficients are adjusted to minimize the mean-squared filter output, which is equal to the prediction mean squared error, denoted by ξ. The mean-squared filter output is given by

$$\xi = E[y_i^2] = E[x_i^2] - 2E[x_i \mathbf{X}_i^T]\mathbf{W} + \mathbf{W}\, E[\mathbf{X}_i \mathbf{X}_i^T]\mathbf{W} \tag{7.13}$$

where $E[\ ]$ is the expected value. The autocorrelation function of x_i, defined by (7.4), can be expressed as

$$E[x_i x_{i+m}] = (S + \sigma_n^2)\delta(m) + J \cos m\Omega T_c \tag{7.14}$$

where σ_n^2 is the thermal noise power, $\delta(m)$ is the Kronecker delta function, S as defined earlier, and $J = V^2/2$ is the power of the interfering tone. Expression (7.14) results from the fact that the signal and noise are independent and spread spectrum signal samples at different taps are approximately uncorrelated. The expected value $E[y_i^2]$ can then be written as

$$E[y_i^2] = E[x_i^2] - 2\mathbf{P}^T\mathbf{W} + \mathbf{W}^T\mathbf{R}\mathbf{W} \qquad (7.15)$$

where

$$\mathbf{P}^T = E[x_i X_i^T] = \{J \cos M\Omega T_c, J \cos(M-1)\Omega T_c, \cdots, \qquad (7.16)$$
$$J \cos \Omega T_c, J \cos \Omega T_c, \cdots, J \cos M\Omega T_c\}$$

and $\mathbf{R}$ is the correlation matrix defined as

$$\mathbf{R} = E[\mathbf{X}_i \mathbf{X}_i^T]$$

$$= \begin{bmatrix} S + J + \sigma_n^2 & J \cos \Omega T_c & \cdots & J \cos 2M\Omega T_c \\ J \cos \Omega T_c & S + J + \sigma_n^2 & \cdots & J \cos(2M-1)\Omega T_c \\ \cdots & & & \\ J \cos 2M\Omega T_c & J \cos(2M-1)\Omega T_c & \cdots & S + J \sigma_n^2 \end{bmatrix} \qquad (7.17)$$

The tap coefficients are obtained by minimizing $E[y_i^2]$. From

$$\frac{\partial E[y_i^2]}{\partial a_k} = 0 \qquad k = -M, \cdots, -1, 1, \cdots, M \qquad (7.18)$$

and (7.15) we obtain

$$-2\mathbf{P} + 2\mathbf{R}\mathbf{W}_o = 0 \qquad (7.19)$$

or

$$\mathbf{W}_o = \mathbf{R}^{-1}\mathbf{P} \qquad (7.20)$$

where $\mathbf{W}_o$ is the optimum tap coefficient vector. Equation (7.20) is the well-known Wiener–Hopf equation. The solution to (7.20) is $a_{k0} = 2\gamma \cos k\Omega T_c$ [6], where

$$\gamma = \frac{J}{2(S + \sigma_n^2) + J\left[2M - 1 + \frac{\sin(2M + 1)\Omega T_c}{\sin \Omega T_c}\right]} \quad (7.21)$$

The minimum power of the filter output can be written as

$$E[y_i^2]_{\min} = S + E[e^2]_{\min} \quad (7.22)$$

where S is the desired signal power and $E[e^2]_{\min}$ is the minimum power of the noise component consisting of the sum of the thermal noise and residual interference. The minimum output noise power is given by [6]

$$E[e^2]_{\min} = J / \left\{ 1 + \frac{J}{2(S + \sigma_n^2)}\left[2M - 1 + \frac{\sin(2M + 1)\Omega T_c}{\sin \Omega T_c} + \sigma_n^2\right]\right\} \quad (7.23)$$

The improvement factor G is defined as the ratio of the SNR with the he suppression filter to the SNR with no suppression filter. The SNR with no suppression filter is given by

$$\text{SNR}_{nf} = \frac{S}{J + \sigma_n^2} \quad (7.24)$$

while the SNR with the suppression filter in place is

$$\text{SNR}_f = \frac{S}{E[e^2]_{\min}} \quad (7.25)$$

where $E[e^2]_{\min}$ is given by (7.23). Taking SNR_f to SNR_{nf} we obtain the improvement factor for the estimation filter shown in Figure 7.2(a), denoted by G_2

$$G_2 = \frac{\text{SNR}_f}{\text{SNR}_{nf}} = (J + \sigma_n^2) \cdot \left\{J / \left\{1 + \frac{J}{2(S + \sigma_n^2)}\left[2M - 1 + \frac{\sin(2M + 1)\Omega T_c}{\sin \Omega T_c} + \sigma_n^2\right]\right\}\right\}^{-1} \quad (7.26)$$

Clearly, G_2 increases either as the number of taps increases or as the ratio of the interference to the signal power J/S increases. The improvement factor for the single-sided error prediction filter shown in Figure 7.2(b), denoted by G_1, is given by [6]

$$G_1 = (J + \sigma_n^2) \cdot \left\{ 2(S + \sigma_n^2) \left[L + 2\frac{S + \sigma_n^2}{J} - \frac{\sin L\Omega T_c}{\sin \Omega T_c} \cos(L + 1)\Omega T_c \right] \right. \tag{7.27}$$

$$\left. / \left[\left(L + \frac{2(S + \sigma_n^2)}{J} \right)^2 - \frac{\sin^2 L\Omega T_c}{\sin^2 \Omega T_c} \right] + \sigma_n^2 \right\}^{-1}$$

Figure 7.3 shows the improvement factor for both two-sided and single-sided transversal filters with $\sigma_n = 0$ and the interference to signal power ratio $J/S = 100$. The transfer function of the two-sided transversal filter is given by [6]

$$H_2(\omega) = 1 - J \left[\left(\sin\frac{(2M+1)(\omega+\Omega)T_c}{2} \Big/ \sin\frac{(\omega+\Omega)T_c}{2} \right) \right.$$

$$+ \left(\sin\frac{(2M+1)(\omega-\Omega)T_c}{2} \Big/ \sin\frac{(\omega-\Omega)T_c}{2} \right) - 2 \right]$$

$$\left. \cdot \left\{ 2(S + \sigma_n^2) + J \left[2M - 1 + \frac{\sin(2M+1)\Omega T_c}{\sin \Omega T_c} \right] \right\}^{-1} \tag{7.28}$$

Figure 7.4 shows an example of filter frequency response $H(\omega)$ for $M = 5$, $\sigma_n^2 = 0$, $J/S = 100$, and $\Omega T_c = 60°$. It can be seen that $H(\omega)$ has a typical shape of a notch filter transfer function.

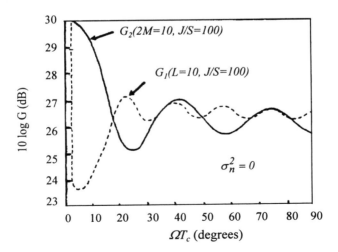

Figure 7.3 Output SNR improvement versus ΩT_c.

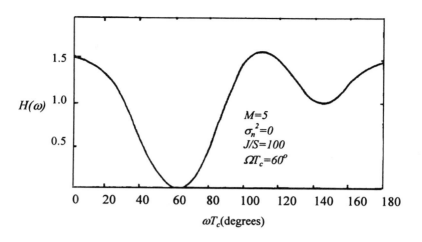

Figure 7.4 Frequency response of the transversal filter. (© 1988 IEEE. *From:* [14].)

7.2 MODELING OF NARROWBAND INTERFERENCE BY AUTOREGRESSIVE PROCESSES

In the previous section it was shown that the optimum filter coefficients depend on the input signal correlation matrix. So, if the interfering signal correlation matrix is specified, the closed-form solution for the SNR improvement factor can be obtained. This will be illustrated in this section by modeling the interference as a narrow band first-order autoregressive process. In this model, the present sample value consists of a linear combination of past sample values plus an error term. At the sampling moments kT_c, after ideal frequency down conversion, the normalized filter input signal can be represented as

$$x(k) = b \cdot C(k) + I(k) + n(k) \quad (7.29)$$

where $b = \pm 1$ is the transmitted binary data, $C(k)$ is the code sequence, $I(k)$ is the narrowband interference sample, and $n(k)$ is the noise sample. We assume that instead of the chip-matched filter in Figure 7.1, only a low-pass filter of bandwidth proportional to $1/T_c$ is used to limit the noise. The interfering signal $I(k)$ is assumed to be a wide sense stationary stochastic process with zero mean and covariance sequence $\rho_i(k)$. At this point we introduce notation $\Re(\alpha, \beta)$ to be a set of integers between α and β including α and β and $\Re_0(\alpha, \beta)$ the same set excluding zero. The filter output signal can be represented as

$$z(k) = \sum_{l \in \Re} h(l) x(k - l) \quad (7.30)$$

where $\Re$ is $\Re(0, M)$ for a one-sided filter, $\Re(-M, M)$ for a two-sided filter, and $h(l)$ is the filter impulse response given by

$$h(l) = \begin{cases} -a_l & l \neq 0 \\ 1 & l = 0 \end{cases} \qquad (7.31)$$

where a_l is defined by (7.11). Substituting (7.29) into (7.30) we obtain for the filter output

$$z(k) = C_0(k) + I_0(k) + n_0(k) \qquad (7.32)$$

where $C_0(k)$, $I_0(k)$, and $n_0(k)$ are the filtered versions of the useful signal, interference, and noise, respectively. Decision variable U used in the decision device in Figure 7.1 is formed by multiplying the filter output signal by the code and can be resolved into three components

$$\begin{aligned} U &= \sum_{k=1}^{N} z(k)C(k) \\ &= \sum_{k=1}^{N} C_0(k)C(k) + \sum_{k=1}^{N} I_0(k)C(k) + \sum_{k=1}^{N} n_0(k)C(k) \\ &= U_1 + U_2 + U_3 \end{aligned} \qquad (7.33)$$

where N is the code sequence length, which is assumed to be equal to the spreading gain.

Under the assumption that signal noise and narrowband interference are mutually independent we have for the average values

$$E[U_1] = b \cdot N \qquad E[U_2] = E[U_3] = 0 \qquad (7.34)$$

Bearing in mind that $b^2 = 1$ we get for the variances

$$\begin{aligned} \text{var}[U_1] &= N \sum_{m \in \Re_0} h^2(m) \\ \text{var}[U_2] &= N \sum_{m_1, m_2 \in \Re} h(m_1)h(m_2)r_i(m_2 - m_1) \\ \text{var}[U_3] &= N \sum_{m_1, m_2 \in \Re} h(m_1)h(m_2)r_n(m_2 - m_1) \end{aligned} \qquad (7.35)$$

where $\rho_i(\ )$ and $\rho_n(\ )$ are covariance functions of the interfering and noise signals, respectively. For the covariance functions we have

$$\text{cov}\{U_i, U_j\} = 0 \quad i \neq j \tag{7.36}$$

The SNR at the filter output can be expressed as [4]

$$(\text{SNR})_0 \triangleq \frac{E^2[U]}{\text{var}[U]} = \frac{N}{\sum_{m \in \Re_0} h^2(m) + \sum_{m_1, m_2 \in \Re} h(m_1)h(m_2)[\rho_i(m_2 - m_1) + \rho_n(m_2 - m_1)]} \tag{7.37}$$

When no suppression filter is used, $h(0) = 1$, and $h(l) = 0$ for $l \neq 0$ so that we get for the signal-to-noise ratio

$$(\text{SNR})_{n0} = \frac{N}{\rho_i(0) + \rho_n(0)} \tag{7.38}$$

The performance improvement factor when the use of the filter is used is given by

$$G = \frac{\rho_i(0) + \rho_n(0)}{\sum_{m \in \Re_0} h^2(m) + \sum_{m_1, m_2 \in \Re}^{M} \sum^{M} h(m_1)h(m_2)[\rho_i(m_2 - m_1) + \rho_n(m_2 - m_1)]} \tag{7.39}$$

7.3 EXAMPLES OF THE INTERFERING SIGNAL

For the signal $x(k)$ given by (7.29), the covariance function $\rho(k)$ can be expressed as

$$\rho(k) = \delta_c(k) + \rho_i(k) + \rho_n(k) \tag{7.40}$$

where $\delta_c(k)$, the Kronecker delta function, is the covariance sequence of the PN code. For $\rho_n(k)$ and $\rho_i(k)$ we will assume

$$\rho_n(k) = \sigma_n^2 \delta_n(k)$$

$$\rho_i(k) = \sigma_n^2 \alpha^{|k|} \quad 0 < \alpha < 1 \tag{7.41}$$

where σ_n^2 and σ_i^2 are the noise and narrowband interference variance, respectively.

The interference power spectral density, denoted by $\phi(\omega)$, is obtained as the Fourier transform of $\rho_i(k)$

$$\phi_i(\omega) = \frac{(1 - \alpha^2)\sigma_i^2/2\pi}{|1 - \alpha \exp(j\omega)|^2} \quad -\pi \leq \omega \leq \pi \tag{7.42}$$

$$= \frac{(1 - \alpha^2)\sigma_i^2/2\pi}{1 + \alpha^2 - 2\alpha \cos \omega}$$

and parameter α will characterize the shape of the spectra. The larger the α, the narrower the spectra and vice versa. By substituting (7.41) into (7.40) we get for the signal covariance

$$\rho(k) = (1 + \sigma_n^2)\delta(k) + \sigma_i^2 \alpha^{|k|} \tag{7.43}$$

It is possible to show that the Wiener-Hopf equation (7.20) for this case becomes

$$a_k(1 + \sigma_n^2) + \sigma_i^2 \sum_{m \in \Re_0} a_m \alpha^{|k-m|} = \sigma_i^2 \alpha^{|k|} \tag{7.44}$$

Solving the filter coefficients from this system of equations and then using (7.41) and (7.39) we get for the filter improvement factor

$$G_{1SF} = (\sigma_n^2 + \sigma_i^2)/[\sigma_n^2 + \sigma_i^2(1 - \alpha^2)]$$
$$\cdot [((1 - \alpha\beta) + (\alpha - \beta)\beta^{2M+1})/((1 - \alpha\beta)^2 - (\alpha - \beta)^2 \beta^{2M})]] \tag{7.45}$$

$$\beta = \gamma - \sqrt{\gamma^2 - 1}$$

$$\gamma = \frac{1}{2\alpha}\left[(1 + \alpha^2) + \frac{\sigma_i^2(1 - \alpha^2)}{1 + \sigma_n^2}\right]$$

for the 1SF and

$$G_{2SF} = (\sigma_n^2 + \sigma_i^2) \tag{7.46}$$
$$/\left[\sigma_n^2 + \sigma_i^2(1 - \alpha^2)\frac{(1 - \alpha\beta) + (\alpha - \beta)\beta^{2M+1}}{(1 - \alpha\beta)(1 + \alpha^2 - 2\alpha\beta) - (\alpha - \beta)(2\alpha - \beta - \alpha^2\beta)\beta^{2M}}\right]$$

for the 2SF, where β and γ are the same as in (7.45). As an illustration, several curves for the filter improvement factor G with the given set of the signal and filter parameters. The improvement factor is plotted versus the SNR per chip with no filtering (given by $10 \log[1/(\sigma_n^2 + \sigma_i^2)]$) and shown in Figure 7.5. Curve D is an upper bound on G, for $M \to \infty$. It can be seen that with only five taps ($M = 2$) the performance of a 2SF is close to the upper bound.

For strong interference, the performance is almost independent of the number of taps. For the analysis of the mutual influence of CDMA and narrowband communications networks we will assume that the interfering signal consists of J_b frequency bands with the power spectral density represented as

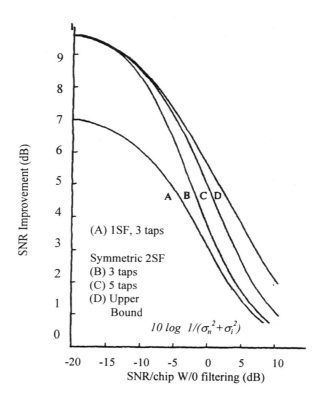

Figure 7.5 Improvement factor for a first-order autoregressive interference with $\alpha = 0.9$, $\sigma_n^2 = 0$.

$$\phi_i(\omega) = \begin{cases} \sigma_i^2/2\pi p & \omega \in A_j, j = 1, \cdots, J_b \\ 0 & \omega \notin \bigcup_{j=1}^{J_b} A_j \end{cases} \quad (7.47)$$

where the intervals A_j's are disjoint and their total length $\Sigma_{j=1}^{J_b}|A_j| = 2\pi p$ for some $0 < p < 1$. The jammer occupies a pth fraction of the signal band. Using the same procedure as in the previous example, the improvement factor bounds for multiband interference are computed and shown in Figure 7.6.

7.4 BIT ERROR RATE

The average probability of error is a better measure of the system performance than the SNR improvement factor.

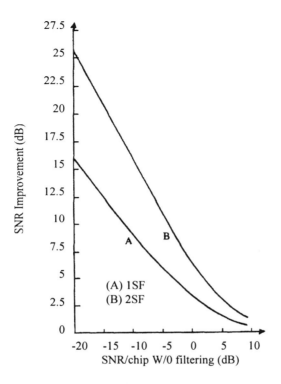

Figure 7.6 Upper bounds on improvement factors for a multiband interference with 20% bandwidth occupancy. (© 1985 IEEE. *From:* [15].)

Assuming that the performance of this system is the same as that of binary PSK signaling corrupted only by white Gaussian noise with variance equal to the total noise due to white noise, interference, and self-noise, at the output of the PN correlator, the bit error rate is given by

$$P_b = \frac{1}{2} \cdot P(U < 0) + \frac{1}{2} P(U > 0) = P(U < 0)$$

since $P(U < 0) = P(U > 0)$. Thus

$$P_b = P(U < 0) = \int_{-\infty}^{0} \frac{1}{\sqrt{2\pi}\sigma} e^{-(U-\mu)^2/2\sigma^2} dU \qquad (7.48)$$

where

$$\sigma^2 = \text{var}(U) = \text{var}(U1) + \text{var}(U2) + \text{var}(U3)$$
$$\mu = E(U) = E(U_1) \tag{7.49}$$

Thus,

$$P_b = Q\left(\sqrt{\frac{\mu^2}{\sigma^2}}\right) = Q(\sqrt{\text{SNR}_0}) \tag{7.50}$$

where SNR_0 is given by (7.39). Some examples of BER curves are given in Figures 7.7 to 7.9 for a given set of parameters.

7.5 ADAPTIVE ESTIMATION FILTERS

Note that all the results presented until now are based on known statistical properties of interference. In practice, the parameters of interference are rarely known to the

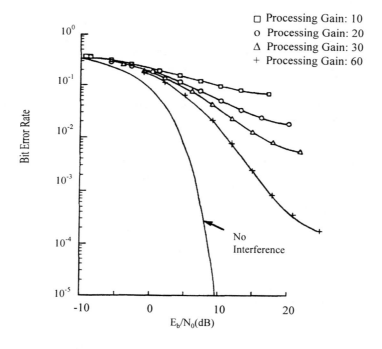

Figure 7.7 Bit error probability under the Gaussian assumption for a four-tap predictor with no matched filter. Filter coefficients are computed using exact autocorrelation coefficients with −20-dB SIR.

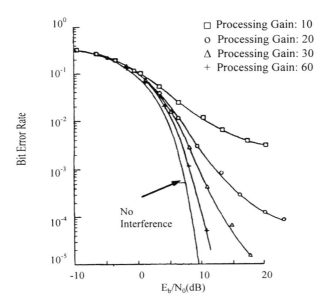

Figure 7.8 Bit error probability under the Gaussian assumption for a four-tap predictor with matched filter. Filter coefficients are computed using exact autocorrelation coefficients with −20-dB SIR.

receiver. Also, interference can have statistics that vary with time. Therefore, an effective suppression filter should be able to adapt itself to variations of interference statistics. There are a variety of adaptive algorithms that can be used [4,5,12,16].

The requirement that an adaptive filter has to meet is to solve the tap coefficients from the Wiener–Hopf equation (7.20). This equation could be solved analytically if and only if all parameters were accurately known. Although conceptually straightforward, this is a computation-intensive operation that requires a matrix inversion. For practical applications a form of recursive algorithm is preferred.

Another option is to build up a recursive algorithm that will evaluate an improved set of filter coefficients in each step. Within this section we will discuss the method of steepest descent and the LMS algorithm. The method of steepest descent uses gradients of the performance surface in seeking its minimum.

7.5.1 The Gradient and the Wiener Solution

The MSE for an estimation filter, as shown in Section 7.1.2, can be written as

$$\xi = E[x_i^2] - 2\mathbf{P}^T\mathbf{W} + \mathbf{W}^T\mathbf{R}\mathbf{W} \qquad (7.51)$$

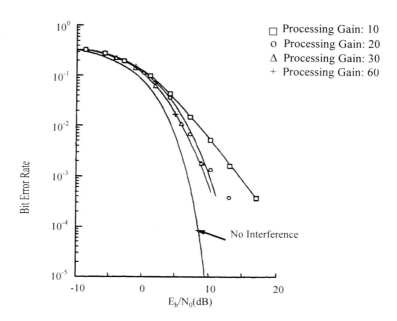

Figure 7.9 Bit error probability from simulation results for a four-tap predictor with matched filter.

We may visualize the MSE function ξ dependence on the elements of the tap vector **W** as a bowl-shaped surface with a unique minimum. We refer to this surface as the error-performance surface of the estimation filter.

It is required to design a filter that operates at the bottom of the error performance surface. The adaptive algorithm continually seeks the bottom or the minimum point of this surface, at which point the tap coefficients take on the optimum values $\mathbf{W}_o$, as defined by the Wiener–Hopf equation (7.20)

$$\mathbf{RW}_o = \mathbf{P} \qquad (7.52)$$

The gradient of the (MSE) function defined by (7.51), denoted as ∇, is given by

$$\nabla = -2\mathbf{P} + 2\mathbf{RW} \qquad (7.53)$$

When we set the gradient to zero, we get the optimal Wiener–Hopf solution defined by (7.20). Substituting (7.20) into (7.15) gives the MMSE

$$\xi_{\min} = E[x_i^2] - \mathbf{P}^T \mathbf{W}_o \qquad (7.54)$$

The MSE in (7.15) can now be represented as

$$\xi = \xi_{min} + (\mathbf{W} - \mathbf{W}_o)^T \mathbf{R}(\mathbf{W} - \mathbf{W}_o) \tag{7.55}$$

This can be further expressed as

$$\xi = \xi_{min} + \mathbf{V}^T \mathbf{R} \mathbf{V} \tag{7.56}$$

where

$$\mathbf{V} \underline{\underline{\Delta}} \mathbf{W} - \mathbf{W}_o \tag{7.57}$$

is the difference between $\mathbf{W}$ and the optimum values $\mathbf{W}_o$. Differentiation of (7.56) gives another form of the gradient

$$\nabla = 2\mathbf{R}\mathbf{V} \tag{7.58}$$

If $\mathbf{Q}$ is an orthonormal modal matrix of a symmetric and positive definite correlation matrix $\mathbf{R}$ and if Λ is its diagonal matrix of eigenvalues given by

$$\Lambda = \text{diag}[\lambda_1, \lambda_2, \cdots, \lambda_n] \tag{7.59}$$

then we can write

$$\mathbf{R} = \mathbf{Q}\Lambda\mathbf{Q}^{-1} = \mathbf{Q}\Lambda\mathbf{Q}^T \tag{7.60}$$

Now (7.56) becomes

$$\xi = \xi_{min} + \mathbf{V}^T \mathbf{Q} \mathbf{L} \mathbf{Q}^{-1} \mathbf{V} \tag{7.61}$$

If we use notation

$$\mathbf{V}' \underline{\underline{\Delta}} \mathbf{Q}^{-1} \mathbf{V} \rightarrow \mathbf{V} = \mathbf{Q}\mathbf{V}' \tag{7.62}$$

(7.61) can be expressed as

$$\xi = \xi_{min} + \mathbf{V}'^T \Lambda \mathbf{V}' \tag{7.63}$$

The primed coordinates are the principal axes of the quadratic surface. In the same way we may apply transformation (7.62) to vector $\mathbf{W}$ itself to get

$$\mathbf{W}' = \mathbf{Q}^{-1}\mathbf{W} \rightarrow \mathbf{W} = \mathbf{Q}\mathbf{W}' \tag{7.64}$$

7.5.2 The Steepest Descent Algorithm

The method of steepest descent is one of the oldest adaptive methods [17]. It updates the filter coefficients at time $i + 1$ by using the simple recursive relation

$$\mathbf{W}_{i+1} = \mathbf{W}_i + \mu(-\nabla_i) \tag{7.65}$$

where μ is a positive real-valued constant that controls stability and rate of adaptation and ∇_i is the gradient at the ith iteration. Substituting (7.58) into (7.64) and (7.65) we have

$$\mathbf{V}'_{i+1} = (\mathbf{I} - 2\mu\Lambda)\mathbf{V}'_i \tag{7.66}$$

where $\mathbf{I}$ is the identity matrix. After successive iterations for $\mathbf{V}'_i$ becomes

$$\mathbf{V}'_i = (\mathbf{I} - 2\mu\Lambda)^i \mathbf{V}'_{in} \tag{7.67}$$

where $\mathbf{V}'_{in}$ is the initial difference between $\mathbf{W}'$ and $\mathbf{W}'_o$

$$\mathbf{V}'_{in} = \mathbf{W}'_{in} - \mathbf{W}'_o \tag{7.68}$$

From (7.67) one can see that each component k of the vector $\mathbf{V}'$ represents a geometric series with the geometric ratio

$$r_k = (1 - 2\mu\lambda_k) \tag{7.69}$$

For convergence it is necessary that the magnitude of this geometric ratio is less than 1

$$|r_{max}| = |1 - 2\mu\lambda_{max}| < 1 \tag{7.70}$$

where λ_{max} is the largest eigenvalue of the correlation matrix $\mathbf{R}$. That is,

$$-1 < 1 - 2\mu\lambda_{max} < 1 \tag{7.71}$$

which results in

$$1/\lambda_{max} > \mu > 0 \tag{7.72}$$

The components of the vector $\mathbf{V}'$ form a geometric series, and thus they decay with time following an exponential envelope.

In order to determine its time constant an exponential envelope is fitted to the geometric series. If the time is normalized to the iteration cycle, the time constant τ_k can be determined from

$$r_k = (1 - 2\mu\lambda_k) \cong \exp\left(-\frac{1}{\tau_k}\right) = 1 - \frac{1}{\tau_k} + \frac{1}{2!\tau_k^2} - \frac{1}{3!\tau_k^3} + \cdots \cong 1 - \frac{1}{\tau_k} \quad (7.73)$$

leading to

$$\tau_k \cong \frac{1}{2\mu\lambda_k} \quad (7.74)$$

Based on this, the time constant can be defined as the maximum value of parameter τ_k

$$\tau = \max_k \tau_k = \frac{1}{2\mu\lambda_{\min}} \quad (7.75)$$

7.5.3 The LMS Algorithm

In practice ∇_i is not known and has to be estimated. So, the algorithm defined by (7.65) becomes

$$\mathbf{W}_{i+1} = \mathbf{W}_i + \mu(-\hat{\nabla}_i) \quad (7.76)$$

where $-\hat{\nabla}_i$ is an estimate of the true gradient ∇_i. When the gradient estimate is obtained as the gradient of the square of a single error sample, we get the LMS algorithm. Taking the derivative of the square of (7.9) we get

$$\hat{\nabla}_i = 2y_i y_i' = -2y_i \mathbf{X}_i \quad (7.77)$$

and (7.76) becomes

$$\mathbf{W}_{i+1} = \mathbf{W}_i + 2y_i \mathbf{X}_i \quad (7.78)$$

In literature y_i is denoted as ϵ_i and the previous equation is written in the form

$$\mathbf{W}_{i+1} = \mathbf{W}_i + 2\epsilon_i \mathbf{X}_i \quad (7.79)$$

The conditions defined by (7.70) to (7.72) are necessary and sufficient for convergence of the LMS algorithm. However, in practice (7.70) to (7.72) are not of much use because the individual eigenvalues are rarely known. Since $tr\ \mathbf{R}$ is the total input power, which is generally known, and $tr\ \mathbf{R} > \lambda_{max}$ because $\mathbf{R}$ is positive definite, the condition expressed by (7.70) to (7.72) can be replaced by

$$1/tr\mathbf{R} > \mu > 0 \tag{7.80}$$

A two-sided transversal filter with the LMS adaptive algorithm is shown in Figure 7.10.

The error probability performance for a system with an unknown interference statistics is shown in Figure 7.11 [13]. Curve B shows the performance when a two-sided transversal filter is used in a system with known interference statistics. Curve C clearly illustrates the degradation caused by the lack of knowledge of the interference statistics in a system with the same filter structure and an LMS adaptive algorithm to update filter coefficients. The results refer to the filter steady state.

It is well known that the convergence rate of the LMS algorithm in combination with a transversal filter structure is very low.

A two-sided filter has a lower convergence than a one sided filter [18]. Lattice structures have also been investigated for narrowband interference suppression [19,20]. The adaptive version of a lattice filter can have a much faster convergence than transversal filters because each section of the lattice converges individually, independent of the remaining sections. A typical lattice filter structure is shown in Figure 7.12.

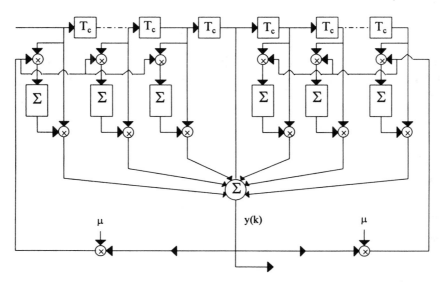

Figure 7.10 A two-sided transversal filter with the LMS adaptive algorithm.

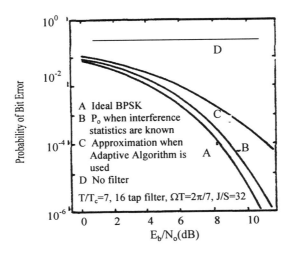

Figure 7.11 Performance of an adaptive receiver. (© 1988 IEEE. *From:* [14].)

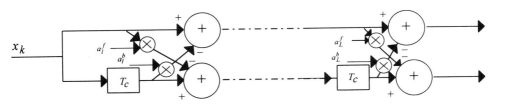

Figure 7.12 Lattice filter.

7.5.4 Decision Feedback Interference Suppression Filter

Another possible filter structure for narrowband interference suppression is a *decision feedback* (DF) filter [21–23]. The block diagram of a decision feedback receiver is shown in Figure 7.13.

The operation of transversal estimation filters corresponds to whitening, that is, making the output samples of the entire received signal uncorrelated. However, whitening just the noise and interference portion of the received signal gives better performance. As the received signal consists of the desired signal plus noise and interference, it is necessary to remove the desired signal. The estimate of the data signal from the output of the receiver can be used instead of the actual transmitted waveform to be subtracted from the received signal. If this estimate of the data symbol is correct, the result of the subtraction operation will give the sum of the thermal noise and interference only, which can be used as the filter input. However,

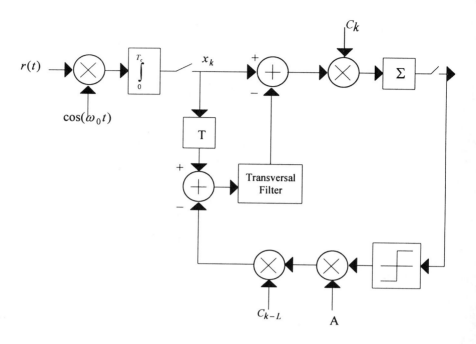

Figure 7.13 Decision feedback receiver.

if the estimate of the data symbol is incorrect, the current error will propagate, since it will affect the subsequent symbol. The error probability curves for an ideal DF receiver with errorfree decisions for a single tone interferer are shown in Figure 7.14 [14]. Clearly, a perfect DF structure is superior to a linear transversal filter.

Other nonlinear filters, such as *approximate conditional mean* (ACM) filter [24,25] and Kalman–Bucy filter [16,19] have also been used to reject narrowband interference modeled as an autoregression process.

7.6 TRANSFORM DOMAIN FILTERS

Transform domain filters suppress narrowband interference by filtering the received signal in the frequency domain [4,7,8,9,14,26,27]. The receiver with a transform domain filter is shown in Figure 7.15.

The basis for this method is the fact that the spectrum of the desired spread spectrum signals is flat and the spectrum of the interference is highly peaked.

This can be explained by Figure 7.16, which shows the spectrum of the received signal and a notch filter frequency response. The received signal spectrum consists of a wideband relatively flat spectrum of the transmitted signal to which a highly

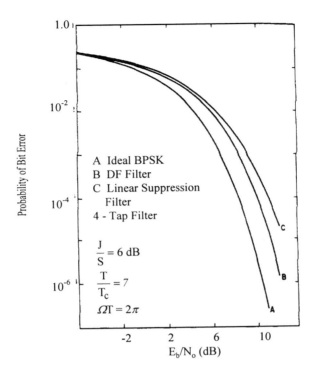

Figure 7.14 Decision feedback filter performance. (© 1988 IEEE. From: [14].)

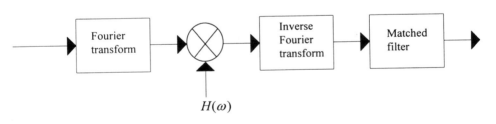

Figure 7.15 Receiver block diagram for transform domain interference suppression.

peaked spectrum of the narrowband interference is superimposed (Figure 7.16(a)). The spectrum of the notch filter, which is designed to suppress the interference peaks, is located in a relatively narrow frequency band $2\delta\omega$. It is shown in Figure 7.16(b).

In the receiver, the Fourier transform of the received signal is estimated and multiplied by the notch filter transfer function $H(\omega)$. The inverse transform of the

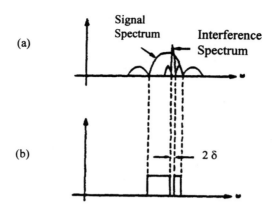

Figure 7.16 Spectrum of the received signal and notch filter transfer function. (© 1988 IEEE. From: [14].)

product is calculated, and the resulting waveform is passed to the detection matched filter.

Clearly, multiplying the received signal spectrum by the notch filter frequency response should suppress considerably the interference power while only slightly reducing the desired signal power.

The first step in a transform domain interference suppression algorithm is the estimation of the power spectral density of the received signal. The spectral estimate can be obtained by any of the well-known spectral analysis techniques [28]. Alternatively, SAW devices can be used to estimate the Fourier transform in real time [14,9,29].

Once the power spectral density of the received signal is estimated, a filter can be designed with notches in the frequency range occupied by the interference. A possible method to design a transversal filter in the discrete time domain is to select its *discrete Fourier transform* (DFT) to be the reciprocal of the square root of the power spectral density at equally spaced frequencies. That is, if the transversal filter consists of N taps, its DFT is given by

$$H(n) = \sum_{k=0}^{N} h(k) e^{-j2\pi nk/N} \quad n = 0, 1, \cdots, N-1 \quad (7.81)$$

where $\{h(k)\}$ are the tap coefficients.

The DFT $H(n)$, $n = 0, 1, \cdots, N-1$, is selected as

$$H(n) = \frac{1}{\sqrt{P(nR_s/N)}} e^{-j\frac{2\pi}{N}\frac{(N-1)}{2}n} \quad (7.82)$$

where $P(f)$, $0 \leq f \leq R_s$ denotes the power spectral density and R_s denotes the sampling rate [28].

The expressions for the average error probability for this technique are presented in [10]. Figure 7.17 illustrates the error probability performance of transform domain processing. It compares a conventional DS spread spectrum receiver with no suppression filter and a transform domain filter receiver. The curves are labeled with the system processing gains and receiver acronyms, where MF stands for the matched filter conventional receiver and TDP for transform domain processing receiver. The curves were obtained for the interference level $\alpha = 20$ and the interference frequency offset $\Omega = 2\pi$. The results show that a conventional system should have processing gain between 255 and 511 to have the same error probability as a system with transform domain processing with processing gain of 31. Thus the performance improvement is about 10 dB.

The drawback of this method is the inherent time-limited nature of the signal processing operation, resulting in significant sidelobes, which effectively spreads the energy of interference over a wide frequency range. Suppressing a large percentage of the interferer's energy would cause significant distortion of the desired signal.

It is proposed to concentrate the interference energy in a small fraction of the spectrum by using windowing functions [26,30].

The windows to be used are the Hamming window [31] or a weighting function that approximates a cosine square function or the overlap and save scheme [26].

The transform domain notch filters can be made adaptive in a similar way as the time domain notch filters [8].

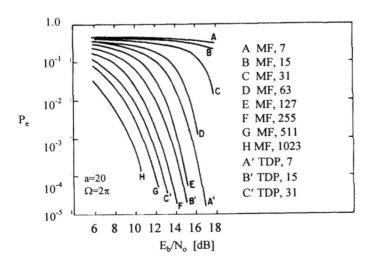

Figure 7.17 Performance of transform domain processing receiver. (© 1988 IEEE. *From:* [14].)

7.7 ADAPTIVE RECEIVERS

As we showed in Chapter 5 adaptive receivers can be used to suppress MAI. The receiver consists of a linear adaptive digital filter followed by a decision device. The filter is trained by a known reference sequence prior to data transmission to determine the optimum initial filter coefficients. This receiver differs from the receiver shown in Figure 7.1 in that it does not require a separate despreading device in addition to a notch filter.

In the data transmission mode the filter coefficients can be continually adjusted by an adaptive LMS algorithm taking the receiver output data symbols as the reference. The receiver does not need to know the spreading sequence of the desired or other signals or timing of other receiver signals. It is important to note that the same receiver can be used to simultaneously suppress both narrowband and multiple-access interference. The performance of the adaptive CDMA receiver in the presence of narrowband interference is presented in [10]. It is shown that in the absence of MAI the performance of the adaptive receiver shown in Figure 7.18 is comparable to the performance of the receiver shown in Figure 7.1. However, the adaptive receiver is able to withstand both narrowband interference of 10% of the signal bandwidth and MAI caused by 33% of the maximum number of users without significant degradation, while the conventional receiver from Figure 7.1 fails to operate. The simulated error probability results are shown in Figure 7.19 for the DS spread spectrum receiver with a two-sided suppression filter from Figure 7.2(a) and an adaptive CDMA receiver in the presence of narrowband and multiple access interference. The curves are labeled with the type of receiver, where AR stands for the adaptive receiver and CR for the conventional receiver from Figure 7.1, the percentage of the *narrowband interference* (NBI) bandwidth relative to the spread spectrum bandwidth and the signal-to-interference power ratio S/J.

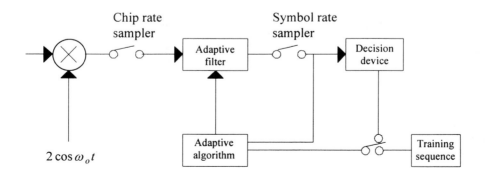

Figure 7.18 Adaptive receiver.

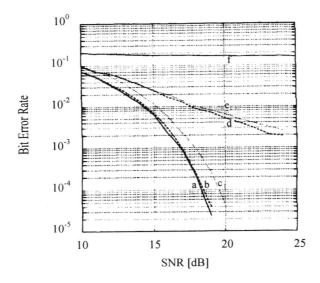

Figure 7.19 Bit error rate of a QPSK CDMA system in the presence of narrowband and multiple-access interference: (a) no NBI; (b) AR, S/NBI = −20dB, 3% NBI bandwidth; (c) AR, S/NBI = −20 dB, 15% NBI bandwidth; (d) CR, S/NBI = −10 dB, 10% NBI bandwidth; (e) CR, S/NBI = −10 dB, 15% NBI bandwidth; and (f) CR, S/NBI = −20 dB, 3% and 15% NBI bandwidth. (©1994. IEEE. From: [11].)

7.8 CDMA OVERLAY FOR PERSONAL COMMUNICATION NETWORKS

7.8.1 CDMA Overlay

Broadband code division multiple access (B-CDMA) has been proposed as a means to provide more efficient use of the overcrowded radio spectrum by overlaying existing narrowband cellular systems [32] or fixed microwave services [1]. The proposed B-CDMA system is a broadband DS spread spectrum personal communication network in the same frequency band with conventional narrowband systems. The receiver contains a notch filter to suppress interference from and to the existing colocated narrowband users [3].

Analog cellular networks are particularly inefficient in radio spectrum utilization and cannot satisfy service demands in some areas. This is especially critical in certain hot spots, such as shopping malls, CBD areas, and transportation centers, where traffic demands are unusually high. Additional capacity could be supplied by overlaying existing analog macrocellular networks by the B-CDMA microcells [3].

In macrocellular systems a large service area is divided into smaller areas, each of which is served by a fixed cell site. The cell sites are arranged in an approximately regular geometric pattern. The basic cells, called macrocells, have radius of the order of up to several kilometers to cover low-density areas. Most of existing cellular systems are based on the frequency division multiple access concept. The available frequency band is divided into seven subbands, and the cells are organized into seven-cell clusters. The cells in a cluster are designated as A to G (Figure 7.20). Seven different frequency subbands are used in each cluster, one subband in each cell. At a sufficient distance from any cell, the same subband is used simultaneously. On the basis of this design, system capacity can be increased by reducing the cell size. However, this incurs an increase in network complexity, particularly in relation to handovers between cells.

Smaller cells, called microcells, with coverage of a few hundred meters, can be arranged in densely populated downtown areas.

Hence, there is a limit to how much the cell areas can be reduced before system complexity becomes unwieldy. An alternative is to overlay the large macrocell network with microcells to serve local "hot-spots."

This design is based on overlaying the existing analog FDMA macrocells by CDMA microcells. The concept is based on the property of spread spectrum formats,

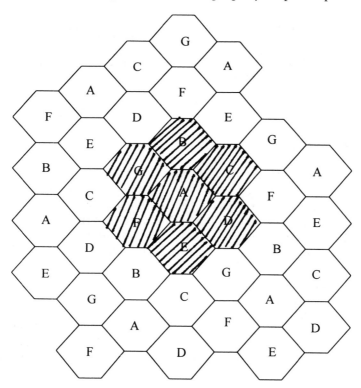

Figure 7.20 Cellular frequency reuse pattern.

as demonstrated in Chapter 1 and Sections 7.1 and 7.2, to operate in parts of electromagnetic spectrum already occupied by other narrowband users without undue degradation of the existing services. Analysis of such a system is the subject of the following sections.

7.8.2 Overlay Geometry

The macrocell is assumed to have a circular shape. In order to reduce cochannel interference it is subdivided into six equal sectors with a radius R_A. Each overlaying microcell is assumed to be circular with radius R_c and centered at a distance D from the macrocell base, as shown in Figure 7.21. It is also assumed that there are maximum 16 analog mobiles randomly distributed within the spread spectrum bandwidth of 10 MHz. Microcells are positioned sufficiently far apart from each other so that the interference from other microcells can be ignored in computing the capacity of each microcell. The macrocell antenna height, denoted by H_A, is much larger than the microcell antenna height, denoted by H_c.

7.8.3 Propagation Model

The propagation mechanism can be represented by a two-ray propagation model [1]. The two-ray model is shown in Figure 7.22.

The signal at the receiver antenna consists of the direct ray and a ray reflected from the ground.

The free space field strength of the direct ray is given by

$$E = \frac{\sqrt{30gP_t}}{d_1} \tag{7.83}$$

where d_1 is the distance shown in Figure 7.22, g is the total antenna gain, and P_t is the transmitted power.

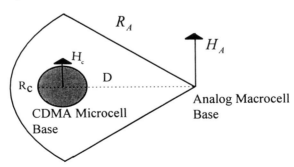

Figure 7.21 Microcell overlay geometry.

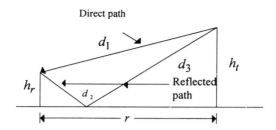

Figure 7.22 Two-ray model.

The electric field of the reflected ray, denoted by E_R, is given by

$$E_R = E \frac{d_1}{d_2 + d_3} R(\phi) e^{j\Delta} \tag{7.84}$$

where $R(\phi)$ is the reflection coefficient, ϕ is the angle of incidence, Δ is the phase difference between the direct and reflected wave, and d_1 and d_2 are the distances shown in Figure 7.22.

The total received electric field strength, denoted by E_r, is given by

$$E_r = E + E_R = E\left[1 + \frac{d_1}{d_2 + d_3} R(\phi) e^{j\Delta}\right] \tag{7.85}$$

Assuming $d_1 \approx d_2 + d_3 \approx r$ and $R(\phi) = -1$ for large distances, where r is shown in Figure 7.22, we get for E_r

$$E_r = E(1 - e^{j\Delta}) \tag{7.86}$$

The magnitude of E_r is then

$$|E_r| = |E|^2 \sin^2 \frac{\Delta}{2} \tag{7.87}$$

The phase difference Δ is given by

$$\Delta = \frac{2\pi}{\lambda}\left\{[r^2 + (h_t + h_r)^2]^{\frac{1}{2}} - [r^2 + (h_t - r_r)^2]^{\frac{1}{2}}\right\} \tag{7.88}$$

where λ is the signal wavelength, h_t is the transmit antenna height, and h_r is the receive antenna height.

For $h_t + h_r \ll r$ Δ can be approximated by [33]

$$\Delta \approx \frac{4\pi h_t h_r}{\lambda r} \qquad (7.89)$$

Assuming $g = 1$, the total received power P_r is given by [33]

$$P_r = \left(\frac{|E_r|\lambda}{2\pi\sqrt{120}}\right)^2 \qquad (7.90)$$

Substituting (7.83) and (7.87) into (7.90) we get

$$P_r = P_t\left(\frac{\lambda}{2\pi r}\right)^2 \sin^2 \frac{\Delta}{2} \qquad (7.91)$$

The smallest phase difference for which P_r has a maximum is

$$\frac{\Delta}{2} = \frac{\pi}{2} \qquad (7.92)$$

which corresponds to the largest distance r. This distance is denoted by R_b and is approximately given by

$$R_b = \frac{4h_t h_r}{\lambda} \qquad (7.93)$$

Using (7.90) and (7.93) and setting $\sin^2 \frac{\Delta}{2} = 1$, we get for the propagation loss at the distance R_b, denoted by L_b,

$$L_b = 10 \log_{10}\frac{P_t}{P_r} = 10 \log_{10}\left[\left(\frac{8\pi h_t h_r}{\lambda^2}\right)^2\right] \qquad (7.94)$$

The variation of the propagation loss upperbound with distance can be described by a piecewise linear curve with a single breakpoint [1] as shown in Figure 7.23. The distance R_b is a breakpoint in the upper bound for the signal propagation loss, denoted by L_u. The propagation loss upper bound is proportional to r^2 before and to r^4 after the breakpoint

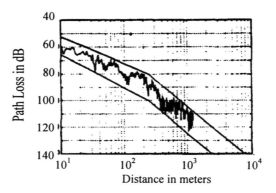

Figure 7.23 Variation of the propagation loss in a suburban area. (© 1992 IEEE. *From:* [1].)

$$L_u = L_b + \begin{cases} 20 \log_{10} \dfrac{r}{R_b} & r \leq R_b \\ 40 \log_{10} \dfrac{r}{R_b} & r > R_b \end{cases} \quad (7.95)$$

In the overlay systems analysis it is assumed, for simplicity, that the microcell radius R_c is

$$R_c = \frac{4 H_m H_c}{\lambda} \quad (7.96)$$

where H_m is the mobile antenna height and H_c is the microcell base station antenna height. Also, assuming that $H_m/\lambda = 5$, we get $R_c = 20 H_c$.

More accurate propagation models incorporate a random log-normal dependence to account for the effects of shadowing [34]. Such effects will be ignored in the subsequent analysis.

7.8.4 Forward Link CDMA Network Capacity

We consider the capacity of the forward link from the base station to mobile in a CDMA microcell without notch filters initially.

7.8.4.1 CDMA Mobile Signal-to-Interference Ratio

In the system model shown in Figure 7.21 we assume that the power of the analog macrocell base station is P_{AB} per channel and the power of the microcell base station is P_{CB} per channel. The power received by a CDMA mobile, from the CDMA base station, in the worst case when it is located at the microcell boundary is [3]

$$Q_{CC}^M = \frac{aP_{CB}}{R_c^2} \qquad (7.97)$$

where a is the propagation constant that applies for distances less than the breakpoint R_c.

The worst case power received by a CDMA mobile from the macrocell analog base station is

$$Q_{CA}^M = \frac{b_A M P_{AB}}{(D - R_c)^4} \qquad (7.98)$$

where $M \leq 16$ is the number of active analog mobiles and b_A is the propagation constant for distances beyond the macrocell breakpoint d_A, given by

$$b_A = a d_A^2 \qquad (7.99)$$

where

$$d_A = \frac{4 H_m H_A}{\lambda} = 20 H_A \qquad (7.100)$$

where H_A is the macrocell base station antenna height. We can express the signal to interference ratio for a CDMA mobile, denoted by Γ_{CM}, as

$$\Gamma_{CM} \cong \frac{g \, Q_{CC}^M}{\alpha K \, Q_{CC}^M + Q_{CA}^M} \qquad (7.101)$$

where g is the processing gain and K is the number of active CDMA mobile users an α is the voice activity factor.

We assume that the error probability of the CDMA mobile is determined by Γ_{CM}; that is, the effect of noise and interference from other microcells is ignored.

7.8.4.2 Analog Mobile Signal-to-Interference Ratio

The lowest signal power of an analog mobile received from the analog macrocell base station is obtained when it is located at the farthest microcell boundary

$$Q_{AA}^M = \frac{b_A P_{AB}}{(D + R_c)^4} \qquad (7.102)$$

The power received by the same mobile from the CDMA base station is

$$Q_{AC}^M = \frac{\alpha K P_{CB} b_C}{R_C^4} \qquad (7.103)$$

where the propagation constant

$$b_C = \alpha (20 H_c)^2 \qquad (7.104)$$

The analog mobile signal-to-interference ratio is given by

$$\Gamma_{AM} = \frac{G_p \cdot Q_{AA}^M}{Q_{AC}^M} \qquad (7.105)$$

where G_p is the spreading gain. Γ_{AM} can be expressed as

$$\Gamma_{AM} = \frac{G_p}{\alpha \rho \delta^2 K (1 + D/R_c)^4} \qquad (7.106)$$

where $\rho = P_{CB}/P_{AB}$ is the microcell-to-macrocell base station power ratio per channel and $\delta = H_c/H_A$ is the microcell-to-macrocell base station antenna height ratio.

7.8.4.3 CDMA Forward Link Capacity

The microcell capacity can be expressed as the number of active CDMA users K

$$K = K_o - \frac{M}{\rho \delta^2 (1 - D/R_c)^4} \qquad (7.107)$$

where $K_o = g/\alpha \Gamma_{CM}$ is the maximum achievable capacity without analog interference. Substituting for $\rho \delta^2$ from (7.106) into (7.107) we can obtain for the CDMA microcell forward link capacity

$$K = K_o \bigg/ \left[1 + \frac{M \Gamma_{AM}}{G_p} \left(\frac{D + R_c}{D - R_c} \right)^4 \right] \qquad (7.108)$$

The microcell forward link capacity is computed using (7.108) for the following conditions

Microcell radius	$R_c = 0.1 R_A$;
Macrocell antenna height	$H_A = 4 H_c$ ($\delta = 0.25$);
CDMA bandwidth	$B_C = 10$ MHz;
Analog bandwidth	$B_A = 15$ KHz;
Spreading gain	$G_p = 666$;
Chip rate	$r_c = 8$ Mcps;
Bit rate	$r_b = 8$ kbps;
Processing gain	$g = 1000$;
Required mobile E_b/N_o	$\Gamma_{CM} = 4.5$ dB;
Required analog mobile signal-to-interference ratio	$\Gamma_{AM} = 17$ dB;
Voice activity factor	$\alpha = 0.75$.

The required CDMA mobile E_b/N_o is based on the assumption that an interleaved half-rate convolutional code of constraint length 7 is used.

The capacity is plotted versus the normalized distance from the macrocell base station $d = D/R_A$ for various numbers of analog users $M = 0, 8$, and 16 and illustrated in Figure 7.24.

The microcell capacity is far less than the maximum capacity of $K_0 = 473$ obtained for no analog interference.

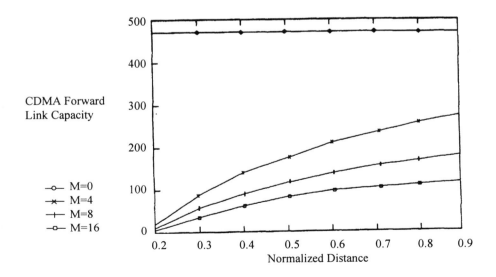

Figure 7.24 CDMA microcell forward link capacity. (© 1994 IEEE. From: [3].)

7.8.4.4 Transmit Notch Filters at Microcell Base Station

In order to reduce interference to the analog macrocell base station, transmit notch filters at the microcell base station are introduced. Assuming a transmit filter notch depth of η_t, the analog signal-to-interference ratio becomes

$$\Gamma_{AM} = \frac{G_p}{\alpha \eta_t \delta^2 \, \rho \, K(1 + D/R_C)^4} \tag{7.109}$$

The capacity equation is then

$$K = K_o / \left[1 + \frac{\eta_t M \Gamma_{AM}}{G_p} \left(\frac{D + R_C}{D - R_C} \right)^4 \right] \tag{7.110}$$

Figure 7.25 shows the capacity of the forward link for a CDMA microcell for a transmit notch filter depth of −30 dB. A significant capacity improvement can be observed relative to the system performance in Figure 7.24 with no notch filters.

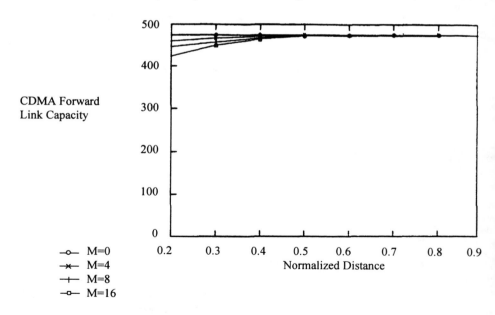

Figure 7.25 CDMA microcell forward link capacity with 30-dB notch filter depth. (© 1994 IEEE. From: [3].)

7.8.5 Reverse Link CDMA Network Capacity

Now we consider the capacity of the reverse link for transmission from the mobile users to the base station. The base station contains the notch filter to suppress narrowband interference originated by the analog network.

7.8.5.1 CDMA Base Station Signal-to-Interference Ratio

The power received by the CDMA base station from a CDMA mobile located at the cell boundary R_c is given by

$$Q_{CC}^B = \frac{R_{CM}a}{R_C^2} \qquad (7.111)$$

where P_{CM} is the maximum CDMA mobile power. Assuming perfect power control, all signals coming from the CDMA mobiles are received at the same power $Q_{CC}^B = Q_0$. In that case the maximum CDMA mobile power is

$$P_{CM} = \frac{Q_0 R_C^2}{a} \qquad (7.112)$$

We will consider the worst case scenario in which all analog mobiles transmit at maximum power P_{AM}, so the interference coming from analog users is maximum. We also assume that the analog mobiles are uniformly distributed over the macrocell excluding the microcell area. Then, the total average analog mobile power received at the CDMA base station positioned at the distance D from the analog macrocell base is given by

$$Q_{CA}^B = \eta_r \left(\frac{3M}{\pi R_A^2}\right) P_{AM} b_C \int_{R_C-\pi}^{\infty} \int \frac{r d\Theta dr}{r^4} = \frac{3\eta_r M P_{AM} b_C}{R_A^2 R_C^2} \qquad (7.113)$$

where η_r is the base station notch filter depth. The CDMA E_b/I_o at the base station is then

$$\Gamma_{CB} = \frac{gQ_0}{\alpha K Q_0 + Q_{CA}^B} \qquad (7.114)$$

7.8.5.2 Analog Base Station Signal-to-Interference Ratio

Now we need to evaluate the impact of CDMA signals on the analog base station. If we assume that there are K CDMA mobiles uniformly distributed over the microcell, the total average power transmitted by all the CDMA mobiles is given by

$$\overline{P}_{CM} = \int_0^{R_c} \int_0^{2\pi} \frac{\alpha K Q_0 r^2}{\pi R_C^2 a} r d\Theta \, dr = \frac{\alpha K Q_0 R_C^2}{2a} \tag{7.115}$$

To simplify the analysis we assume that $D \gg R_C$ and thus we can approximate the total CDMA power received at the macrocell base station as

$$Q_{AC}^B \cong \overline{P}_{CM} \frac{bA}{D^4} \tag{7.116}$$

The worst case power received by the analog base station from an analog mobile, located at the macrocell boundary R_A, is

$$Q_{AA}^B = \frac{P_{AM} b_A}{R_A^4} \tag{7.117}$$

As a result, the worst case carrier-to-interference ratio is given by

$$\Gamma_{AB} = \frac{G_p Q_{AA}^B}{Q_{AC}^B} = \frac{2a G_p P_{AM} D^4}{\alpha K Q_0 R_C^2 R_A^4} \tag{7.118}$$

7.8.5.3 CDMA Reverse Link Capacity

The CDMA capacity can be expressed as (from (7.114))

$$K \cong K_0 - \frac{Q_{CA}^B}{Q_0} \tag{7.119}$$

where K_0 is the maximum achievable capacity obtained with no interference from the analog network. If we solve (7.118) for Q_0 and substitute it into (7.119) we get for the reverse link capacity

$$K = K_0 / \left(1 + \frac{3 \eta_r M \Gamma_{AB} R_A^2 R_C^2}{2 G_p D^4}\right) \tag{7.120}$$

where we have used the equality $b_C = aR_C^2$ from the adopted propagation model.

Figure 7.26 shows the CDMA microcell reverse link capacity versus normalized microcell distance from the base station, $d = D/R_A$, for various numbers of macrocell users, for the case without the notch filters in the base station. The system parameters were the same as in the previous example. The capacity is very high when the microcell is located far from the macrocell base station and drops considerably when the separation is small.

Figure 7.27 presents the capacity when the notch filters are used in the receiver of the base station. The filters had a depth of $\eta_t = -30$ dB. As the figure indicates, the capacity is now independent of distance and microcell usage and the impact of macrocell on the CDMA network is completely eliminated.

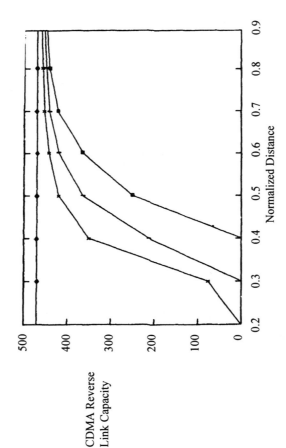

Figure 7.26 CDMA microcell reverse link capacity without notch filters. (© 1994 IEEE. *From:* [3].)

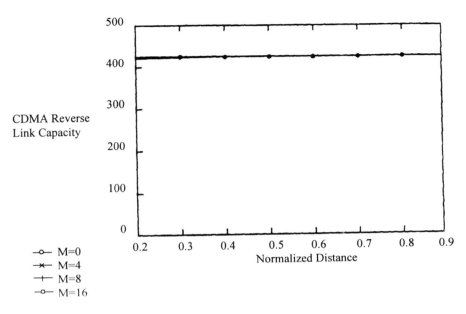

Figure 7.27 CDMA microcell reverse link capacity with 30-dB base receive notch filters. (© 1994 IEEE. From: [3].)

References

[1] Milstein, L. B., et al., "On the Feasibility of a CDMA Overlay for Personal Communication Networks," *IEEE J. Selected Areas in Communications*, Vol. 10, No. 4, May 1992, pp. 655–667.
[2] Wei, P., J. R. Zeidler, and W. H. Ku, "Adaptive Interference Suppression for CDMA Overlay Systems," *IEEE J. Selected Areas in Communications*, Vol. 12, No. 9, Dec. 1994, pp. 1510–1523.
[3] Grieco, D., "The Capacity Achievable with a Broadband CDMA Microcell Underlay to an Existing Cellular Macrosystem," *IEEE J. Selected Areas in Communications*, Vol. 12, No. 4, May 1994, pp. 744–750.
[4] Ketchum, J. W., and J. G. Proakis, "Adaptive Algorithms for Estimating and Suppressing Narrowband Interference in PN Spread Spectrum Systems," *IEEE Trans. on Communications*, Vol. COM-30, May 1982, pp. 913–924.
[5] Hsu, F. M., and A. A. Giordano, "Digital Whitening Techniques for Improving Spread Spectrum Communications Performance in the Presence of Narrowband Jamming and Interference," *IEEE Trans. on Communications*, Vol. COM-26, Feb. 1978, pp. 209–216.
[6] Li, L., and L. B. Milstein, "Rejection of Narrowband Interference in PN Spread Spectrum Systems Using Transversal Filters," *IEEE Trans. on Communications*, Vol. COM-30, May 1982, pp. 925–928.
[7] Milstein, L. B., and P. Das, "Spread Spectrum Receiver Using Acoustic Surface Wave Technology," *IEEE Trans. on Communications*, Vol. 25, Aug. 1977, pp. 841–847.
[8] Milstein, L. B., and P. Das, "An Analysis of a Real-Time Transform Domain Filtering Digital Communication System, Part I: Narrowband Interference Rejection," *IEEE Trans. on Communications*, Vol. 28, June 1980, pp. 816–824.

[9] Milstein, L. B., and P. Das, "An Analysis of a Real-Time Transform Domain Filtering Digital Communication System—Part II: Wideband Interference Rejection," *IEEE Trans. on Communications*, Vol. 31, Jan. 1983, pp. 21–27.
[10] Rapajic, P., and B. Vucetic, "Adaptive Single User Receiver for Asynchronous CDMA Systems in the Narrowband Interference Environment," Adelaide, Australia, Nov. 12–15, 1992, pp. 143–148.
[11] Rapajic, P., and B. Vucetic, "Adaptive Receiver Structures for Asynchronous CDMA Systems," *IEEE J. Selected Areas in Communications*, Vol. 12, No. 4, May 1994, pp. 685–697.
[12] Iltis, R. A., and L. B. Milstein, "Performance Analysis of Narrowband Interference Rejection Techniques in DS Spread Spectrum Systems," *IEEE Trans. on Communications*, Vol. COM-32, Nov. 1984, pp. 1169–1177.
[13] Iltis, R. A., and L. B. Milstein, "An Approximate Statistical Analysis of the Widrow LMS Algorithm with Application to Narrow-band Interference Rejection," *IEEE Trans. on Communications*, Vol. COM-33, Feb. 1985, pp. 121–132.
[14] Milstein, L. B., "Interference Rejection Techniques in Spread Spectrum Communications," *IEEE Proc.*, Vol. 76, No. 6, June 1988, pp. 657–671.
[15] Masry, E., "Closed Form Analytical Result for the Rejection of Narrowband Interference in PN Spread Spectrum Systems—Part II: Linear Interpolation Filters," *IEEE Trans. on Communications*, Vol. COM-33, Jan. 1985, pp. 10–19.
[16] Poor, H. V., "Signal Processing for Wideband Communications," *IEEE Inform. Theory Soc. Newsletter*, Vol. 42, No. 2, June 1992, pp. 1–10.
[17] Murray, W., *Numerical Methods for Unconstrained Optimization*, New York: Academic Press.
[18] Ungerboeck, G., "Theory on the Speed of Convergence in Adaptive Equalizers for Digital Communications," *IBM J. Research and Development*, Vol. 16, Nov. 1972, pp. 546–555.
[19] Sorenson, H. W., and D. L. Alspach, "Recursive Bayesian Estimation Using Gaussian Sums," *Automatica*, Vol. 7, 1971, pp. 465–479.
[20] Guilford, J., and P. Das, "The Use of Adaptive Lattice Filter for Narrowband Jammer Rejection in DS Spread Spectrum Systems," *Proc. IEEE Int. Conf. Communications*, June 22–26, 1985, pp. 822–826.
[21] Li, L., and L. B. Milstein, "Rejection of CW Interference in QPSK Systems Using Decision Feedback Filters," *IEEE Trans. on Communications*, Vol. COM-31, April 1983, pp. 473–483.
[22] Lin, F., and L. M. Li, "Rejection of Finite Bandwidth Interference in QPSK Systems Using Decision-Feedback Filters," *IEEE Int. Conf. Communications*, June 1987, pp. 24.6.1–24.6.5.
[23] Takawira, F., and L. B. Milstein, "Narrowband Interference Rejection in PN Spread Spectrum Systems Using Decision Feedback Filters," *IEEE Military Commun. Conf.*, Oct. 1986, pp. 20.4.1–20.4.5.
[24] Vijayan, R., and H. V. Poor, "Nonlinear Techniques for Interfernce Suppression in Spread Spectrum Systems," *IEEE Trans. on Communications*, Vol. 38, July 1990, pp. 1060–1065.
[25] Garth, L. M., and H. V. Poor, "Narrowband Interference Supression in Impulsive Channels," *IEEE Trans. on Aerospace Electronic Sysems.*, Vol. 28, Jan. 1992, pp. 15–39.
[26] Davidovici, S., and E. Kanterakis, "Narrow-Band Interference Rejection Using Real-Time Fourier Transform," *IEEE Trans. on Communications*, Vol. 37, July 1989, pp. 713–722.
[27] Saulnier, G., "Suppression of Narrowband Jammers in a Spread Spectrum Receiver Using Transform Domain Adaptive Filtering," *IEEE J. Selected Areas in Communications*, Vol. 10, May 1992, pp. 792–749.
[28] Oppenheim, A. V., and R. W. Schafer, *Digital Signal Processing*, Englewood Cliffs, NJ: Prentice-Hall, 1975.
[29] Milstein, L. B., and P. Das, "Surface Acoustic Wave Devices," *IEEE Communications*, Sept. 1979, pp. 25–33.
[30] Gerargiz et al., "A Comparison of Weighted and Nonweighted Transform Domain Processing Systems for Narrowband Interference Excision," *IEEE Military Commun. Conf.*, Oct. 1984, pp. 32.3.1–32.3.4.

[31] Harris, F. J., "On the Use of Windows for Harmonic Analysis Discrete Fourier Transform," *Proc. IEEE*, Vol. 66, 1978, pp. 51–83.
[32] Schilling, D. L., et al., "Broadband CDMA Overlay," *Proc. IEEE 43rd VTS Conf.*, Secaucus, NJ, May 18–20, 1993, pp. 452–455.
[33] Lee, W., *Mobile Communications Design Fundamentals*, New York: Howard W. Sons & Co, 1986.
[34] Gilhousen, K. S., et al, "On the Capacity of a Cellular CDMA System," *IEEE Trans. on Vehicular Technology*, Vol. 40, May 1991, pp. 303–313.

Selected Bibliography

Bouvier, M. J., "The Rejection of Large CW Interferers in Spread Spectrum Systems," *IEEE Trans. on Communications*, Vol. 28, Feb. 1978, pp. 254–256.

Li, Z., H. Yuan, and G. Bi, "Rejection of Multitone Interference in PN Spread Spectrum Systems Using Adaptive Filters," *IEEE Int. Conf. Communications*, June 1987, pp. 24.5.1–24.5.5.

Proakis, J. G., *Digital Communications*, New York, NY: McGraw Hill, 1983.

Saulnier, G. J., K. Yum, and P. Das, "The Suppression of Tone Jammers Using Adaptive Lattice Filtering," *IEEE Int. Conf. Communications*, June 1987, pp. 24.4.1–24.4.5.

Wang, Y. C., and L. B. Milstein, "Rejection of Multiple Narrowband Interference in Both BPSK and QPSK DS Spread Spectrum Systems," *IEEE Trans. on Communications*, Vol. COM-36, Feb. 1988, pp. 159–204.

Widrow, B., et al., "Adaptive Noise Cancelling: Principles and Applications," *Proc. IEEE*, Vol. 63, Dec. 1975, pp. 1692–1716.

CHAPTER 8

MULTIPLE-ACCESS PROTOCOLS IN SPREAD SPECTRUM PACKET RADIO NETWORKS

This chapter examines various multiple-access techniques, protocols, and throughput performance aspects in spread spectrum packet radio networks. We provide the background for packet radio networks and multiple-access technique fundamentals. The bulk of the chapter is dedicated to the role of spread spectrum in packet radio networks. The key features of spread spectrum signaling that enable multiple-access capability, capture, and multipath resistance are considered. Both DS/CDMA and frequency hopping multiple-access techniques are presented in a random access mode of operation. The throughput performance of a number of random access CDMA systems is analyzed. In particular, the use of adaptive receivers in random access CDMA packet radio networks is discussed, including its effect on the throughput performance.

8.1 PACKET RADIO NETWORKS—GENERAL CONCEPTS

Packet radio networks provide communication between distant users via a radio medium based on packet switching.

Examples include mobile packet radio networks [1,2], wireless data communications [3], analog cellular mobile networks [4], satellite data networks [5,6], IS-95 Qualcomm CDMA network [7], GSM cellular mobile network [8], mobile satellite networks such as Iridium [9] and Globalstar [10], and mobile packet data networks including ARDIS by IBM and Motorola, Mobitex by Ericsson and Swedish Telecom, and *cellular digital packet data* (CDPD) by IBM et al. [3].

There are two general approaches, known as *packet switching* and *circuit switching*, that can be used to transmit the information for various services.

A circuit-switched connection is one for which a dedicated communication path or *circuit* is established during each call. The circuit, once established, remains uninterrupted for the entire duration of the call. Circuit switching is used for voice transmission in public telephone networks.

In a packet-switched network, the continuous messages are divided into segments of the fixed length prior to transmission. The segments are referred to as *packets*. Packet switching uses existing circuits in the network to transmit packets that contain the necessary information to direct the packet to its destination. Series of packets move independently through the network. They form queues for processing and transmission in the network nodes through which they pass. The original messages are reassembled at the destination on a packet-by-packet basis. Note that the distinguishing feature of packet switching is that at each switching node each packet is received and then stored in a buffer until the link to the next node becomes available for forward transmission. If transmission lines are congested, the packet can be retained for later transmission or it could be sent by some roundabout route. Thus, a communication link in a packet-switched network is shared *dynamically* among many competing users. In contrast, in a circuit-switched network, a communication link is dedicated to the exclusive use of the two users communicating with each other for the entire duration of their call. Therefore, packet switching enables more efficient use of transmission lines than circuit switching. The use of packet switching is suited to computer communications services in which exchange of data occurs in an unpredictable and bursty fashion. The packet data service is also suitable for low-rate low-volume data transmission in applications such as messaging, paging, and telex. Closed user groups may use this type of transmission for their own private radio networks.

A *protocol* is a set of rules, including formats and procedures that enable one user to communicate, through the hub station or packet-switched node, with the other users. It is implemented in a *controller*.

The controller could be located in the hub station, distributed among users or in only one designated radio station.

In the analysis of packet networks two significant performance indicators are throughput and delay.

The *thoughput* is defined as the number of successfully transmitted packets per packet transmission time. It is computed as the ratio of the number of successfully transmitted packets to the total numbers of packets that were sent. Accordingly, the throughput assumes values within the range 0 to 1.

The *delay* of a packet is the time between its start and successful completion of transmission and delivery to the user.

The following services are possible in packet radio networks:

- Interactive computer to terminal sessions;
- File transfer;
- Streams (packetized voice, facsimile);
- Datagrams;
- Paging.

A datagram is a self-contained packet particularly used in broadcast applications.

A stream is service characterized by volatile periodic traffic for which the maximum acceptable delay is strictly limited. Examples are voice and some forms of facsimile.

A packet radio network typically consists of radio terminals that communicate with each other via broadcast radio channels and packet switching nodes. Each node contains a *radio terminal, digital controller*, and an *antenna*.

The radio terminal acts like a modem providing connectivity between neighboring nodes.

The antenna transforms the electrical signals received from the radio terminal into electromagnetic waves and radiates them into the air.

The digital controller is a processor that provides packet switching. It receives packets from neighboring terminals, makes the routing decisions, performance and error control, and forwards the packets to the next nodes [11].

Typical examples of packet radio networks are satellite and cellular networks. A *satellite network* consists of a geosynchronous or a low Earth orbit satellite and a number of Earth stations that communicate with each other. In a single-beam satellite there is a transponder that operates as a repeater. It receives uplink signals, amplifies, and translates to a downlink frequency band. In a multibeam satellite there are multiple transponders that point to different regions on the Earth. Terminals within the same beam will use the same uplink and downlink frequency bands. A transmission between two terminals in different regions will therefore involve two roundtrips to the satellite, unless on-board switching between beams is provided on the satellite. It may be advantageous to do on-board processing for multibeam satellites or in a network consisting of several single-beam satellites. Then the transponder has to make certain routing decisions for each message.

Another important network topology is a *cellular network* with frequency reuse. It is used in cellular and personal mobile telephone and data communications. A cellular network consists of a large number of mobile users spread over wide geographic areas communicating via base stations. A large service area is divided into smaller areas called cells. Each cell is served by a base station. The cell sites are designed in approximately regular geometric patterns to cover the whole service area with a proper signal strength. Each cell is a centralized network with all communication to and from a mobile user. All cells are connected to the public-switched telephone network. Cellular networks are designed with frequency reuse in order to maximize the overall capacity for given frequency allocations. At a sufficient distance from a cell, the same set of frequencies is used simultaneously in other cells.

In this chapter we will focus on the features of packet radio networks. In packet-switched radio networks packets are multiplexed prior to transmission. In the next section we will discuss fundamentals of various multiplexing and multiple-access techniques.

8.2 MULTIPLEXING AND MULTIPLE-ACCESS TECHNIQUES

Multiplexing and multiple access both refer to sharing a common transmission channel by a number of users. However, multiplexing is related to fixed or slowly changing channel allocations. The channel sharing is usually limited to the local site. Multiple access typically relates to the remote sharing of a resource like satellite or base station by a number of users located over a large geographic region.

8.2.1 Multiplexing

Multiplexing is the transmission of information from more than one source, located at the same site, to more than one destination over the same physical medium. There are several types of multiplexing but three are most common.

Frequency-Division Multiplexing (FDM): Signals that originally occupied the same frequency spectrum are each shifted to a different frequency band and transmitted simultaneously over a single transmission medium. Common examples are AM, FM radio, and TV broadcasting. Another example is the analog public telephone network.

Time-Division Multiplexing (TDM): Signals from various sources are interleaved in the time domain. The most common example is the PCM-TDM digital telephone system in which several voice band channels are sampled, converted to PCM codes, and time-division multiplexed for transmission over the same cable.

Code-Division Multiplexing (CDM): Signals are assigned their own unique signatures before combining them for simultaneous transmission using the full channel bandwidth. An example is the Qualcomm CDMA cellular mobile network.

Space-Division Multiplexing (SDM): Signals are separated by pointing spot beam antennas in different directions. This method allows the frequency band reuse in different areas. Examples are cellular mobile networks.

Polarization Division (PD): Orthogonal polarizations of signals are used for their separation. Examples are INTELSAT V satellite communications systems where each beam is generated in two polarizations to increase bandwidth efficiency.

8.3 MULTIPLE-ACCESS SCHEMES

In packet radio networks there is a need to communicate between a large number of radio terminals and a single hub station. In these systems there are two types of communication channels. A *forward* or *broadcast channel*, which is used to transmit information from the hub station to the terminals, is represented by the dash line in Figure 8.1. The signals in this channel are controlled by the hub station, and it is relatively easy to synchronize them. Accordingly, they are most often combined for transmission by a simple TDM or FDM technique. For example, a combination of FDM and TDM is used in GSM cellular radio networks for mobile voice communications [8].

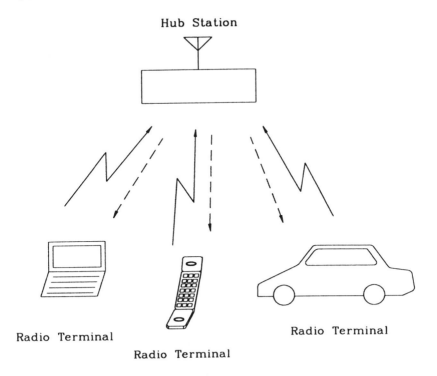

Figure 8.1 A packet radio network model.

The *reverse* or *multiple-access* channel, marked in Figure 8.1 by solid lines, is used to transmit data from remote terminals to a single hub station. The remote terminals are obviously uncoordinated and scattered over a broad geographic region. Hence, the problem of sharing a multiple-access channel is much more challenging than sharing a broadcast channel. The basic problem is how to enable a varying group of radio terminals to share the hub station in such a way to optimize the system capacity, spectrum utilization, the hub station power, cost, and user acceptability [12]. In different systems these requirements will have different priorities, and they should be addressed for each individual network.

There are basically three broad types of multiple-access techniques: *fixed access*, *random access*, and *controlled access*.

In fixed access a fraction of the multiple-access channel is permanently allocated to each radio terminal. In *frequency division multiple access* (FDMA) all terminals transmit at the same time, but each of them is confined to its own unique frequency band. In *time division multiple access* (TDMA) the terminals transmit in turn in their own unique time slots. In *code division multiple access* (CDMA) terminals transmit signals that overlap both in time and frequency. Transmissions are separated by orthogonally coded spread spectrum modulation.

Fixed access schemes are good choices for networks with steady traffic. In many networks the traffic pattern from radio terminals might vary significantly with time of day or season. In other networks there might be a large number of users transmitting in bursts with long inactive periods between two successive transmission bursts. Permanent allocations of channel capacity in such a network tend to be inefficient.

In random access there is no permanent assignment of transmission resources. The multiplexing strategy is based on contention. The entire channel capacity is allocated to one terminal only when it is needed. These schemes have been successfully used in radio and satellite communications to provide services to a large population of bursty users.

Controlled access or reservation schemes improve the capacity of random access techniques by assigning channel resources to users on demand. In these schemes a separate channel, called the *request channel*, is allocated for transmission of the request for capacity when it is needed.

8.3.1 Frequency Division Multiple Access

FDMA is illustrated in Figure 8.2. The multiple-access channel available bandwidth B is divided among N users, which transmit their signals simultaneously.

In an FDM/*Frequency Modulation* (FM) system, a multiplexer takes the baseband signals from many individual users; translates them to nonoverlapping channels in the frequency domain, called *subdivisions*; and combines them. The resulting

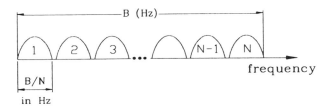

Figure 8.2 Frequency division multiple access scheme.

composite FDM signal frequency-modulates an IF carrier to create an FM multiplex signal.

Subdivisions are also combined into groups, super groups, or master groups. This form of FDM is called *multichannel per carrier* (MCPC). *Single-channel-per-carrier* (SCPC) multiplex systems transmit each baseband signal on its own carrier. This multiplexing method is easy to reconfigure to adapt to varying traffic conditions. Also, the carrier in an SCPC system is transmitted only when the link is active, while in an FDM/FM system the carrier is always present. Thus SCPC offers a saving in the transmitter power. However, SCPC requires more bandwidth than MCPC for the same number of users. The reason is that guard bands are needed between carrier bands; the more carriers there are, the more guard bands are in the signal spectrum. Furthermore, more carriers will cause more intermodulation products in nonlinear systems, so the output power must be backed off. Thus, the system can service fewer users.

An MCPC system is more efficient than a SCPC system under heavy traffic conditions if all subdivisions are almost always filled. However, if some subdivisions within a group are not active, they cannot be turned off. One important feature of SCPC is the ability to utilize voice-activated carriers in voice transmission. That is, the channel carrier is turned on only when active speech is transmitted in the channel. In normal phone transmissions one speaker is silent while the other is talking. There are also silent periods and hesitations in normal single-speaker speech. It has been shown that the average single speaker's speech activity takes only 40% of the total available channel time. In a two-way conversation, by turning off the carrier during silent periods on an SCPC system, one can save the system power or bandwidth as well as reduce intermodulation distortion between various terminals and utilize channel resources more efficiently.

The carrier frequencies in the hub stations for FDMA may be either preassigned to individual users for their exclusive use or demand assigned. In preassigned systems a control mechanism is used to make sure that no two users transmit on the same subdivision at the same time. In demand-assigned systems a control mechanism is used to establish and terminate communication sessions between the transmit and receive terminals. Hence, any of the subdivisions may be used by any of the users at any given time, providing that they are available.

The first FDMA demand assignment system for satellite communication was the *single-channel-per-carrier PCM multiple-access demand assignment equipment* (SPADE) system used on the INTELSAT IV satellites [13].

8.3.2 Time Division Multiple Access

In TDMA a number of users transmit bursts of RF energy in separate nonoverlapping time slots. Accordingly, a carrier frequency can be shared by several terminals. Consider a system where the available bandwidth is B Hz. Assume that the total transmission time is divided into frames of T sec each, and that each frame is further partitioned into N time slots of duration T/N sec. In a TDMA system each slot is permanently allocated to one of the N users. Each user will use the entire system bandwidth of B Hz, and its transmission is confined to its own time. Thus, only one user's carrier is present in the hub station at any given time, thus avoiding collisions with other users carriers. A TDMA scheme is illustrated in Figure 8.3.

The number of time slots per carrier depends primarily on modulation scheme and available bandwidth. For example, in the European cellular mobile system (GSM) the number of voice channels per carrier is eight, while in the American IS-54 system this is three.

The bursts transmitted from various radio terminals must arrive at the hub station in a given order so that signal overlapping does not occur. If the radio terminals are mobile they have to adjust their transmission times to compensate for variations in their respective positions relative to the hub station.

To enable proper interleaving of individual bursts from multiple radio terminals a TDMA signal has a frame structure. An example of a basic TDMA frame is shown in Figure 8.4. A frame consists of N time slots. Each time slot contains a *reference burst* or *preamble*, which provides absolute time for the system. The reference burst is transmitted by a control station on the ground in satellite systems or by the base

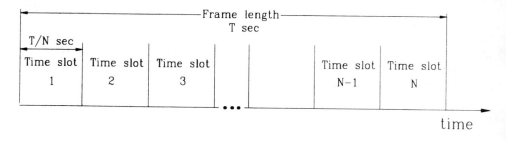

Figure 8.3 Time division multiple access scheme.

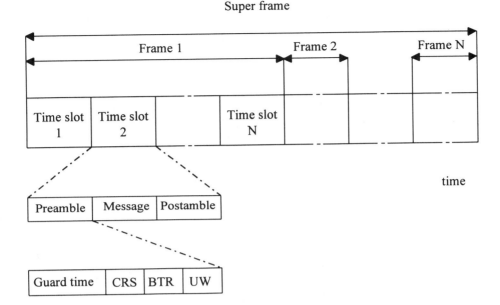

Figure 8.4 A basic TDMA frame.

station in cellular mobile systems. The reference burst contains a *carrier recovery sequence* (CRS), which is used to recover frequency and phase for coherent demodulation in the receiving station.

The *bit recovery sequence* (BTR) is used to acquire bit timing in the demodulator. The *unique word* (UW) transmitted at the end of the reference burst provides a time reference to the radio terminals. This information is used by each radio terminal to determine the transmission time of its burst. The UW is usually a binary sequence chosen for its correlation properties.

At the receiver side, once the synchronization circuit has achieved lock, it starts searching for the unique word. The circuit that does this is a correlator. Once the receiver recognizes the unique word it updates its timing relative to the beginning of the frame and its position in the frame.

It is evident from the description of TDMA systems that they are much more complex to implement than comparable FDMA systems due to synchronization requirements. Another shortcoming of TDMA lies in the fact that the transmissions from radio terminals take the form of bursts. This means that the transmitters will have a high peak power relative to their average transmitted power. This might be undesirable for small radio terminals subject to hardware limitations.

However, TDMA has several advantages over FDMA. In a TDMA system only the signal from one radio terminal is present in the hub station at any given instant, thus reducing intermodulation distortion. In contrast, in a FDMA, each radio terminal must be capable to transmit and receive a multitude of carrier frequencies to achieve multiple accessing capabilities. This results in spectral regrowth and intermodulation distortion in nonlinear systems. In addition, all practical TDMA systems are digital and, therefore, have all advantages of digital transmission over analog FDMA systems.

A fixed assignment TDMA scheme is very efficient when the traffic generated by the radio terminals is steady. However, if the terminal has no data to send in its own slot, that slot is wasted. When the traffic is time varying, there are more efficient schemes based on dynamic assignments. TDMA architectures are widely used in digital satellite and cellular mobile communications for transmission of voice signals [12,14].

8.4 CODE DIVISION MULTIPLE ACCESS

In FDMA and TDMA systems individual users are separated from each other in frequency and time, respectively. CDMA technique can be considered as a combination of FDMA and TDMA. In CDMA different user transmissions overlap in both frequency and time, while using the entire system bandwidth. The properties of CDMA schemes and their advantages over FDMA and TDMA in radio cellular networks were discussed in detail in Chapter 5.

CDMA techniques have been used for a variety of communication systems, such as NASA TDRSS [15], cellular mobile communications [16], and satellite mobile communications [17], and in military applications [18,19].

8.4.1 Random Access Schemes

In packet radio networks composed of a large number of radio terminals transmitting the information in bursts (Figure 8.5), fixed assignment multiple access schemes tend

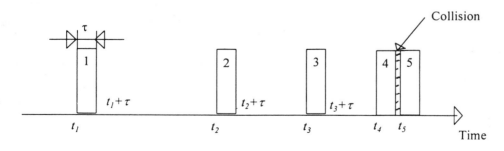

Figure 8.5 A low duty cycle ALOHA channel.

to be inefficient. Examples are digital satellite networks consisting of thousands of small Earth stations, called *very small aperture terminals* (VSATs) [20] and *ultra small aperture terminals* (USATs) [21] transmitting data in bursts. For such networks, random multiple-access schemes are more appropriate. Random access schemes are based on contention strategies. Two examples of random access schemes that can be used to provide communications to a large population of bursty users are pure ALOHA and slotted ALOHA schemes. It is important to note that in a packet radio network the need to apply random access only arises in transmissions over the multiple-access channel, that is, from the users to the hub station.

8.4.1.1 Pure ALOHA

Pure ALOHA is a packet radio system that was first introduced at the University of Hawaii. It was applied to interconnect several university computers via a communication satellite by use of a random access protocol [5]. Such a system could be used to provide communication between distributed users or between users and a hub station. In what follows we will consider the schemes for communication between users and a hub station. The data to be transmitted from the users to the hub station are packetized. Each packet contains the data section, the address of the source, and a *cyclic redundancy check* (CRC) for error detection. The transmitting operation consists of the following steps.

Each terminal transmits a packet immediately upon receiving it from the source, independently from other terminals.

The packets sent from different terminals may overlap in the hub station receiver and cause transmission errors. This situation is called *collision*.

In this case the errors are detected in the hub station and the users receive a negative acknowledgment.

After receiving a negative acknowledgment message the transmitter retransmits the packet. To avoid repeated collisions, the message is retransmitted after a random delay.

If the user does not receive either a positive or negative acknowledgment within a given time after transmitting a packet, the packet is retransmitted following the same procedure until it is finally acknowledged.

Let us assume that the arrival process of new packets is represented by the Poisson distribution. This assumption is valid for unrelated users in a network. That is, the probability of having m new messages arrive during a time interval of τ, denoted by $P_t(m)$, is given by [22]

$$P_t(m) = \frac{(\lambda \tau)^m e^{-\lambda \tau}}{m!} \qquad m \geq 0 \qquad (8.1)$$

where λ is the average packet arrival rate and τ is the packet duration.

Due to collisions, some of the messages will be unsuccessful. Therefore, we define the combined process of arrival of new packets with the rate of λ and retransmissions at the rate of λ_r. The total traffic arrival rate, denoted by λ_t, is then $\lambda_t = \lambda + \lambda_r$; and it is also modeled by a Poisson distribution. The packet transmission operation is shown in Figure 8.6.

A user can successfully transmit a message as long as no other user began one within the previous τ packets or starts one within next τ packets. Thus a space of 2τ packets is needed for each message.

Let P_s be the probability that a given message is successful. That is, the message will not collide with any other message. P_s can be computed by finding the probability that no other messages are transmitted during an interval of 2τ packet time, which is equivalent to the probability that $m = 0$ in (8.1). Assuming that the traffic is Poissonian, we can substitute for the total traffic λ_t and time interval 2τ in (8.1) to compute $P_s = P_t(m = 0)$ as

$$P_s = P_t(m = 0) = \frac{(2\lambda_t \tau)^0 e^{-2\tau\lambda_t}}{0!} = e^{-2\tau\lambda_t} \tag{8.2}$$

The probability of successful packets can be defined as the ratio of the new arrival rate λ and the total arrival rate λ_t

$$P_s = \frac{\lambda}{\lambda_t} \tag{8.3}$$

By combining (8.2) and (8.3) we get

$$\lambda = \lambda_t e^{-2\tau\lambda_t} \tag{8.4}$$

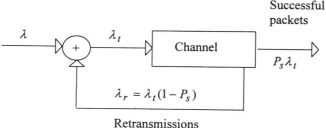

Figure 8.6 Operation of an ALOHA system.

Defining the *normalized throughput* S as the number of successfully transmitted packets per packet time, given by

$$S = \lambda \tau \tag{8.5}$$

and the *total traffic* as the number of packets transmitted per packet time

$$G = \lambda_t \tau \tag{8.6}$$

(8.4) can be written in the form

$$S = Ge^{-2G} \tag{8.7}$$

Equation (8.7) is plotted in Figure 8.7. The maximum possible value for the throughput S is $e/2 = 0.18$ corresponding to the total traffic of $G = 0.5$. That is, for pure ALOHA only about 18% of the available capacity is used. In fact, the system must operate much below its capacity in order to achieve system stability.

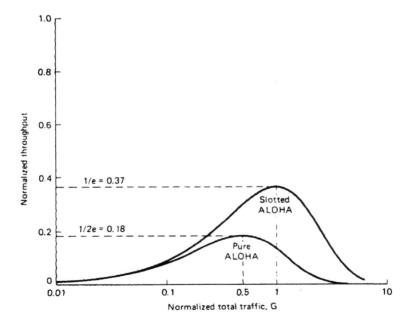

Figure 8.7 Normalized throughout for ALOHA systems.

8.4.1.2 Slotted ALOHA

In the slotted ALOHA the probability of collisions is reduced by introducing a small degree of coordination between the users. The time axis is considered to be universal for all terminals and divided into equally sized slots. A node with a packet scheduled for transmission in a particular slot transmits that packet, synchronizing the start of transmission to coincide with the beginning of the slot. It is assumed that the packet duration is equal to the time slot.

A user can start transmitting a message only at the beginning of a time slot. All the users are assumed to be correctly synchronized so that the slot definitions are the same for the whole network.

A user can successfully transmit a message if no other user transmits within an interval of τ packets. Combining (8.1) and (8.2) the probability that a message is successful in a slotted ALOHA system is

$$P_s = e^{-\tau \lambda_t} \qquad (8.8)$$

The packet arrival rate and normalized throughput are given by

$$\lambda = \lambda_t e^{-\lambda_t \tau} \qquad (8.9)$$

$$S = Ge^{-G} \qquad (8.10)$$

The maximum throughput of the slotted ALOHA scheme is $1/e$ or about 36% of the available capacity, and it is obtained for the load of $G = 1$. The capacity of the slotted ALOHA is doubled relative to the capacity of the pure ALOHA system. Figure 8.7 shows the relationship between the normalized throughput and the total traffic for pure ALOHA and slotted ALOHA.

Both pure ALOHA and slotted ALOHA are susceptible to instability at high traffic load. As the total traffic load increases, the probability of collisions goes up and eventually no packet can be successfully transmitted.

The slotted ALOHA system is used in the European cellular radio mobile system [8] to provide initial access of mobile users to the base station. Surprisingly, the propagation effects of mobile radio, which are detrimental to error performance, greatly enhance the capacity of the slotted ALOHA. The most important effect is multipath propagation. It occurs in mobile radio due to signal scattering and various propagation delays experienced by different scattered signals. The scattered signals with approximately the same propagation delay undergo Rayleigh distribution. The total signal, which is the sum of Rayleigh distributed signals with various propagation delays, is frequency selective. The second significant propagation effect is the near-far phenomenon. It manifests in unequal power levels of the signals received at the base station originated by mobiles at various positions in the cell. The combination of these two effects has a beneficial impact on the capacity of the slotted ALOHA. In a standard slotted ALOHA scheme we assume that the power levels of the signals received by the base station, denoted by W, are the same. Hence, when a collision occurs, the received signal in the base station will be subject to high interference at the power level of at least W. Thus any collision will cause destruction of all messages involved and they have to be repeated.

On the other hand, in mobile cellular networks it is possible that a signal received at the base station is much stronger than other received signals coming from distant users. If the total interference of other signals is low, the base station might receive successfully the strongest signal. This successful reception in the presence of other interfering signals is called *capture*.

The capture effect increases the system capacity relative to the standard slotted ALOHA since a signal can be successfully received even if a collision occurs.

Assume that the power level of the desired signal is W and the total interference level is I. The capture will occur if the signal power is higher than the total interference level I by a certain threshold R_0.

$$\frac{W}{I} > R_0 \qquad (8.11)$$

Assume the slotted ALOHA protocol, Rayleigh fading environment, and the near-far effect in a cell expressed as

$$\overline{W} = \left(\frac{r}{r_{max}}\right)^{-\mu} \qquad 2 \leq \mu \leq 5.5 \qquad (8.12)$$

where $\overline{W}$ is the mean signal power, r is the distance of the mobile from the base station, r_{max} is the maximum distance of the mobile, and μ is a parameter describing signal attenuation with distance [7]. The analysis of the throughput for the slotted ALOHA system in this environment was made by Arnbak et al. [14]. The throughput curves are shown in Figure 8.8.

8.4.2 Controlled Access Schemes

We have shown that pure ALOHA and slotted ALOHA random access schemes can only provide relatively low capacity. Higher capacity can be obtained with more complex contention resolution methods.

One example is the *carrier sense multiple access* (CSMA). In this scheme each terminal monitors the status of the channel before it starts transmitting. If the channel

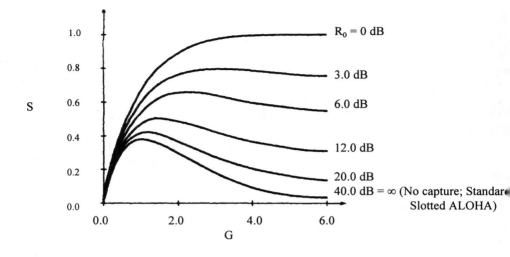

Figure 8.8 Throughput curves for various capture thresholds. (© 1987 IEEE. *From:* [14].)

is idle, the terminal transmits; otherwise it waits. The information about the status of the channel is acquired by carrier sensing. The propagation delay has a substantial effect on the throughput in CSMA systems. Another shortback of CSMA is a problem of *hidden terminals*, which occurs when the network is not fully connected. Assume that there is no direct connection between nodes i and j. If they both try to transmit to node k, i, and j cannot hear each other and a collision will occur. This scheme is suitable for cable networks and has been used in the Ethernet system [23]. The maximum throughput of CSMA exceeds 80% if the propagation delay is negligible compared to the packet transmission time [24]. In satellite systems this scheme is not practical since the propagation delay is too long. In mobile radio systems, channel monitoring is affected by the near-far effect.

We will now consider a simple way of improving the throughput of random access schemes. If we transmit long packets in ALOHA or slotted ALOHA they are often wasted in collision or waiting. It would be more efficient to send very short packets in a contention mode to reserve noncontending slots for transmission of actual data. In this way, the packets wasted in collisions or by idles are all short, so the overall throughput becomes much higher than in random access schemes.

8.4.2.1 Reservation ALOHA

Reservation ALOHA (R-ALOHA) scheme can be looked upon as a combination of TDMA and slotted ALOHA. The channel data is structured in frames and slots as in TDMA. Each slot within a frame can be declared as available for connection or reserved. Terminals contend for access to the idle slots before transmissions. Once the terminal is successful it gets a time slot in the subsequent frames until it has no more packets to send. The time frame is divided into $M + 1$ slots whenever a reservation is made. The first slot is subdivided into subslots to be used for reservation requests. The other M slots are used for actual data transmission. The organization of frames and slots in an R-ALOHA system is shown in Figure 8.9.

This type of multiple access is used for the European digital cellular mobile radio system GSM in which request messages are sent based upon a random access slotted ALOHA system and the actual data are transmitted in reservation-based TDMA, as explained [8]. The request message consists of only eight bits. Three of them indicate the reason for access, which could be the answer to paging, emergency call, call re-establishment, or user request. The remaining five are chosen randomly. They serve as a random discriminator to enable the user to reduce ambiguity in getting the reply message from the network after sending the initial request for capacity. This is done by correlating the response from the network with the user request message.

8.4.2.2 Demand Assignment Multiple Access

The *demand assignment multiple access* (DAMA) scheme is similar to the R-ALOHA scheme. The only difference is that the R-ALOHA is based on the TDMA approach, while DAMA is based on FDMA. The structure of channel subdivisions for a DAMA scheme is shown in Figure 8.10.

The entire available bandwidth is partitioned into M_r request channels and M_m data channels. A user with data to transmit first sends a request packet to the hub station.

The request channels are accessed by a random access scheme such as pure or slotted ALOHA. After getting an assignment from the hub station, the user transmits in one of data channels based on FDMA. The protocol for the American mobile

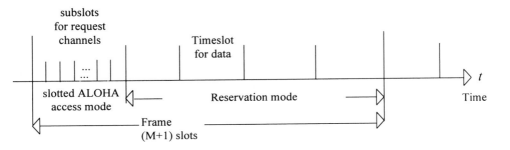

Figure 8.9 Organization of frames and slots in a reservation ALOHA system.

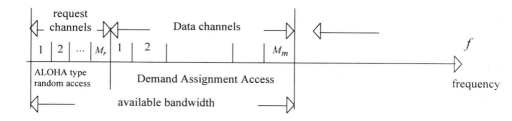

Figure 8.10 Channel subdivisions in a DAMA system.

satellite network MSAT-X [25] and the protocol for the Australian mobile satellite network Mobilesat [9] are variations of the DAMA scheme.

MSAT-X and Mobilesat provide mobile satellite communication voice and data services for rural areas.

MSAT-X multiple-access protocol [25] is also based on a DAMA scheme. The satellite channel bandwidth is divided into M channels. There are M_r reservation channels, M_v voice channels, and M_d data channels, so that $M = M_r + M_v + M_d$. There is a *network management center* (NMC) that is responsible for the management of the network, including the channel assignment to the terminals. NMC periodically estimates the data and voice call arrival rates and accordingly calculates the number of reservation, voice, and data channels in order to optimize the network performance. That is, the NMC might dynamically adjust the number of these channels according to the offered traffic. The identities of the newly assigned channels are broadcast on the same status channel that is monitored by all active terminals in the network when a terminal has a message to retransmit; it monitors the status channel to get the information about the up-dated reservation channels. It then sends a request on one of the reservation channels to NMC using the slotted ALOHA protocol. The request consists of the origin and destination addresses; whether it is a voice or data request; and, in the case of data request, the length of the data message. After transmitting the request, the mobile terminal waits for a channel assignment from NMC. When the channel is assigned, it transmits its message. If the channel assignment is not received within a specified time-out period, the mobile will retransmit the request after a random delay.

The DAMA scheme can support a large number of bursty users because of the random access scheme used for the request. It can also achieve a high throughput efficiency because of the DAMA mode of data transmission.

8.4.3 CDMA Slotted ALOHA Schemes

In digital cellular mobile [8,16] and satellite mobile [9,25,10], commercial networks requiring high-efficiency use of channel resources, controlled access schemes are most frequently used. In these systems capacity is allocated on demand in response to user requests. Upon receiving a request, the network allocates channel capacity for transmission of actual data on the basis of FDMA, TDMA [22], or CDMA schemes [16]. For transmission of request signals a separate request channel is needed. The request channel is shared among various users in these networks on the basis of random access slotted ALOHA protocols.

There are some other digital packet radio networks in which the traffic is basically random and bursty. For such networks, highly efficient use of channel resources can be achieved by selecting a random access protocol for all the network traffic rather than for a request channel only. Examples of such networks are satellite networks composed of a large number of VSAT terminals [20] providing shared databases, reservation, and financial services.

The operation of such a random access network will depend on the channel signaling method, which could be *narrowband* or *spread spectrum*. There are several properties of spread spectrum that can improve the networks performance.

The main advantage of spread spectrum is its inherent resistance to multipath propagation, narrowband interference, and jamming.

In spread spectrum networks it is possible to exploit the capture effect to increase the throughput. The capture effect is the ability of a receiver to receive correctly a packet despite the presence of time overlapping packets. In narrowband networks the overlap of two packets with approximately the same signal power will result in the destruction of both packets. A capture might occur only if the packets have distinctly different signal powers.

In spread spectrum random access networks packet overlapping does not necessarily result in packet loss. It is possible to correctly receive a packet if several of them overlap with each other even if each packet uses the same code sequence. Capture capability in spread spectrum systems refers to the ability of the receiver to receive a packet in the presence of overlapping signals with the same code sequences. In order to receive the first of overlapped packets, it should arrive several chips to several symbols before other packets [1] and the interference from other packets should not be too high.

Spread spectrum enables discrimination between a number of users in the code domain. The channel resource is shared by transmitting orthogonal codes that overlap in time without affecting each other. A spread spectrum receiver can discriminate between packets coming from various users on the basis of code sequences. This is referred to as the *multiple access capability* of spread spectrum and is called CDMA.

These properties make spread spectrum signaling in combinations with random access schemes attractive for many packet radio networks with bursty traffic and mobile terminals. These combined schemes are referred to as CDMA ALOHA schemes when different codes are assigned to terminals or spread ALOHA when one code sequence is assigned to all terminals. The rules that govern the selection of the code sequence for a particular packet are called *spread spectrum protocols*.

There are two forms of CDMA, based on *frequency hopping* (FH) and *direct sequence* (DS) signaling.

The operation and performance of a CDMA ALOHA network depend on the selected channel access and spread spectrum transmission protocols. There are some common properties of DS and FH spread spectrum systems that influence the choice of the protocols, and we will consider these in the next section.

8.4.3.1 Spread Spectrum Transmission Protocols

Discrimination between various users' transmissions is based on spreading sequences, which are referred to as signatures or chip sequences for DS systems or frequency hopping patterns for FH systems.

Multiple-access, capture, and antimultipath properties as well as network throughput performance are intimately related to the rule for spreading sequence assignment. In a CDMA system a number of orthogonal or quasi-orthogonal signals are transmitted simultaneously in order to share the common channel. However, collisions are possible in spread spectrum systems as in narrowband ALOHA systems. There are two reasons for collisions in CDMA ALOHA systems.

Primary collisions occur due to two or more users simultaneously transmitting packets using the same spreading sequence. The packets involved in the collision will be lost and must be retransmitted. However, note that in CDMA ALOHA it is possible that two packets collide and get destroyed and a third packet transmitted with a different spreading sequence can be successfully received at the same time. This is an important difference from narrowband ALOHA in which collisions refer to individual time slots or receivers, while in CDMA ALOHA they refer to individual packets.

Secondary collisions in CDMA ALOHA occur due to high interference coming from other spreading codes that are not exactly orthogonal to the given spreading code for the desired signal. We assume that the hub station multiple-access receiver can recover a given received packet free of errors if the number of multiple-access interferers at a given time is smaller than a specified maximum value. We also assume that all these interferers use spreading sequences that are different from those of the packet under consideration. If the number of interferers is larger than the maximum number, all packets are destroyed.

A packet will be lost if it is involved in any primary or secondary collisions.

A designer can choose one of the following ways to assign spreading sequences to active users in CDMA ALOHA systems.

Transmitter-Receiver-Oriented Protocol

Every pair of nodes is assigned a distinct code sequence. That is, if terminal i has a message to send to terminal j, it will use a code sequence assigned to this pair, denoted by C_{ij}. In a network with N terminals, $N(N-1)$ code sequences are needed.

This protocol is suitable for point-to-point communication. In order to successfully receive a given spreading sequence, the receiver must know the code sequence and monitor it at the right time. This spread spectrum protocol can be used in combinations with random access protocols only if the receiver can monitor the code sequences of all terminals in the network at all times. Otherwise if the receiver can accept and monitor only one signal at a time, reservation-based access protocols are necessary to enable the receiver to know the code sequence for the given transmitter that wants to send a message.

The transmitter-receiver protocol is clearly not appropriate for broadcast transmission since the transmitter does not know the identity of receivers.

Transmitter-Oriented Protocol

Every transmitter is assigned its own individual code sequence. For example, terminal i will always use code sequence C_i for transmission regardless of the identity of receiver and transmission time. For a network of N terminals, N code sequences are needed. A receiver in this system must monitor all code sequences in the system.

This protocol can be used in broadcast mode since the transmitter does not need to know the identity of the receiver. The protocol can be used with random access schemes for receivers that can monitor all code sequences in the system. Clearly, the protocol enables multiple-access capability if the receivers are implemented as multiple-access receivers. The protocol must be used with reservation-based access schemes if the receivers can monitor only one code sequence at a time.

The advantage of the transmitter-oriented protocol is that primary collisions do not occur since different transmitters have different code sequences and packets coming from different transmitters will not collide even if they are transmitted at the same time. Hence, this system has good capture capabilities. That is, if the receiver can receive only one code sequence at a time, it will successfully receive a packet even if it overlaps with a number of other packets, providing that MAI of other packets is not high.

The disadvantage of this system is that the receiver must search over the whole set of spreading sequences if random access protocols are used. That means the receiver must generate all spreading sequences in the system. These receivers are complex to implement and have poor acquisition properties.

The main source of errors with this protocol is a secondary collision caused by MAI.

Receiver-Oriented Protocols

Every receiver is assigned its own distinct code regardless of the identity of the transmitter. Node i always monitors code sequence C_i. If terminal C_j wants to transmit to terminal i, it must use code sequence C_i. A total of N code sequences are required for a network with N terminals.

Receiver-oriented protocols are not suitable for broadcasting since the code sequences are related to the identity of individual receivers and a separate transmission is required for each receiver. The advantages of these protocols are fast acquisition time of the receiver and simultaneous transmissions by several terminals to different receivers over the same channel. However, primary collisions are possible if two or more terminals transmit to the same receiver simultaneously. Hence, this protocol has poor capture properties.

Common Code Protocol

All terminals are assigned a common spreading sequence. Any terminal with a message to transmit to any other terminal uses this sequence. All terminals monitor the common sequence all the time. Only one code sequence is required in the system.

This protocol is suitable for broadcast operation since all receivers are tuned to the common sequence all the time. Discrimination between incoming packets is based on the time difference in their arrival and time offsets between the received code sequences. If two or more packets are transmitted at the same time and have the same propagation delay, they will be lost due to primary collision.

This protocol enables a simpler design of terminals relative to other protocols. The implementation of the hub station is particularly simplified relative to the transmitter-oriented protocols. Instead of a different matched filter for each user, a single matched filter can be used to detect all signals on a multiple-access channel. Synchronization at the chip level could be enabled by feedback pilot tone from the hub station [6].

Adaptive receivers have been proposed for spread spectrum systems to increase capacity and simplify synchronization [26]. The receiver requires a training sequence known to both the transmitter and the receiver in order to determine the spreading sequence and the channel parameters and to enable synchronization. Hence, a training sequence needs to be assigned to a code sequence. All spread spectrum protocols described could be used for adaptive receivers as well, with an additional training sequence that is assigned to a particular code sequence for a given protocol. The training sequences are chosen to have good autocorrelation properties in order to obtain fast acquisition and good crosscorrelation properties to reduce MAI.

When the number of active terminals at a given time is much smaller than the total number of terminals in the network, the number of code sequences in transmitter-oriented protocols can be reduced. That is, the total number of code sequences can be less than the total number of terminals in the network. The code sequences could be assigned on demand to active terminals only or one sequence per a group of terminals.

8.4.3.2 Channel Access Protocols

Channel access protocols define the rules for sharing the channel. The choice of a protocol depends on the network type, traffic patterns, spread spectrum protocol, and the channel properties. In general, for spread spectrum networks it is possible to use controlled access and random access schemes. Controlled access schemes include CSMA, R-ALOHA, and DAMA techniques.

The R-ALOHA access scheme achieves high throughput but requires higher complexity than random access schemes. It is used for the transmission of actual data in the digital cellular radio networks [16].

In CSMA schemes for spread spectrum systems, the hub station typically monitors the channel load, expressed by the number of ongoing transmissions [27,28]. If the channel load is below a certain threshold, then the packets from terminals are accepted. Otherwise, when the channel load is above the threshold, packet accesses are rejected. Channel load sensing schemes are sensitive to propagation delays, near-far, and hidden terminals effects and are not suitable for spread spectrum radio systems.

Random access protocols include various types of ALOHA schemes. Pure ALOHA protocol in a spread spectrum system is easy to implement as it requires no time coordination.

In spread spectrum systems due to spreading sequences used for modulation and demodulation, there is a need for system synchronization. Therefore, these systems have the inherent capability for slotted operation such as that used in the slotted ALOHA random access schemes. Hence, slotted ALOHA is often considered as the channel access protocol in combination with CDMA [29,30].

The operation of a slotted ALOHA scheme in spread spectrum systems is similar to that in narrowband systems.

8.4.3.3 Direct Sequence CDMA Slotted ALOHA Schemes

One way of achieving spread spectrum operation consists of using DS pseudonoise modulation. We consider a slotted ALOHA DS/CDMA scheme in which the data sequence is spread by a spreading sequence consisting of G_p chips per bit. The code length might be G_p or higher. We assume that the spreading sequences can be approximated by random sequences. We also assume that the total number of terminals in the network is large enough to get a Poisson distribution function for the offered traffic. Hence, the probability $P_t(m)$ that m packets are generated within one time slot for the slotted ALOHA scheme is

$$P_t(m) = \frac{G^m}{m!} e^{-G} \qquad (8.13)$$

where G is the average number of transmitted packets per packet time. The throughput of the system is defined as the average number of successfully received packets per packet time and is given by

$$S = \sum_{m=1}^{K_{max}} m P_t(m) P_s(m) \qquad (8.14)$$

where K_{max} is the maximum number of terminals that the system can handle and $P_s(m)$ is the probability that a packet is received correctly. A packet is received correctly when there are no errors in it. In that case the receiver sends an acknowledgment to the transmitter.

The probability that a packet is received correctly on a Gaussian channel with no error control coding is given by

$$P_s = 1 - (1 - P_b)^{N_p} \qquad (8.15)$$

where N_p is the number of bits per packet and P_b is the bit error probability. The bit error probability depends on the modulation type and the type of the receiver. For BPSK modulation, Gaussian channel, and the matched filter receiver, the bit error probability can be approximated by [31]

$$P_b \cong Q\left\{\left[\frac{K-1}{3G_p} + \frac{N_o}{3E_b}\right]^{-1/2}\right\} \qquad (8.16)$$

where K is the number of active terminals in the system, E_b is the transmitted energy per bit, and N_o is the one-sided noise power spectral density. The throughput for a matched filter receiver can be obtained by substituting (8.16) into (8.15) and then (8.15) into (8.14).

For adaptive receivers we can assume that

$$P_s(m) = 1 \quad \text{for} \quad m \leq K_c$$
$$P_s(m) = 0 \quad \text{for} \quad m > K_c$$

where K_c is the critical number of users equal to 70% of the spreading gain [26]. Therefore, for these receivers the throughput is given by

$$S = \sum_{m=1}^{K_c} m P_t(m) \qquad (8.17)$$

In order to make a fair comparison with narrowband slotted ALOHA systems that require a bandwidth G_p times less than the corresponding CDMA slotted ALOHA systems, the throughput S must be normalized by dividing it with the spreading gain G_p

$$S_N = \frac{S}{G_p} \qquad (8.18)$$

The curves showing the normalized throughput of the narrowband slotted ALOHA system of the slotted ALOHA CDMA system with matched filter and adaptive filter receivers on Gaussian channels are shown in Figure 8.11. As the figure indicates, the normalized throughput of slotted ALOHA CDMA with a matched filter receiver and no error control is less than that of slotted ALOHA; while adaptive receivers, due to their MAI suppression properties, achieve much higher throughput. The results are achieved for a spreading gain of $G_p = 50$.

If the system is operating on fading channels, then the probability of error is computed using formulae derived for fading channels in Chapter 5. The normalized throughput results for narrowband slotted ALOHA and CDMA slotted ALOHA with matched filter receivers on a satellite mobile fading channel are shown in Figure 8.12. The satellite mobile channel is modeled by a shadowed line-of-sight signal with a log-normal envelope distribution plus a sum of multipath signals with

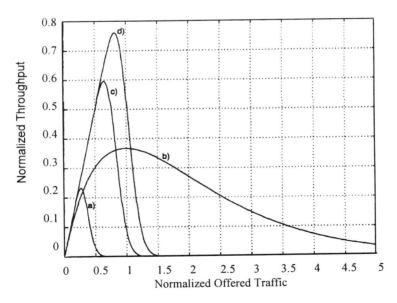

Figure 8.11 Normalized throughput of slotted ALOHA DS/CDMA on Gaussian channels.

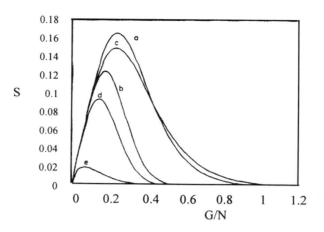

Figure 8.12 Normalized throughput of slotted ALOHA CDMA on satellite mobile fading channels. (© 1995 IEEE. From: [32].)

a Rayleigh distributed envelope. The corresponding results for the narrowband slotted ALOHA on the same channel for comparison are given in Figure 8.13.

The results for both Gaussian and fading channels indicate that a slotted ALOHA DS/CDMA system with no error control makes poorer use of channel bandwidth than using narrowband slotted ALOHA systems. However, if error control is used to correct some of the packet errors caused by MAI, the network capacity

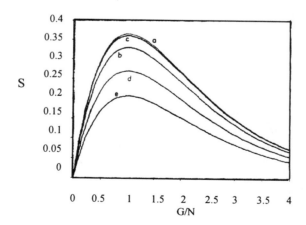

Figure 8.13 Normalized throughput of narrowband ALOHA on satellite mobile fading channels. (© 1995 IEEE. From: [32].)

peaks beyond that of corresponding narrowband systems with the same bandwidth efficiency [29].

8.4.3.4 Frequency Hop Slotted ALOHA Schemes

We will consider a standard slow FH system as described in Chapter 1. The total available bandwidth is divided into q frequency subbands. Each transmitter-receiver pair hops randomly among all q frequency subbands with probability of $1/q$ for each subband independent of previous hop frequencies. The FH pattern is determined by a binary code. Each terminal is assigned a different FH pattern, and it uses the entire spectrum but never occupies more than a small portion of the entire spectrum at any given time. We will assume that the duration of one frequency hop is equal to the symbol interval. *Frequency shift keying* (FSK) is the most often used modulation scheme with FH. At any given time, a terminal transmits one of the two frequencies, representing either mark or space, in a particular frequency subband in which it is transmitting.

We consider a FH system with slotted ALOHA random access. That is, the time axis, which is universal for all terminals, is divided into time slots; and each transmission must take place within a time slot. We assume that the packet length is equal to the time slot.

A *hit* occurs whenever two or more symbols from different terminals are transmitted simultaneously in the same frequency subband.

We will consider a situation in which the receiver can determine the hit by using side information. There are several methods by which to obtain side information [33–35]. The method based on test symbols proved to be most practical for packet radio networks. A known sequence of symbols is included in each hop interval. The number of symbols that are received correctly within each hop interval is used as a statistic to estimate the reliability of the symbol in that hop interval.

We assume that when a hit occurs the receiver erases the received symbol.

The probability that the other symbol is transmitted at the same frequency at any time during the interval occupied by that symbol is called the probability of a hit and is denoted by P_h. Typical models for the random hopping patterns are such that [36]

$$P_h \leq \frac{2}{q} \qquad (8.19)$$

If m packets are transmitted simultaneously by terminals within range of a given receiver, the probability that a particular symbol in a received packet is hit by at least one of the other $m - 1$ symbols is $1 - (1 - P_h)^{m-1}$. As each symbol that is hit is erased, the conditional probability of symbol erasure, given there are $m - 1$ other simultaneous transmissions is given by [37]

$$P_e(m) = 1 - (1 - P_h)^{m-1} \tag{8.20}$$

An efficient scheme consists of erasing the symbols that have been hit and using Reed–Solomon (RS) codes to correct the erased symbols. We assume that each packet consists of an (n, k) RS codeword and erasure-only [38] decoding is employed. Since an (n, k) RS code can correct up to $e = n - k$ erasures, a packet containing more than $(n - k)$ symbol erasures must be retransmitted. The conditional probability that the transmission of a packet from a given terminal is successful given that there are m simultaneous transmissions, $P_s(m)$, is

$$P_s(m) = \sum_{j=0}^{n-k} \binom{n}{j} P_e^j(m)[1 - P_e(m)]^{n-j} \tag{8.21}$$

We assume that the receiver SNR is high enough so that if the symbol is not hit, it is correctly demodulated. That is, the effect of thermal noise is ignored.

Assuming that the transmitted traffic process is Poisson, the throughput, defined as the average number of successful packets per packet time is given by [36]

$$S = \overline{m} \cdot P_s = \overline{m}(1 - P_E) \tag{8.22}$$

where $\overline{m}$ is the average number of transmissions per slot, P_s is the average probability that a packet is successful, and P_E is the probability that a packet is in error when the average number of packets per slot is $\overline{m}$.

We consider a slotted ALOHA fully connected network with Poisson traffic. The throughput is a function of $\overline{m}$.

For narrowband slotted ALOHA systems the maximum value of the throughput is e^{-1} obtained for $\overline{m} = 1$, and the resulting packet error probability from (8.22) is $1 - e^{-1}$, which is approximately 0.632.

For frequency hop radio networks, in order to obtain the throughput per unit time and unit bandwidth for comparison with narrowband systems without error correction, we need to normalize the throughput S

$$S_N = \frac{k}{nq} S \tag{8.23}$$

where S_N is the normalized throughput, k is the number of information symbols in a packet that consists of n symbols. S_N depends on $\overline{m}$, and the maximum value is denoted $S_{N\max}$.

Some values of the maximum normalized throughput, $S_{N\max}$, for a FH slotted ALOHA system are given in Table 8.1 for two values of P_E [36]. The codes are extended RS codes of block length $n = 64$, the number of frequency subbands

Table 8.1
S_{nmax} for RS Codes, $n = 64$, $q = 100$

k	S_{nmax}	
	$P_E = 10^{-2}$	$P_E = 10^{-6}$
32	0.0917	0.0372
24	0.1056	0.0503
20	0.1078	0.0545
16	0.1056	0.0561
12	0.0978	0.0541

Table 8.2
S_N for Narrowband Slotted ALOHA

$P_E = 10^{-2}$	$P_E = 10^{-6}$
0.0099	$1.000e^{-6}$

$q = 100$, and the traffic has Poisson distribution. The number of information symbols in a block varies from 12 to 32. The normalized thoughput for a narrowband slotted ALOHA system for the same values of the packet error probability is given in Table 8.2.

Computations of throughput for DS spread spectrum packet radio networks are more complex due to the statistical dependence between symbol errors in these systems. Some comparisons between the throughput for DS and frequency hop packet radio networks are given in [39].

References

[1] Davis, D., and S. Gonmeyer, "Performance of Slotted ALOHA Random Access with Delay Capture and Randomized Time of Arrival," *IEEE Trans. on Communications*, Vol. COM-28, No. 5, May 1980, pp. 703–710.
[2] Jubin, J., and J. Tornow, "The DARPA Packet Radio Network Protocols," *Proc. IEEE*, Vol. 75, No. 1, 1987, pp. 21–32.
[3] Pahlavan, K., and A. H. Levesque, "Wireless Data Communications," *Proc. IEEE*, Vol. 82, Sept. 1994, pp. 1398–1430.
[4] Ehrich, N., "The Advanced Mobile Phone Service," *IEEE Commounications Mag.*, Vol. 17, March 1979, pp. 9–15.
[5] Abramson, N., "The ALOHA System," *Computer Communication Networks*, N. Abramson and F. Kuo (eds.), Englewood Cliffs, NJ: Prentice-Hall Inc., 1973.
[6] Abramson, N., "Fundamentals of Packet Multiple Access for Satellite Networks," *IEEE J. Selected Areas in Communications*, Vol. 10, No. 2, Feb. 1992, pp. 309–316.

[7] Gilhousen, K. S., I. M. Jacobs, R. Padovani, and L. A. Weaver, Jr., "Increased Capacity Using CDMA for Mobile Satellite Communication," *IEEE J. Selected Areas in Communications*, Vol. 8, No. 4, May 1990, pp. 503–514.
[8] Mouly, M., and M. B. Paulet, *The GSM System for Mobile Communications*, Palaiseau, France: Mouly and Paulet, 1992.
[9] Miller, M., B. Vucetic, and L. Berry, *Satellite Communications—Mobile and Fixed Services*, Kluwer, 1993.
[10] Description of the Globalstar System, Globalstar, L. P., Palo Alto, CA, 1993.
[11] Tobagi, F. A., "Modelling and Performance Analysis of Multihop Packet Radio Networks," *IEEE Proc.*, Vol. 75, Jan. 1987, pp. 135–155.
[12] Puente, W. G. Schmidt, and A. M. Werth, "Multiple Access Techniques for Commercial Satellites," *Proc. IEEE*, Vol. 59, Feb. 1971, pp. 218–229.
[13] Puente, J. G., and A. M. Werth, "Demand Assigned Service for the Intelsat Global Network," *IEEE Spectrum*, Jan. 1971.
[14] Arnbak, D. C., and W. Van Blitterswijk, "Capacity of Slotted ALOHA in Rayleigh Fading Channels," *J. Selected Areas in Communications*, Vol. 5, Feb. 1987, pp. 261–269.
[15] Stampfl, R. A., and A. E. Jones, "Tracking and Data Relay Satellite Systems," *IEEE Trans. on Aerospace Electronic Systems*, Vol. AES-6, May 1970, pp. 276–287.
[16] Viterbi, A. J. "Principles of Spread Spectrum Multiple Access Communications," Qualcomm Inc., 1994.
[17] Stiglitz, I. G., "Multiple Access Considerations—A Satellite Example," *IEEE Trans. on Communications*, Vol. COM-21, May 1973, pp. 577–582.
[18] Ince, A. N., "Code Division Multiplexing for Satellite Systems," *Communication Systems and Random Theory*, J. K. Skwirzynski (ed.), Darlington, UK, 1978, pp. 821–857.
[19] JTIDS/TIES Consolidated Tactical Communications, Electronic Warfare, Sept. 1977, pp. 45–53.
[20] Raychaudhuri, D., and K. Joseph, "Channel Access Protocols for Ku Band VSAT Networks: A Comparative Evaluation," *IEEE Communications Mag.*, Vol. 26, No. 5, May 1988, pp. 34–44.
[21] Polydoros, A., and J. Silvester, "Slotted Random Access Spread-Spectrum Networks: An Analytical Framework," *IEEE J. Selected Areas in Communications*, Vol. 5, July 1987, pp. 989–1001.
[22] Kleinrock, L., *Queuing Systems, Vol. 1, Theory*, New York: John Wiley & Sons, Inc., 1975.
[23] Metcalf, R. M., and D. R. Boggs, "Ethernet: Distributed Packet Switching for Local Computer Networks," *Communications ACN*, 1976, pp. 395–404.
[24] Kleinrock, L., and F. Tobagi, "Random Access Techniques for Data Transmission over Packet Switched Radio Channels," *Proc. NCC'75*, May 1975, pp. 187–201.
[25] Li, V. O. K., and T. Y. Yan, "An Integrated Voice and Data Multiple-Access Scheme for a Land Mobile Satellite System," *Proc. IEEE*, Vol. 72, No. 11, Nov. 1984, pp. 1611–1619.
[26] Rapajic, P., and B. Vucetic, "Adaptive Receiver Structures for Asynchronous CDMA Systems," *IEEE J. Selected Areas in Communications*, Vol. 12, No. 4, May 1994, pp. 685–697.
[27] Abdelmonem, A. H., and T. N. Saadawi, "Performance Analysis of Spread Spectrum Packet Radio of Spread Spectrum Packet Radio Networks with Channel Load Sensing," *IEEE J. Selected Areas in Communications*, Vol. 7, No. 1, Jan. 1989, pp. 161–166.
[28] Toshimitsu, T., T. Yamazato, M. Katayama, and A. Ogawa, "A Novel Spread Slotted ALOHA System with Channel Load Sensing Protocol," *IEEE J. Selected Areas in Communications*, Vol. 12, No. 4, May 1984, pp. 665–672.
[29] Morrow, R., and J. Lehnert, "Packet Throughput in Slotted ALOHA DS/SSMA Radio Systems with Random Signature Sequences," *IEEE Trans. on Communications*, Vol. 40, No. 7, July 1992, pp. 1223–1230.
[30] Makrakis, D., and K. M. S. Murthy, "Spread Slotted ALOHA Techniques for Mobile and Personal Satellite Communication Systems," *IEEE J. Selected Areas in Communications*, Vol. 10, No. 6, Aug. 1992, pp. 985–1002.

[31] Pursley, M., "Performance Evaluation for Phase Coded Spread Spectrum Multiple Access Communications—Part I: System Analysis," *IEEE Trans. on Communications*, Vol. COM-25, No. 8, Aug. 1977, pp. 795–799.
[32] van Nee, R. D. J., R. N. van Wolfswinkel, and R. Prasad, "Slotted ALOHA and Code Division Multiple Access Techniques for Land Mobile Satellite Personal Communications," *IEEE J. Selected Areas in Communications*, Vol. 13, No. 2, Feb. 1995, pp. 382–388.
[33] Pursley, M., and H. B. Russel, "Network for Frequency-Hop Packet Radios with Decoder Side Information," *IEEE J. Selected Areas in Communications*, Vol. 12, No. 4, May 1994, pp. 612–621.
[34] Pursley, M. B., "The Derivation and Use of Side Information in Frequency Hop," *IEICE Trans. on Communications, Special Issue on Spread Spectrum Techniques and Application*, Aug. 1993, pp. 814–824.
[35] Wang, Q., "Frequency-Hopped Multiple Access Communications with Coding and Side Information," *IEEE J. Selected Areas in Communications*, Vol. 10, No. 2, Feb. 1992, pp. 317–327.
[36] Pursley, M., "The Role of Spread Spectrum in Packet Radio Networks," *IEEE Proc.*, Vol. 75, Jan. 1987, pp. 116–134.
[37] Pursley, M., "Frequency-Hop Transmission for Satellite Packet Switching and Terrestrial Packet Radio Networks," *IEEE Trans. on Information Theory*, Vol. IT-32, No. 5, Sept. 1986, pp. 652–667.
[38] Lin, S., and D. Costelo, *Error Control Coding*, Englewood Cliffs, NJ: Prentice-Hall, 1983.
[39] Pursley, M., and D. J. Tapale, "Error Probabilities for Spread Spectrum Packet Radio with Convolutional Codes and Viterbi Decoding," *IEEE Trans. on Communications*, Vol. 35, Jan. 1987, pp. 1–12.

Selected Bibliography

Balston, D. M., "Pan-European Cellular Radio," *Electron. Communications Engr. J.*, Jan./Feb. 1989, pp. 7–13.

Parker, E. B., "Micro Earth Stations as Personal Computer Accessories," *Proc. IEEE*, Vol. 72, No. 11, Nov. 1984, pp. 1526–1531.

Pratt, T., and C. W. Bostian, *Satellite Communications*, New York: John Wiley & Sons, 1986.

Pritchard, W. L., and J. A. Sciulli, *Satellite Communications System Engineering*, Englewood Cliffs, NJ: Prentice-Hall, 1986.

Yacoub, M. D., *Foundations of Mobile Radio Engineering*; Boca Raton, FL: CRC Press, 1993.

ABOUT THE AUTHORS

Savo Glisic is professor of electrical engineering at the University of Oulu, Finland and vice president of Globalcomm Institute for Telecommunications. He was visiting scientist at the Cranfield Institute of Technology, Cranfield, England (1976–1977) and at the University of California at San Diego (1986–1987). He has been active in the field of spread spectrum for 20 years and has published four books and a number of papers. He is currently doing consulting in this field throughout Europe, the United States, and Australia. Professor Glisic served as technical program chairman of the Third IEEE International Symposium on Spread Spectrum Techniques and Applications (ISSSTA'94) and is technical program chairman of the Eighth IEEE International Symposium on Personal, Indoor and Mobile Radio Communications (PIMRC'97).

Branka Vucetic received the Ph.D. degree in electrical engineering from the University of Belgrade, Belgrade in 1982. In 1986 she joined the University of Sydney, where she is currently associate professor in the Department of Electrical Engineering and director of the Communications Science & Engineering Laboratory. Her research interests include digital communications, coding, modulation, and channel modeling.

INDEX

Acoustic charge transport (ACT), 188
Acquisition detectors, 62
Active correlation, 62
Adaptive cancellation, 266
Adaptive control algorithm, 286–87
    LMS, 287
    performance, 287
    RLS, 287
Adaptive estimation filters, 306–14
    gradient and Wiener solution, 307–9
    LMS algorithm, 311–13
    requirement, 307
    steepest descent algorithm, 310–11
    See also Estimation filters
Adaptive MMSE receivers, 283–86
    advantages, 286
    illustrated, 285
    implementation of, 284
    requirements, 284
    timing recovery, 284
    See also Suboptimum DS/CDMA receivers
Adaptive receivers, 318–19
    filter coefficients, 318
    illustrated, 318
Additive white Gaussian noise (AWGN), 217
    superimposed modulated signals in, 217
    with two-sided spectral density, 295
Algorithms
    adaptive control, 286–87
    CFAR, 85, 88–91
    CML tracking system, 155, 167–68
    eML tracking, 144–45
    ITS, 86, 88
    LMS, 287, 311–13
    MF tracking system, 146

    MITS, 91, 93
    steepest descent, 310–11
    transform domain, 316
    Viterbi, 270, 274
ALOHA
    normalized throughput, 349
    pure, 347–49
    reservation, 352–53
    slotted, 350–51
    system operation, 348
Analog-analog correlators, 185–86
Analog-binary correlators, 186
Analog shift register, 178
Antijam communications, *xiii*, 2–4
Aperiodic correlation function, 177
Approximate conditional mean (ACM) filter, 314
Approximations, 153–54, 171–74
Autocorrelation function, 18, 25
    of chip pulse, 189
    multipath spread, 232
    periodic, 177
    spatial, 249
    See also Correlation functions
Automatic decision threshold level control (ADTLC), 84–91
    CFAR algorithm, 85, 88–91
    circuitry block diagram, 87
    ITS algorithm, 86, 88
    See also Code acquisition
Automatic gain control (AGC), 243
Autoregressive (AR) processes, 158

Bandwidth
    coherence, 230

Bandwidth (continued)
  loop-noise, 115
Bang-bang control loop, 243
Baseband code tracking loop, 120–21
  defined, 120
  illustrated, 121
  purpose of, 120–21
  See also Tracking loops
Binary arithmetic, 37
Binary field, 19, 37
Binary irreducible polynomials, 40–41
Binary phase shift keying (BPSK), 10
  bit error probability for, 240
  coherent digital demodulation of, 190–91
  modulation, 190–91
  receivers, 11
  receiver structure, 191
  transmitter, 11
Binary sequences, composite, 29–30
Bit error probability, 220–23
  average, 220
  for BPSK modulation, 240
  for four-tap predictor, 306, 307, 308
  on Gaussian channel, 222
  lower bound, 222
  variance from thermal Gaussian noise, 221
Bit error rate, 304–6
Bit recovery sequence (BTR), 345
Blocking probability, 252, 253, 254
  in cellular system, 256
  in cellular system with imperfect power control, 257
Broadband code division multiple access (B-CDMA), 319
Broadcast channel, 341
Broken/center z search, 80–82
  defined, 80
  mean acquisition time improvement factor, 81

Carrier recovery sequence (CRS), 345
Carrier sense multiple access (CSMA), 351–52
CCD
  attributes, 178
  charge transfer in, 179
  components, reported signal processing based on, 185–88
  CTI effects on PNMF performance, 183
  defined, 178
  FIR filter, 204
  MDAC approach, 186
  MDAC-based PTF, 205
  pipe-organ structure, 187
  PN-matched filter imperfections, 178–85
  samples, 182
  technology, 175–78
CCD-matched filters
  architecture illustration, 176
  CD zero offset, 178–80
  charge transfer inefficiency, 180–84
  dark current, 184–85
  extending, 177
  illustrated, 176
  impairments, 184–85
  imperfections, 178–85
  input circuit noise, 184
  odd/even structure of, 188
  pulse propagation, 181
  reset noise, 185
CDMA, 6–7
  base station, 329
  blocking in, 252
  broadband (B-CDMA), 319
  capacity advantage, 212
  code sequences, 212
  controlled access schemes, 351–54
  defined, 342
  forward link capacity, 326–27
  mobile power, 329
  mobile signal-to-interference ratio, 324–25
  multiple-access techniques vs., 212
  as multiple purpose scheme, *xiv*
  near-far effect, 214–15
  networks, capacity of, 7–8
  overlay, 319–21
  overlay type networks, 4
  random access schemes, 346–51
  receivers, 212
  reverse link capacity, 330–31
  self-jamming, 214
  single-channel demodulation, 215–20
CDMA slotted ALOHA schemes, 354–65
  channel access protocols, 359
  direct sequence, 359–62
  frequency hop, 363–65
  primary collisions, 356
  secondary collisions, 356
  spread spectrum transmission protocols, 356–58
CDMA systems, 211–61
  asynchronous, 215, 268
  authorized users of, 215
  B-CDMA, 319
  bit error probability, 220–23

capacity of, 223–24
cellular, 211–15
conventional capacity, 243
disadvantages of, 214
DS/CDMA, 215–18
multipath propagation, 225–36
multiple-cell, 241–61
Rake receiver, 236–40
single-cell, 223–24
synchronous, 217
Cells
   defined, 58
   definition of, 60
   hexagonal cellular structure, 246
   observation time, 66
Cellular CDMA systems, 211–15
   advantages of, 212
   basic structure, 213
   blocking probability, 256
   clusters, 212, 213
   defined, 246
   disadvantages, 214
   Erlang capacity of reverse link in, 255–58
   forward link capacity, 258–61
   hexagonal structure cells, 246
   with imperfect power control, 256–58
   near-far effect, 214
   number of users and, 214
   performance, 214
   reuse factor, 212
   reverse link capacity, 246–52
   reverse link outage probability, 251
   self-jamming, 214
   soft handoff, 214
   voice activity monitoring, 213–14
   See also CDMA systems
Cellular digital packet data (CDPD), 338
Cellular networks, 339–40
Central Limit Theorem, 227
Channel access protocols, 359
Channel model, 156–58
Channels
   broadcast, 341
   coefficients, 157
   delays, 157
   fading, 154–63, 229–30
   flat fading, 230
   Gaussian, 215–20
   multipath response, 230
   multiple-access, 342
   paging, 263

request, 342
time-dispersive, 229
traffic, 263
wideband models, 230–36
Characteristic polynomial, 21
Charge coupled devices. *See* CCD
Charge transfer
   inefficiency (CTI), 180–84
   mechanism, 180
Chi-square distribution, 238
Circuit switching, 338
Circular state diagram, 74–76
   branch subscripts, 75
   for continuous/center z serial search, 78
   illustrated, 75
Closed-loop power control, 243
Clusters
   defined, 212
   illustrated, 213
   seven-cell, 213
   See also Cellular CDMA systems
CML tracking system algorithm, 146
   correlation functions for, 155, 167–68
   defined, 146
   mutual signal positions in system with, 147
   tracking error probability, 161
Cochannel interference, 212–13
Code acquisition, 49–99
   ADTLC, 84–91
   lock detectors, 95–99
   ML-motivated code synchronization, 53–60
   performance analysis tools, 63–84
   performance measures, 60–63
   randomness sources, 60–61
   sequential detection, 91–95
   time variance, 102
   See also Code synchronization
Code division multiple access. *See* CDMA
Code division multiplexing (CDM), 340
Code synchronization
   coarse, 58
   coherent, 55
   ML-motivated, 53–60
   noncoherent, 55
   steps for, 49
Code tracking, 58, 105–63
   in fading channels, 154–63
   loops, 120–54
   PLL fundamentals, 105–20
   See also Code synchronization
Coherence bandwidth, 230

Coherent code synchronization, 55
Coherent digital demodulators, 190–91
Coherent PAM, 197–98
Common code protocol, 358
Complex-number field, 41
Composite binary sequences, 29–30
   correlation function, 29–30
   crosscorrelation function, 30
Concentrated sync group (CSG), 137
   approximations, 153–54
   CML tracking system algorithm, 146
   eML tracking system algorithm, 144–45
   MF tracking system algorithm, 146
   performance, 154
   sequence windowing, 148–49
   state transition probabilities, 149–53
   sync frame, 144
   systems with, 143–54
   tracking system parameters, 146–48
Constant false alarm rate (CFAR)
   algorithm, 85, 88–91
   defined, 85, 88–89
   threshold pdf of, 91, 92
Continuous/center z serial search, 76–79
   cell numbering, 77
   equivalent circular-state diagram for, 78
   flow graph for, 79
   See also Serial searches
Continuous crosscorrelation function, 18
Controlled access schemes, 351–54
   DAMA, 353–54
   reservation ALOHA, 352–53
   See also Multiple-access schemes
Controllers
   digital, 339
   location of, 338
Convolution function, 198–200
Correlation functions, 18–19
   aperiodic, 177
   autocorrelation, 18, 25
   for CML system, 155
   for composite binary sequences, 29
   continuous crosscorrelation, 18
   discrete, 19
   sequence, 174
   spaced-frequency spaced-time, 233
Correlators
   analog-analog, 185–86
   analog-binary, 186
   high-speed, 188
   SAW memory, 205

Crosscorrelation function
   between two sequences, 30
   of binary m-sequences, 27–28
   of code sequences, 219
   of signature waveforms, 269, 271–72
   See also Correlation functions
Cumulative probability, 153
Cumulative probability distribution function
   (CPDF), 61
Cyclic redundancy check (CRC), 347
Cyclic sequence windowing, 166–67

Dark current, 184–85
DC zero offset, 178–80
Decision feedback filter, 313–14
   performance, 315
   receiver block diagram, 314
Decorrelating receiver, 277–81
   average probability of error, 280
   decision feedback, 283
   defined, 277
   low complexity of, 278
   performance, 281
   for synchronous DS/CDMA system, 279
Delay-locked loop (DLL)
   defined, 121
   double dither, 125–28
   See also Tracking loops
Demand assignment multiple-access
   (DAMA), 353–54
   bursty user support, 354
   channel subdivisions in, 354
   R-ALOHA vs., 353
Demodulators
   coherent digital, 190–91
   multipath, 237
   noncoherent amplitude, 197
   sum and difference, 195
Detectors
   DS structures, 55–56
   FH structures, 56–57
   lock, 95–99
   passive/active correlation, 62
   verification modes, 62
Differently coherent demodulation of
   QPSK, 193–94
   defined, 193
   illustrated, 194
Digital PLL (DPLL), 117–20
   block diagram, 119
   equivalent baseband model, 119
   See also Phase-locked loop (PLL)

Digital signal processing (DSP), 175, 292
Direct method, 70–73
    defined, 70
    mean synchronization time, 70–72
    parameters, 70
    See also Performance analysis tools
Direct sequence spread spectrum (DSSS), 8–13
    BPSK, 10, 11
    detector structures, 55–56
    hybrid system, 14, 16–17
    QPSK, 10, 12
Discrete correlation function, 19
Discrete Fourier transform (DFT), 316
Discrete tracking systems, 131–54
    with CSG, 143–54
    with DSG, 137–43
    process, 131–37
    for slow FH, 137–54
    See also Tracking loops
Distributed sync group (DSG)
    (h, 1)-ML system, 137–43
    defined, 137
    systems with, 137–43
DLL tracking loop, 86
Doppler offset, 55, 56, 59
Doppler shift
    defined, 225
    of ith component, 226
    maximum, 227, 228
    relative approximation, 54
    for single tone transmission, 225–29
Double dither DLL, 125–28
    average S-curve for, 127
    block-diagram, 128
    See also Delay-locked loop (DLL)
Double sampling, 186–87
DS/CDMA system
    asynchronous, 218, 269–70, 288
    decorrelating receiver for, 279
    maximum likelihood detector, 270
    in multipath environment, 235
    multiuse adaptive receiver, 288
    single-cell, 220
    slotted ALOHA, 359–62
    suboptimum receivers, 276–89
    synchronous, 217, 273, 276, 279
    tapped delay line for, 273
    transmitter block diagram, 216
    transmitter model, 215
    two-user synchronous, 274
    See also CDMA systems

DS spread spectrum receiver, 293–94
Dual space, 39–40
Early late tracker (ELT)
    defined, 53
    structure, 52
Electrical/acoustic (E/A) wave transducer, 198
EML tracking algorithm, 144–45
    defined, 144
    signal waveforms in system with, 145
    See also Concentrated sync group (CSG)
Energy-to-noise spectral density ratio, 238
Erlang-B formula, 258
Erlang capacity, 252–58
    defined, 252
    of FDMA system, 258
    occupancy parameter, 253
    outage probability and, 258
    of reverse link in cellular CDMA, 255–56
    of reverse link in cellular CDMA with
        imperfect power control, 256–58
    of reverse link in single-cell CDMA, 252–55
Estimation filters, 293–300
    adaptive, 306–14
    coefficients, 296
    in DS spread spectrum systems, 293–94
    error-performance surface of, 308
    MSE for, 307
    narrowband interference suppression
        by, 294–300
    one-sided structure, 294
    output, 296
    output minimum power, 298
    structures, 293–94
    two-sided structure, 294
Euclidian distance, 276
Expanding-window search, 63
    advantages of, 84
    mean acquisition time vs. number of partial
        windows, 84, 85, 86
    strategies, 82–84
    technique definitions, 83
Extended Kalman filter (EKF)
    defined, 159
    PN code delay and multipath using, 158–63
    theory, 159

Fade rate, 227
Fading channels
    code tracking in, 154–63
    narrowband ALOHA on satellite mobile, 362
    Rayleigh, 240, 244

False alarm (FA)
  absorbing, 62
  catastrophic, 61
  probability per cell, 62
Fast FH (FFH), 56
Fibonacci configuration, 20
Field-effect transistor (FET) bridge circuit, 186
Fingers, 237
Finite fields. *See* Galois fields
Flat fading channels, 230
Forward CDMA channel structure, 264
Forward link capacity, 258–61, 324–29
  bit error rate support, 261
  CDMA, 326–27
  illustrated, 261
  microcell, 326–28
  power control, 258–59
  See also Reverse link capacity
Fourier transform, 200–201
  discrete (DFT), 316
  inverse, 201–2
  real-time, 200, 201
Frames, 344–45
  illustrated, 345
  time slots, 344
  See also Time division multiple access (TDMA)
Frequency division multiple access (FDMA), 211
  cochannel interference, 212–13
  defined, 342
  Erlang capacity, 258
  frequency reuse in, 212–13
  guard times/guard bands, 241
  handoff, 214
  "hard blocking" condition, 214
  illustrated scheme, 343
  macrocells, 320
  MCPC system, 343
  SCPC system, 343
  SPADE system, 344
  subdivisions, 342, 343
  voice activity monitoring and, 214
  See also CDMA; Time division multiple access (TDMA)
Frequency division multiplexing (FDM), 340
Frequency hop (FH) slotted ALOHA
  schemes, 363–65
Frequency hopping spread spectrum (FHSS), 13–14
  code tracking loop for, 128–29
  detector structures, 56–57
  hybrid system, 14, 16–17

modem illustration, 15
Frequency-selective fading, 230
Frequency shift keying (FSK), 13, 363
Full-time early-late noncoherent tracking loop, 121–24
  discriminator S-curve, 132
  FHSS, 129
  frequency plane, 131
  signal waveforms, 131
  filter output noise density, 123
  illustrated, 123
  input signal, 121, 123
  signal-to-noise power ratio, 124
  See also Tracking loops

GaAs semiconductor material, 188
Galois fields, 41–46
  addition, 43–45
  defined, 41, 42
  division, 43
  extension, 23–24
  multiplication, 42
  primitive polynomial, 45
  subtraction, 45
Gaussian channel, 215–20
  bit error probability on, 222
  correctly received packet on, 360
Gaussian noise, 154
  additive white (AWGN), 217
  one-sided power spectral density, 222
  sample, 219
  thermal, 8, 221
  white, 268
Generating function, 65
  flow diagram, 67
  moment, 68
  second derivative of, 66
Global positioning system (GPS), 188
Gold sequences, 30–31
GSM system, 344

Handoff, 214
High-speed correlators, 188
Hybrid DS/FH, 14, 16–17
  defined, 14
  illustrated, 16–17
Hypotheses, 59–60

Immediate-rejection logic, 62
Input circuit noise, 184
Instantaneous threshold setting (ITS)
  algorithm, 86, 88
  defined, 88

threshold probability density
    function, 89, 90, 91
Intercell interference
    defined, 246
    factor, 249, 255
    normalized power, 249
    power, 247
Interference
    autoregressive, 304
    intercell, 246, 249
    intracell, 246
    low-level, 7
    multiple-access (MAI), 13, 265, 266
    narrowband (NBI), 318, 319
    power spectral density, 302
    signal examples, 302–4
    suppression, 4
Interference canceler, 281–83
    block diagram, 282
    complexity of, 281
    operation elements, 281
    performance, 283
Interference-limited systems, 8, 40
Intracell interference, 246
Inverse Fourier transform, 201–2
Irreducible polynomials, 40
    primitive, 40
    root of, 41
IS-95 standard, 262–64

Kalman-Bucy filter, 314
Kasami sequences, 31

Laplace approximation, 172
Laplace transform, 110
Lattice filter, 312, 313
Least mean square (LMS) algorithm, 287, 311–13
    convergence of, 311, 312
    two-sided transversal filter with, 312
Least mean square error (LMSE) transversal
    filter, 292
Likelihood function (LF), 55
Linear combination, 39
Linear feedback shift register (LFSR)
    circuit implementation, 22
    maximum number of states, 24
    minimum memory, 22
    sequence generation with, 22–23
Linear recursions, 19–22
Linear span, 22
Lock detectors, 95–99
Loops

code tracking, 120–54
filter transfer function, 108, 115
gain, 107
noise bandwidth, 115
parameter specification, 111
transfer function, 108
transient second-order response, 112
transient third-order response, 113
See also Delay-locked loop (DLL);
    Phase-locked loop (PLL)

Macrocell base station, 325
    signal-to-interference ratio, 330
    See also Microcell base station
Macrocells, 319
    FDMA, 320
    shape of, 321
    See also Microcells
Markov absorbing chain, 89, 96, 97
    fundamental matrix for, 98
    standard results from, 98
M-ary frequency shift keying (MFSK), 13
Matched filter (MF), 146
Maximum aposterior probability (MAP)
    estimate, 50
Maximum likelihood (ML)
    estimation, 50–51
    structure, 52
    See also ML-motivated code synchronization
Maximum likelihood sequence detection, 270
Maximum likelihood tracker (MLT), 51, 52
Mean square error (MSE), 284
    for estimation filter, 307
    for finite-length filter, 286
Metal-oxide-semiconductor (MOS) transistor
    multiplier, 186
MF tracking system algorithm, 146, 152
Microcell base station, 324
    transmit notch filters at, 328
    See also Macrocell base station
Microcells
    B-CDMA, 319
    defined, 320
    forward link capacity, 326–28
    overlay geometry, 321
    positioning of, 321
    radius, 324
    reverse link capacity, 331–33
    See also Macrocells
Minimum polynomials, 46–47
Minimum-shift-keyed (MSK) waveform, 204
MITS algorithm, 91, 93

ML/IT subalgorithm, 150
ML-motivated code synchronization, 53–60
    DS signal detector structures, 55–56
    FH signal detector structures, 56–57
    parallel vs. serial implementation, 57–58
    problem definition, 53–54
    serial search, 59–60
    signal parameters uncertainty range, 54–55
    two-dimensional search, 58–59
    See also Code acquisition
Mobile signal-to-interference ratio, 324–26
    analog, 325–26
    CDMA, 324–25
Modulation
    BPSK spread spectrum, 190–91
    noncoherent pulse amplitude, 195–97
    OQPSK spread spectrum, 191–92
    QPSK data, 193–94
Modulo-2 addition, 37
Modulo-2 multiplication, 37
MSAT-X, 354
m-sequences, 24–26
    with arbitrary shift, 26
    balance property, 25
    binary, 25
    binary, crosscorrelation of, 27–29
    correlation properties, 25
    cyclically distinct, 24
    defined, 24
    delayed version, 26
    shift and add property, 25
    trace representation, 26
Multichannel per carrier (MCPC), 343
Multipath channels
    in frequency domain, 232
    impulse response time, 231
    output, 232
    response, 230
    spaced-frequency spaced-time correlation function, 233
    tapped delay line model, 235
    wideband models, 230–36
Multipath delay spread, 234
Multipath intensity profile, 232
Multipath propagation, 225–36
    fading and Doppler shifts, 225–29
    frequency selective fading channel, 229–30
    illustrated, 4
    operation with, 4–6
    result of, 4
    typical environment, 226

Multipath spread, 232
Multiple access capability, 355
Multiple-access channel, 342
Multiple-access interference (MAI), 265
    defined, 13
    immunity against, 266
Multiple-access schemes, 341–65
    CDMA, 346–65
    controlled, 351–54
    FDMA, 342–44
    random, 346–51
    TDMA, 344–46
Multiple-access techniques, 342
Multiple-cell CDMA systems, 241–62
    Erlang capacity, 252–58
    forward link capacity, 258–61
    mobile receiver, 245
    power control, 242–46
    propagation loss model, 241–42
    reverse link capacity, 246–52
    See also CDMA systems
Multiplexing, 340–41
    code-division (CDM), 340
    defined, 340
    frequency-division (FDM), 340
    polarization division (PD), 341
    space-division (SDM), 340
    time-division (TDM), 340
Multistage receivers, 281, 282, 283
    illustrated, 282
    for two users, 283
Multiuser CDMA receivers, 265–89
    adaptive, 288
    optimal, 269–76
    suboptimum, 276–89
    system model, 267–69

Narrowband interference (NBI), 318–19
    modeling by autoregressive processes, 300–302
    suppression by estimation filters, 294–300
    suppression techniques, 292
Narrowband signal(s_n), 1
Near-far effect, 214–15
    counteracting, 265
    defined, 214
    power control and, 215
    See also CDMA systems; Cellular CDMA systems
Network management center (NMC), 354
Noise
    additive model, 114

additive white Gaussian (AWGN), 217
filter output density, 123
Gaussian, 154
input circuit, 184
loop bandwidth, 115
PLL operation in presence of, 110–15
reset, 185
thermal, 8, 253
Noncoherent code synchronization, 55
Noncoherent pulse amplitude
    modulation, 195–97
Non-immediate-rejection logic, 62
Nonuniform serial searches, 76–84
    broken/center z, 80–82
    continuous/center z, 76–79
    expanding-window strategies, 82–84
    See also Serial searches
Notch filters, 316
    microcell forward link capacity with, 328
    microcell reverse link capacity with, 333
    transform domain, 317
    transmit, 328
Nyquist sampling interval, 156

Occupancy parameter, 253
One-channel noncoherent code tracking loop
    frequency plane, 135
    illustrated, 134
    signal waveforms, 135
Open-loop configuration, 51
Open-loop power control, 243
Open-loop transfer function, 108
Optimum receivers, 269–76
    for asynchronous DS/CDMA systems, 271
    bit sequence, 269
    decision making, 270
    error probability, best/worst cases, 277
    for synchronous DS/CDMA system, 276
OQPSK modulation, 191–92
    differently coherent demodulation of, 193–94
    receiver structure for, 191
Organization, this book, *xiv–xv*
Outage probability
    of cellular CDMA system reverse link, 251
    defined, 250
    Erlang capacity and, 258
    propagation model, 251
    for reverse link system with power
        control, 257
    upper bound, 250
Outer loop power control, 244

Packet radio networks, 337–40
    antenna, 339
    defined, 337
    digital controller, 339
    elements of, 339
    examples, 338, 339
    illustrated model, 341
    radio terminal, 339
    services, 339
Packets
    defined, 338
    delay of, 339
    overlapping, 355
Packet switching, 338
Passive correlation, 63
Performance, 154
    adaptive algorithms, 287
    decision feedback filter, 315
    decorrelating receiver, 281
    interference canceler, 283
    measures, 60–63
    power loop control, 256
    Rake receiver, 237–40
    transform domain processing receiver, 317
Performance analysis tools, 63–84
    direct method, 70–73
    modified search strategies, 73–84
    transform domain analysis, 64–69
Periodic autocorrelation function, 177
Phase-locked loop (PLL), 105–20
    additive noise model, 114
    analog, 105–7
    baseband model, 107
    digital (DPLL), 117–20
    illustrated, 106
    linearized baseband model, 108
    linear model, 107–8
    loop gain, 107
    nonlinear model of, 115–17
    operation in the presence of noise, 110–15
    steady-state phase error, 108–10
    trajectories, 117, 118
    transient response of, 110
Poisson distribution, 254
Polarization division (PD), 341
Power control, 242–46
    closed-loop, 243
    commands for, 245
    example, 244
    forward links and, 258–59
    imperfect, 256–58

Power control (continued)
  log-normal approximation, 245
  open-loop, 243
  outer loop, 244
  perfect blockage probability, 255
  See also Multiple-cell CDMA systems
Primary collisions, 356
Primitive elements, 46
Primitive polynomials
  defined, 40
  Galois fields, 45
  list of, 41
Probability
  bit error, 220–23
  blocking, 252, 253, 254
  cumulative, 153
  of detection, 61
  FA per cell, 62
  of handover, 69
  of missing the code, 61
  outage, 250, 251
  state transition, 149–52
Probability distribution function (pdf), 73
  CFAR, 91, 92
  ITS, 81, 89, 90
  MITS algorithm, 93
Probability mass function (pmf), 153, 170–71
Programmable transversal filters (PTFs), 185, 204
Propagation delay, 54
Propagation loss
  model, 241–42
  power variations due to, 265
Protocols
  channel access, 359
  common code, 358
  defined, 338
  receiver-oriented, 357–58
  spread spectrum transmission, 356–58
  transmitter-oriented, 357
  transmitter-receiver-oriented, 356–57
Pseudonoise (PN) sequence, 8
Pure ALOHA, 347–49

Quadratype phase shift keying (QPSK), 10
  data modulation, 193–94
  receiver, 12
  transmitter, 12

Rake receivers, 5, 157, 236–40
  bit-energy-to-noise spectral density ratio, 238
  defined, 236
  fingers, 237
  with L branches, 237
  multipath demodulator in, 237
  performance of, 237–40
Random access schemes, 346–51
  pure ALOHA, 347–49
  slotted ALOHA, 350–51
  See also Multiple-access schemes
Rayleigh distribution, 238
Rayleigh fading, 240, 244, 258
Real-number field, 41
  extension, 42
  subfield, 41
Receiver-oriented protocols, 357–58
Receivers
  adaptive, 318–19
  adaptive MMSE, 283–86
  for BPSK data, 191
  CDMA, 212
  configurations, 188–98
  decision feedback, 314
  decorrelating, 277–81
  front end, 189
  MMSE, 267
  multistage, 281, 282, 283
  multiuser CDMA, 265–89
  for OQPSK modulation, 191
  Rake, 5, 157, 236–40
  suboptimum DS/CDMA, 276–89
Recursive least square (RLS) algorithm, 287
Reed-Solomon (RS) codes, 364
Request channel, 342
Reservation ALOHA, 352–53
Reset noise, 185
Reverse CDMA channel structure, 262
Reverse link capacity, 246–52, 329–33
  CDMA, 330–31
  in cellular CDMA, 255–56
  in cellular CDMA with imperfect power control, 256–58
  microcell, 331–33
  outage probability, 251
  in single-cell CDMA, 252–55
  user support, 251
  See also Forward link capacity
Rician distribution, 228–29, 236

Satellite networks, 339
SAW/FET devices, 205
  capability of, 206
  filter response, 205
  hybrid, 205
Scalar multiplication, 37–38

Searches
  expanding-window, 63, 82–84
  modified strategies, 73–84
  sequential state estimation, 63
  serial, 59–60
  straight, 63
  two-dimensional, 58–59
  Z, 63
Secondary collisions, 356
Self-jamming, 214
Sequence generation, 18–31
  characteristic polynomial, 21
  correlation functions, 18–19
  with elements of extension Galois field, 23–24
  with LFSR long division, 22–23
  linear recursions, 19–22
  linear span, 22
  m-sequences, 24–26
Sequences
  composite binary, 29–30
  correlation function, 174
  cyclic shifting, 166
  decimation of, 27
  Gold, 30–31
  Kasami, 31
Sequence windowing, 148–49
  cyclic, 166–67
  defined, 148
Sequential detection, 91–95
  acquisition circuit, 94
  defined, 92
  integrator plus threshold output, 95
  theory, 92–93
Sequential state estimation, 63
Serial searches, 59–60
  broken/center z, 80–82
  continuous/center z, 76–79
  defined, 59
  likelihood-ratio code acquisition receiver, 60
  nonuniform, 76–84
  start of, 59
  strategy for, 49
  strategy structure, 74
  types of, 63
  See also Searches
Shadowing, 242, 265
Signal flow graph
  illustrated, 64
  method, 64–69
Simultaneous/parallel search, 187–88
Single-cell CDMA systems, 223–24, 252–55

Single channel per carrier (SCPC), 343
Single-channel-per-carrier PCM multiple-access demand assignment equipment (SPADE) system, 344
Slotted ALOHA, 350–51
  capacity, 350
  CDMA schemes, 354–65
  collision probability, 350
  DS/CDMA, 359–62
  FH, 363–65
  narrowband, 364, 365
  throughput, 351
Slow FH (SFH), 56
  discrete tracking systems for, 137–54
  one-channel noncoherent code tracking loop for, 134
  time-shared two-channel noncoherent code tracking loop, 132
  tracking loop for, 129–31
Soft handoff, 214
Spaced-frequency spaced-time correlation function, 233
Space-division multiplexing (SDM), 340
Spread spectrum
  antijamming capabilities, 2–4
  defined, *xiii*
  direct sequence, *xiv*
  key elements, *xiii*
  See also Spread spectrum systems
Spread spectrum protocols, 356–59
  common code, 358
  defined, 355
  receiver-oriented protocol, 357–58
  transmitter-oriented protocol, 357
  transmitter-receiver-oriented protocol, 356–57
  See also Protocols
Spread spectrum systems
  concept of, 1–8
  DSSS, 8–13
  examples of, 8–18
  FHSS, 13–14
  hybrid DS/FH schemes, 14, 16–17
  illustrated concept, 2
  packet overlapping, 355
  technology, 175–206
  throughput and, 355
State transition probabilities, 149–52
Steady-state phase error, 108–10
Steepest descent algorithm, 310–11
Straight search, 63

Subdivision, 342, 343
Suboptimum DS/CDMA receivers, 276–89
   adaptive algorithm, 286–87
   adaptive MMSE, 283–86
   decorrelating, 277–81
   interference cancellers, 281–83
Subspace, 39, 40
Sum and difference demodulator, 195
Surface acoustic wave (SAW), 175, 198–206
   chirp filters, 206
   convolution function, 198–200
   convolver block diagram, 200
   device fabrication complexity, 205
   Fourier transform, 200–201
   hybrid PTFs, 206
   inverse Fourier transform, 201–2
   memory correlators, 205
   programmable transversal filter (PTF), 204
   reported devices, 203–6
   signal propagation, 199
   tapped delay line, 204
   transform domain signal filtering, 202–3
   See also SAW/FET devices

Tapped delay lines, 178
   multipath channels, 235
   SAW, 204
   for synchronous DS/CDMA system, 273
   trellis diagram, 275
Tau dither early-late noncoherent tracking loop, 124–25
   advantage of, 124–25
   defined, 124
   illustrated, 126
   operation of, 125
   See also Tracking loop
Technology
   CCD, 175–78
   SAW, 175, 198–203
   spread spectrum system, 175–206
Thermal noise
   Gaussian, 8
   power, 8, 260, 297
   power spectral density, 259
   ratio, 253, 259
Throughput
   for ALOHA systems, 349
   for capture thresholds, 352
   computation of, 338–39
   defined, 338
   for narrowband ALOHA on satellite mobil fading channels, 362

   slotted ALOHA, 351
   slotted ALOHA DS/CDMA, 361, 362
   in spread spectrum networks, 355
Time dispersion, 225
Time-dispersive channels, 229
Time division multiple access (TDMA), *xiv*, 131
   advantages/disadvantages of, 345–46
   BTR, 345
   cochannel interference, 212–13
   CRS, 345
   defined, 342
   fixed assignment, 346
   frame, 345
   frequency reuse in, 212–13
   guard times/guard bands, 241
   handoff, 214
   "hard blocking" condition, 214
   illustrated scheme, 344
   in mobile radio systems, 211–12
   voice activity monitoring and, 214
   See also CDMA; Frequency division multiple access (FDMA)
Time-division multiplexing (TDM), 340
Time field, 19
Time-shared two-channel noncoherent tracking loop
   discriminator S-curve, 134
   frequency plane, 133
   illustrated, 132
   signal waveforms, 133
Trace function, 26, 33–36
   linear, 35–36
   properties, 34–36
Trace polynomial, 26, 33–34
Tracking CML/IT error probability, 161
Tracking error power density, 114–15
Tracking loops, 120–54
   baseband full-time early-late, 120–21
   discrete tracking process and, 131–54
   double dither DLL, 125–28
   for FHSS, 128–29
   full-time early-late noncoherent, 121–24
   for slow FHSS systems, 129–31
   tau dither early-late noncoherent, 124–25
Tracking systems. *See* Discrete tracking systems
Transform domain analysis, 64–69
Transform domain filters, 314–17
   defined, 314
   interference suppression algorithm, 316
   notch, 317
   performance, 317

receiver block diagram, 315
Transform domain signal filtering, 202–3
   adaptive receiver, 203
   defined, 202
   nonadaptive receiver, 203
Transmission functions, 64
Transmitter-oriented protocol, 357
Transmitter-receiver-oriented protocol, 356–57
Transversal filters, 294
   DFT, 316
   frequency response, 300
   with LMS adaptive algorithm, 312
   two-sided transfer function, 299
Two-dimensional searches, 58–59
Two-ray propagation model, 321–24
   illustrated, 322
   propagation loss in suburban area, 324
   propagation loss upper bound, 323
   ray free space field strength, 321
   received electric field strength, 322
   reflected ray electric field, 322

Ultra small aperture terminals (USATs), 347
Uniformly most powerful test (UMPT), 60

Vectors
   addition, 38
   all-zero, 38
   defined, 38
   inner product of, 39
   linear combination of, 39
   orthogonal, 39
Vector space, 38
Vector space over, 37–38
Verification modes, 62
Very small aperture terminals (VSATs), 347
Viterbi algorithm, 270, 274
Voice activity monitoring, 213–14
Voltage-controlled oscillator (VCO), 106

Wald sequential test, 62
Wiener-Hopf equation, 297, 303, 307

Z-search, 63

The Artech House Telecommunications Library
Vinton G. Cerf, Series Editor

Advanced High-Frequency Radio Communications, Eric E. Johnson, Robert I. Desourdis, Jr., et al.

Advanced Technology for Road Transport: IVHS and ATT, Ian Catling, editor

Advances in Computer Systems Security, Vol. 3, Rein Turn, editor

Advances in Telecommunications Networks, William S. Lee and Derrick C. Brown

Advances in Transport Network Technologies: Photonics Networks, ATM, and SDH, Ken-ichi Sato

An Introduction to International Telecommunications Law, Charles H. Kennedy and M. Veronica Pastor

Asynchronous Transfer Mode Networks: Performance Issues, Second Edition, Raif O. Onvural

ATM Switching Systems, Thomas M. Chen and Stephen S. Liu

Broadband: Business Services, Technologies, and Strategic Impact, David Wright

Broadband Network Analysis and Design, Daniel Minoli

Broadband Telecommunications Technology, Byeong Lee, Minho Kang, and Jonghee Lee

Cellular Mobile Systems Engineering, Saleh Faruque

Cellular Radio: Analog and Digital Systems, Asha Mehrotra

Cellular Radio: Performance Engineering, Asha Mehrotra

Cellular Radio Systems, D. M. Balston and R. C. V. Macario, editors

CDMA for Wireless Personal Communications, Ramjee Prasad

Client/Server Computing: Architecture, Applications, and Distributed Systems Management, Bruce Elbert and Bobby Martyna

Communication and Computing for Distributed Multimedia Systems, Guojun Lu

Community Networks: Lessons from Blacksburg, Virginia, Andrew Cohill and Andrea Kavanaugh, editors

Computer Networks: Architecture, Protocols, and Software, John Y. Hsu

Computer Mediated Communications: Multimedia Applications, Rob Walters

Computer Telephone Integration, Rob Walters

Corporate Networks: The Strategic Use of Telecommunications, Thomas Valovic

Digital Beamforming in Wireless Communications, John Litva, Titus Kwok-Yeung Lo

Digital Cellular Radio, George Calhoun

Digital Hardware Testing: Transistor-Level Fault Modeling and Testing, Rochit Rajsuman, editor

Digital Switching Control Architectures, Giuseppe Fantauzzi

Digital Video Communications, Martyn J. Riley and Iain E. G. Richardson

Distributed Multimedia Through Broadband Communications Services, Daniel Minoli and Robert Keinath

Distance Learning Technology and Applications, Daniel Minoli

EDI Security, Control, and Audit, Albert J. Marcella and Sally Chen

Electronic Mail, Jacob Palme

Enterprise Networking: Fractional T1 to SONET, Frame Relay to BISDN, Daniel Minoli

Expert Systems Applications in Integrated Network Management, E. C. Ericson, L. T. Ericson, and D. Minoli, editors

FAX: Digital Facsimile Technology and Applications, Second Edition, Dennis Bodson, Kenneth McConnell, and Richard Schaphorst

FDDI and FDDI-II: Architecture, Protocols, and Performance, Bernhard Albert and Anura P. Jayasumana

Fiber Network Service Survivability, Tsong-Ho Wu

A Guide to the TCP/IP Protocol Suite, Floyd Wilder

Implementing EDI, Mike Hendry

Implementing X.400 and X.500: The PP and QUIPU Systems, Steve Kille

Inbound Call Centers: Design, Implementation, and Management, Robert A. Gable

Information Superhighways Revisited: The Economics of Multimedia, Bruce Egan

Integrated Broadband Networks, Amit Bhargava

International Telecommunications Management, Bruce R. Elbert

International Telecommunication Standards Organizations, Andrew Macpherson

Internetworking LANs: Operation, Design, and Management, Robert Davidson and Nathan Muller

Introduction to Document Image Processing Techniques, Ronald G. Matteson

Introduction to Error-Correcting Codes, Michael Purser

An Introduction to GSM, Siegmund Redl, Matthias K. Weber, Malcom W. Oliphant

Introduction to Radio Propagation for Fixed and Mobile Communications, John Doble

Introduction to Satellite Communication, Bruce R. Elbert

Introduction to T1/T3 Networking, Regis J. (Bud) Bates

Introduction to Telephones and Telephone Systems, Second Edition, A. Michael Noll

Introduction to X.400, Cemil Betanov

LAN, ATM, and LAN Emulation Technologies, Daniel Minoli and Anthony Alles

Land-Mobile Radio System Engineering, Garry C. Hess

LAN/WAN Optimization Techniques, Harrell Van Norman

LANs to WANs: Network Management in the 1990s, Nathan J. Muller and Robert P. Davidson

Minimum Risk Strategy for Acquiring Communications Equipment and Services, Nathan J. Muller

Mobile Antenna Systems Handbook, Kyohei Fujimoto and J.R. James, editors

Mobile Communications in the U.S. and Europe: Regulation, Technology, and Markets, Michael Paetsch

Mobile Data Communications Systems, Peter Wong and David Britland

Mobile Information Systems, John Walker

Networking Strategies for Information Technology, Bruce Elbert

Packet Switching Evolution from Narrowband to Broadband ISDN, M. Smouts

Packet Video: Modeling and Signal Processing, Naohisa Ohta

Personal Communication Networks: Practical Implementation, Alan Hadden

Personal Communication Systems and Technologies, John Gardiner and Barry West, editors

Practical Computer Network Security, Mike Hendry

Principles of Secure Communication Systems, Second Edition, Don J. Torrieri

Principles of Signaling for Cell Relay and Frame Relay, Daniel Minoli and George Dobrowski

Principles of Signals and Systems: Deterministic Signals, B. Picinbono

Private Telecommunication Networks, Bruce Elbert

Radio-Relay Systems, Anton A. Huurdeman

RF and Microwave Circuit Design for Wireless Communications, Lawrence E. Larson

The Satellite Communication Applications Handbook, Bruce R. Elbert

Secure Data Networking, Michael Purser

Service Management in Computing and Telecommunications, Richard Hallows

Smart Cards, José Manuel Otón and José Luis Zoreda

Smart Highways, Smart Cars, Richard Whelan

Super-High-Definition Images: Beyond HDTV, Naohisa Ohta, Sadayasu Ono, and Tomonori Aoyama

Television Technology: Fundamentals and Future Prospects, A. Michael Noll

Telecommunications Technology Handbook, Daniel Minoli

Telecommuting, Osman Eldib and Daniel Minoli

Telemetry Systems Design, Frank Carden

Teletraffic Technologies in ATM Networks, Hiroshi Saito

Toll-Free Services: A Complete Guide to Design, Implementation, and Management, Robert A. Gable

Transmission Networking: SONET and the SDH, Mike Sexton and Andy Reid

Troposcatter Radio Links, G. Roda

Understanding Emerging Network Services, Pricing, and Regulation, Leo A. Wrobel and Eddie M. Pope

Understanding GPS: Principles and Applications, Elliot D. Kaplan, editor

Understanding Networking Technology: Concepts, Terms and Trends, Mark Norris

UNIX Internetworking, Second edition, Uday O. Pabrai

Videoconferencing and Videotelephony: Technology and Standards, Richard Schaphorst

Wireless Access and the Local Telephone Network, George Calhoun

Wireless Communications in Developing Countries: Cellular and Satellite Systems, Rachael E. Schwartz

Wireless Communications for Intelligent Transportation Systems, Scott D. Elliot and Daniel J. Dailey

Wireless Data Networking, Nathan J. Muller

Wireless LAN Systems, A. Santamaría and F. J. López-Hernández

Wireless: The Revolution in Personal Telecommunications, Ira Brodsky

Writing Disaster Recovery Plans for Telecommunications Networks and LANs, Leo A. Wrobel

X Window System User's Guide, Uday O. Pabrai

For further information on these and other Artech House titles, contact:

Artech House
685 Canton Street
Norwood, MA 02062
617-769-9750
Fax: 617-769-6334
Telex: 951-659
email: artech@artech-house.com

Artech House
Portland House, Stag Place
London SW1E 5XA England
+44 (0) 171-973-8077
Fax: +44 (0) 171-630-0166
Telex: 951-659
email: artech-uk@artech-house.com

WWW: http://www.artech-house.com